Artificial Intelligence of Things (AIoT) for Productivity and Organizational Transition

Sajad Rezaei
University of Worcester, UK

Amin Ansary
University of the Witwatersrand, South Africa

A volume in the Advances in
Computational Intelligence and
Robotics (ACIR) Book Series

Published in the United States of America by
   IGI Global
   Business Science Reference (an imprint of IGI Global)
   701 E. Chocolate Avenue
   Hershey PA, USA 17033
   Tel: 717-533-8845
   Fax: 717-533-8661
   E-mail: cust@igi-global.com
   Web site: http://www.igi-global.com

Library of Congress Cataloging-in-Publication Data

Names: Rezaei, Sajad, 1984- editor. | Ansary, Amin, 1979- editor.
Title: Artificial intelligence of things (AIoT) for productivity and
   organizational transition / edited by Sajad Rezaei, Amin Ansary.
Description: Hershey, PA : Business Science Reference, [2024] | Includes
   bibliographical references and index. | Summary: "This book aims to
   contribute high-quality and original works highlighting the most
   important research topics, concepts, trends, issues, practices, and
   implications of Artificial Intelligence of Things (AIoT) for
   productivity and organization transitions"-- Provided by publisher.
Identifiers: LCCN 2023053010 (print) | LCCN 2023053011 (ebook) | ISBN
   9798369309933 (hardcover) | ISBN 9798369309940 (ebook)
Subjects: LCSH: Artificial intelligence--Industrial applications. |
   Internet of things. | Organizational effectiveness.
Classification: LCC TA347.A78 A87 2024 (print) | LCC TA347.A78 (ebook) |
   DDC 006.3--dc23/eng/20240116
LC record available at https://lccn.loc.gov/2023053010
LC ebook record available at https://lccn.loc.gov/2023053011

This book is published in the IGI Global book series Advances in Computational Intelligence and Robotics (ACIR) (ISSN: 2327-0411; eISSN: 2327-042X)

British Cataloguing in Publication Data
A Cataloguing in Publication record for this book is available from the British Library.

All work contributed to this book is new, previously-unpublished material.
The views expressed in this book are those of the authors, but not necessarily of the publisher.

For electronic access to this publication, please contact: eresources@igi-global.com.

Advances in Computational Intelligence and Robotics (ACIR) Book Series

ISSN:2327-0411
EISSN:2327-042X

Editor-in-Chief: Ivan Giannoccaro, University of Salento, Italy

MISSION

While intelligence is traditionally a term applied to humans and human cognition, technology has progressed in such a way to allow for the development of intelligent systems able to simulate many human traits. With this new era of simulated and artificial intelligence, much research is needed in order to continue to advance the field and also to evaluate the ethical and societal concerns of the existence of artificial life and machine learning.

The **Advances in Computational Intelligence and Robotics (ACIR) Book Series** encourages scholarly discourse on all topics pertaining to evolutionary computing, artificial life, computational intelligence, machine learning, and robotics. ACIR presents the latest research being conducted on diverse topics in intelligence technologies with the goal of advancing knowledge and applications in this rapidly evolving field.

COVERAGE

- Machine Learning
- Artificial Intelligence
- Cognitive Informatics
- Natural Language Processing
- Brain Simulation
- Pattern Recognition
- Evolutionary Computing
- Computer Vision
- Robotics
- Artificial Life

IGI Global is currently accepting manuscripts for publication within this series. To submit a proposal for a volume in this series, please contact our Acquisition Editors at Acquisitions@igi-global.com or visit: http://www.igi-global.com/publish/.

Titles in this Series

For a list of additional titles in this series, please visit:
http://www.igi-global.com/book-series/advances-computational-intelligence-robotics/73674

Innovative Machine Learning Applications for Cryptography
J. Anitha Ruth (SRM Institute of Science and Technology, India) G.V. Mahesh Vijayalakshmi (BMS Institute of Technology and Management, India) P. Visalakshi (SRM Institute of Science and Technology, India) R. Uma (Sri Sai Ram Engineering College, India) and A. Meenakshi (SRM Institute of Science and Technology, India)
Engineering Science Reference • copyright 2024 • 294pp • H/C (ISBN: 9798369316429) • US $300.00 (our price)

The Ethical Frontier of AI and Data Analysis
Rajeev Kumar (Moradabad Institute of Technology, India) Ankush Joshi (COER University, Roorkee, India) Hari Om Sharan (Rama University, Kanpur, India) Sheng-Lung Peng (College of Innovative Design and Management, National Taipei University of Business, Taiwan) and Chetan R. Dudhagara (Anand Agricultural University, India)
Engineering Science Reference • copyright 2024 • 456pp • H/C (ISBN: 9798369329641) • US $365.00 (our price)

Empowering Low-Resource Languages With NLP Solutions
Partha Pakray (National Institute of Technology, Silchar, India) Pankaj Dadure (University of Petroleum and Energy Studies, India) and Sivaji Bandyopadhyay (Jadavpur University, India)
Engineering Science Reference • copyright 2024 • 314pp • H/C (ISBN: 9798369307281) • US $300.00 (our price)

Computational Intelligence for Green Cloud Computing and Digital Waste Management
K. Dinesh Kumar (Amrita Vishwa Vidyapeetham, India) Vijayakumar Varadarajan (The University of New South Wales, Australia) Nidal Nasser (College of Engineering, Alfaisal University, Saudi Arabia) and Ravi Kumar Poluru (Institute of Aeronautical Engineering, India)
Engineering Science Reference • copyright 2024 • 405pp • H/C (ISBN: 9798369315521) • US $300.00 (our price)

For an entire list of titles in this series, please visit:
http://www.igi-global.com/book-series/advances-computational-intelligence-robotics/73674

701 East Chocolate Avenue, Hershey, PA 17033, USA
Tel: 717-533-8845 x100 • Fax: 717-533-8661
E-Mail: cust@igi-global.com • www.igi-global.com

Table of Contents

Chapter 10

 Surjit Singha, Kristu Jayanti College (Autonomous), India
 Ranjit Singha, Christ University, India

Chapter 11

 V. Santhi, Vellore Institute of Technology, India
 Yamala N. V. Sai Sabareesh, Vellore Institute of Technology, India
 Ponnada Prem Sudheer, Vellore Institute of Technology, India
 Villuri Poorna Sai Krishna, Vellore Institute of Technology, India

Detailed Table of Contents

Chapter 1

    Helen Watts, University of Worcester, UK
    Abdulmaten Taroun, University of Worcester, UK
    Richard Jones, University of Worcester, UK

With the rise of big data analytics in recent years, organisations now have more data than ever before to make decisions. Whilst AI solutions are developing to aid big data decision making, employees continue to analyse big data introducing bias and variability into the process. This chapter details a case study examining the efficacy of the big data analytic capability (BDAC) model, and how it can be augmented to account for cognitive bias to improve the model's organisational value. Qualitative semi-structured interviews were conducted within a global data-driven manufacturing organisation. Thematic analysis elucidated that cognitive bias impacts decision making when analysing big data. This case study yields recommendations for a modified big data analytics capability model to recognise an additional sub-dimension under the 'intangibles' dimension of 'objective decision-making'. Implications for manufacturing organisations and the role of AI in both removing and adding bias are discussed.

Chapter 2

    Ranjit Singha, Christ University, India
    Surjit Singha, Kristu Jayanti College (Autonomous), India

Integrating AIoT technologies provide businesses with increased productivity, cost savings, data-driven insights, and enhanced consumer interactions. Nevertheless, difficulties include data privacy, ethics, regulatory compliance, and technical complexities. The recommendations include transparent practices, accountability, bias mitigation, data minimization, informed consent, and ethical design. Policymakers must develop adaptable regulations, place a premium on privacy and security, and involve stakeholders. A user-centric approach and training in data ethics are essential. AIoT offers enormous potential but requires a delicate balance between innovation and responsibility, with ethics, privacy, and policy compliance at the forefront.

Chapter 3

Qasim Hamakhurshid Hamamurad, Management Information System, UTM-Skudai, Malaysia

Normal Mat Jusoh, Azman Hashim International Business School, Malaysia

Farahakmar Anorsalim, Azman Hashim International Business School, Malaysia

Sathiswaran Uthamaputhran, Azman Hashim International Business School, Malaysia

Muslim Amin, Azman Hashim International Business School, Malaysia

Artificial intelligence and technological advancements impact human life and activities. Through structured surveys to find the effects of ChatGPT use among postgraduate students in Arab countries, we collected data on students' experiences and thoughts about using ChatGPT in their academic pursuits. In this study, one hundred postgraduate students from various academic subjects and universities in Arab countries participated. We used inferential statistics (including regression analysis and t-tests) to investigate the relationships between ChatGPT usage and academic achievement and descriptive statistics to summarise demographic data. We assessed the collected data using the Statistical Package for the Social Sciences (SPSS). The findings, which show that ChatGPT usage positively affects learning outcomes, support past research outlining the potential benefits of AI-assisted learning technology.

Chapter 4

Tarun Kumar Vashishth, IIMT University, India
Vikas Sharma, IIMT University, India
Kewal Krishan Sharma, IIMT University, India
Bhupendra Kumar, IIMT University, India
Sachin Chaudhary, IIMT University, India
Rajneesh Panwar, IIMT University, India

The integration of artificial intelligence and the internet of things (AIoT) in education has ushered in a paradigm shift, transforming traditional learning environments and revolutionizing educational technology. This chapter explores the multifaceted impact of AIoT on education, encompassing personalized learning experiences, intelligent tutoring systems, and the creation of smart educational spaces. By leveraging AI algorithms and IoT-connected devices, educators can tailor instructional content to individual student needs, fostering a dynamic and adaptive learning environment. The interconnectivity of devices enables real-time data collection, facilitating insights into student performance, engagement, and preferences. Moreover, AIoT contributes to the development of smart classrooms equipped with interactive technologies, enhancing collaboration and active participation. As we navigate this transformative era, the potential of AIoT in education holds promise for optimizing educational outcomes and shaping a future where technology seamlessly integrates with the learning experience.

Chapter 5

   C. V. Suresh Babu, Hindustan Institute of Technolgy and Science, India
   M. Sowmi Saltonya, Hindustan Institute of Technology and Science,
       India
   Suresh Ganapathi, Hindustan Institute of Technology and Science, India
   A. Gunasekar, Hindustan Institute of Technology and Science, India

This chapter provides a comprehensive exploration of networking challenges in today's evolving landscape. It introduces the concept of artificial intelligence of things (AIoT) and demonstrates the synergy between AI and IoT in addressing these challenges. Through a holistic AIoT strategy, the chapter guides readers in designing and implementing proactive network monitoring, predictive maintenance, enhanced security measures, quality of service management, automation, and data-driven decision-making. It emphasizes the tangible benefits of AIoT, including optimized networking performance, cost reduction, and improved user experiences. Real-world case studies highlight successful AIoT implementations, offering valuable lessons. The chapter also delves into emerging trends, ethical considerations, and concludes by emphasizing the implications and the need for a forward-looking approach in networking productivity.

Chapter 6

   Mukesh Chaware, Indian Institute of Management, Kozhikode, India

Artificial intelligence (AI) and internet of things (IoT) have been combined to create the artificial intelligence of things (AIoT), which could revolutionize education.

Education has opportunities and difficulties from this growth. The benefits include tailored learning, real-time feedback, and immersive learning. Data privacy, security, and accessibility are problems beyond the digital divide. Strategic decision-making, communication, and stakeholder engagement are needed to maximise productivity and results using AIoT in education. The chapter provides a plan for ethical and equitable usage of the technology while avoiding hazards. This study investigates the role of AIoT in education to provide a much-needed understanding of future perspectives and ramifications. Responsible and ethical AIoT implementation improves learning and outcomes, according to the chapter.

Chapter 7

At the end of the 1950s, the possibility of machine learning in computers began to be discussed. After more than half a century, thanks to technological advances, machine learning is more than a reality; it is a tool that supports different fields, such as education, business, science, and management. This chapter explores how machine learning takes place in decision-making, especially in the managerial field, and the advantages that it offers when studying scenarios, choosing strategies, exploring the possible consequences of our actions, and/or predicting the likely responses of the parties involved in a specific situation.

Chapter 8

This chapter delves into the impact of artificial intelligence (AI) on the educational systems, spanning both schools and universities. The primary components of an educational system include students, teachers, monitoring and evaluation (M&E), and administrative organization. While there are undoubtedly other components, these mentioned items are the most crucial. Artificial intelligence significantly influences all of them. In this chapter, the authors thoroughly examine and analyze the effect of AI on each of the four educational components, employing a systematic review and library methodology. Consequently, after the introduction, the chapter is organized into four parts, corresponding to each component. The main finding indicates that AI technology has revolutionized education by offering personalized and adaptive learning, flexible teaching, as well as smart administration and swift assessment systems. This chapter also highlights the drawbacks associated with it.

Saumya Ketan Jhaveri, School of Liberal Studies, Pandit Deendayal Energy University, India

Anshika Chauhan, School of Liberal Studies, Pandit Deendayal Energy University, India

Nausheen Nizami, School of Liberal Studies, Pandit Deendayal Energy University, India

This chapter assesses the correlation between AI expansion and its influence on diverse economic factors like labor supply, wage rates, and working hours. In the subsequent segment, the authors explore the connection between the utility function and the derived hours equation. Unlike the direct utility indicators of leisure and consumption, indirect utility functions are represented through income/wage rates or prices. These models and diagrams are developed by the authors to illustrate the correlation between income/consumption and its association with the balance between working hours and leisure hours. This chapter encompasses diverse models predicated on specific assumptions, contingent on market demand and supply scenarios.

Surjit Singha, Kristu Jayanti College (Autonomous), India

Ranjit Singha, Christ University, India

AIoT, or artificial intelligence of things, is a transformative combination of artificial intelligence and the internet of things (IoT) that has far-reaching ramifications across multiple domains. This chapter examines the theories and models underlying its development and implementation. Businesses can assess the costs, benefits, and competitive advantages of AIoT by using economic models and market dynamics. Understanding human behaviour and trust is crucial for user acceptance, while ethical considerations underpin the development of accountable AIoT applications. Data management, security, and interoperability are technical facets that architectural frameworks address. The alignment of AIoT with human needs is enhanced by cognitive models and user experience, thereby fostering well-being. Change management and organizational learning are essential for effective implementation, which fosters innovation. AIoT promotes innovation and efficiency in manufacturing, healthcare, and smart cities.

Chapter 11

V. Santhi, Vellore Institute of Technology, India
Yamala N. V. Sai Sabareesh, Vellore Institute of Technology, India
Ponnada Prem Sudheer, Vellore Institute of Technology, India
Villuri Poorna Sai Krishna, Vellore Institute of Technology, India

Smart homes, smart buildings, and smart cities increase the quality of living by creating a world that is better and more secure, reducing dependency on human needs and efforts. This chapter investigates the complex landscape of security threats resulting from the integration of AIoT (artificial intelligence of things) and cloud platforms within the domains of smart homes, smart buildings, and smart cities. It examines the distinctive challenges and vulnerabilities that arise when these technologies converge. These technologies hold the potential to significantly enhance the quality of life, reinforce home security, and facilitate elderly care. However, their effectiveness hinges on access to extensive data about users' homes and private lives. This chapter underscores the critical need to address the ensuing security and privacy concerns, which present substantial barriers to the widespread adoption of these promising technologies.

Preface

The integration of Artificial Intelligence (AI) and the Internet of Things (IoT) has led to the advent and nascent of Artificial Intelligence of Things (AIoT) as an emerging technology that provides tremendous opportunities and imperative challenges through which organizations are struggling to implement into their core business activities. AIoT and related technologies are becoming a strategic tool to enhance the firm's value chain process in several business domains and disciplines such as education, IS, IT, marketing, hospitality, tourism, logistics, human resource management, and psychology. This emerging technology is valuable for overcoming several technical issues in organizations while posing concerns and risks for organizations and societies in general. AIoT empowers organizations' productivity by contributing a wider stance that is related to firms' products and services. AIoT strategic decisions in an organization involve business leaders' approach in executing AIoT at the organization's level and functional levels to create and enhance the communications channels, the flow, and distribution of information, to sustain and conduct the relationships with stakeholders that are aiming for productivity and strategic results. Accordingly, successful management and implementation of AIoT are believed to enhance corporate productivity, KPIs, and strategic outcomes in the long run. As a result, this book discusses trends, opportunities, issues, risks, and implications of AIoT development, implementation, and management as new phenomena in organizations and for society in general.

This book aims to contribute high-quality and original works highlighting the most important research topics, concepts, trends, issues, practices, and implications of Artificial Intelligence of Things (AIoT) for productivity and organization transitions. The book serves as a source of knowledge for researchers, educators, students, and industry practitioners around the globe. Despite its significance, the integration of AI into IoT and its impact on a firm's productivity and transition have not been extensively addressed in the literature. The present book addresses this issue and advances a framework that distinguishes between applications of AI in IoT systems within several key organizational contexts across several disciplines. Therefore, this book aims to address AIoT development, implementation, and management in transiting organizations across departments and industries. This book would be useful for researchers, business leaders, instructors, e-commerce managers, business analysts, and undergraduate/post-graduate and doctorate students in a variety of disciplines including IS, IT, marketing, tourism, and human resource management. In addition to academicians and researchers, entrepreneurs and technology developers would be the potential target audience for this book.

ORGANIZATION OF THE BOOK

The book is organized into 11 chapters. A brief description of each of the chapters follows:

1. **Theories and Models in AIoT: Exploring Economic, Behavioral, Technological, Psychological, and Organizational Perspectives.** This chapter discusses AIoT, the fusion of Artificial Intelligence (AI) and the Internet of Things (IoT), enhances IoT devices with advanced data processing and decision-making capabilities through AI algorithms and machine learning. This convergence enables autonomous decision-making, adaptive analysis, and intelligent responses without human intervention. AIoT systems seamlessly integrate with IoT devices, sensors, and networks to comprehend context, learn from data, and evolve over time. This transformative technology has vast implications across various sectors like smart cities, healthcare, manufacturing, and agriculture by optimizing processes, predicting faults, automating tasks, and enabling real-time informed decision-making. It drives progress by reducing operational costs, improving resource efficiency through predictive maintenance and energy management, and offering customized services like remote patient monitoring in healthcare. AIoT also addresses environmental challenges by promoting sustainability in smart cities and industries. While unlocking economic growth and innovation opportunities, AIoT raises concerns about data privacy, security, ethics, and potential job displacement. Its multidisciplinary nature underscores the need for collaboration across diverse fields to maximize its potential for societal transformation and productivity enhancement.

2. **Trends and Challenges in AIoT Implementation for Smart Homes, Smart Buildings, and Smart Cities in Cloud Platforms.** This chapter discusses that smart homes and cities, fueled by IoT and AI, have evolved from luxury amenities to fundamental elements of modern living. Connected devices enable automated, adaptive, and intelligent solutions, simplifying daily routines and optimizing resources. Smart homes feature interconnected appliances

responding to user preferences and delivering remote access, alerts, and notifications. Meanwhile, smart cities utilize IoT and AI to foster innovation, promote well-being, and ensure sustainability, leveraging cloud platforms for data-driven decision-making. This symbiotic relationship between technology and lifestyle improvements continues to shape urban landscapes and daily experiences.

3. **Theoretical Impact of AI on Working Hours and Wage Rate.** This chapter discusses that throughout history, each industrial revolution has propelled progress yet introduced novel challenges. The ongoing Fourth Industrial Revolution, centered around information and communication technologies, encompasses AI, Automation, Digitalization, and Machine Learning. Like past eras, this revolution brings concerns such as unemployment, job displacement, market shifts, and fluctuating prices. Historically, doubts about technology's impact on labor demand trace back to ancient times, exemplified by Plato's concern about writing replacing human memory. Similar anxieties surfaced during the Industrial Revolution concerning machinery taking jobs. Presently, AI and robotics raise questions about job loss, although research indicates they may eventually create new employment opportunities despite initial creative destruction.

4. **AIoT Concepts and Integration: Exploring Customer Interaction, Ethics, Policy, and Privacy.** This chapter discusses that AIoT infuses intelligence into IoT devices, empowering them to self-learn, adapt, and make autonomous decisions. This synergy enables IoT devices to gather, analyze, and apply data, driving operational efficiencies, predictive maintenance, and resource optimization. AIoT offers a competitive edge through enhanced customer experiences, accelerated response times, and innovative services. It reduces operational costs, promotes energy efficiency, and minimizes waste generation. AIoT's multidisciplinary nature spans sensor integration, data analytics, machine learning, and cybersecurity. Ethical considerations, policy developments, and privacy safeguards are critical for successful AIoT implementation. The chapter examines AIoT's principles, impacts, and challenges, highlighting the necessity for holistic approaches to fully realize AIoT's potential.

5. **A Case Study of Big Data Analytics Capability and the Impact of Cognitive Bias in a Global Manufacturing Organisation.** This chapter discusses that the 4th Industrial Revolution (Industry 4.0) is revolutionizing manufacturing through data-driven optimization, with the Industrial Internet of Things (IIoT) playing a pivotal role in enhancing efficiency, safety, and predictive maintenance in manufacturing processes. The abundance of big data generated by interconnected sensors and devices empowers organizations to drive competitive performance gains, improve decision-making, adopt agile practices, and enhance supply

chain sustainability. Big data is essential for smart manufacturing, enabling continuous monitoring, simulation, and optimization of production activities while reducing costs. Integrating Artificial Intelligence (AI) with big data allows businesses to make accurate decisions at scale efficiently, leading to improved decision quality, efficiency, and competitive advantage. The fusion of AI with the Internet of Things (IoT) in Artificial Intelligence of Things (AIoT) enables real-time analysis across multiple devices for proactive decision-making and forecasting, potentially offering superior returns on investment through quicker and more precise decision-making processes.

6. **AIoT-Education: The Effects of ChatGPT Use Among Postgraduate Students in Arab Countries.** This chapter discusses that the rapid evolution of AI technology, exemplified by OpenAI's ChatGPT, is reshaping education by offering college students a transformative learning tool. AI, mimicking human cognition, is advancing rapidly, with ChatGPT showcasing AI's progress and potential in education. This innovative language model enhances the learning experience by providing instant access to information, fostering critical thinking, and aiding in complex problem-solving. While ChatGPT's benefits are significant, its ethical use and impact on student performance must be carefully considered. By facilitating access to new knowledge and technological advancements, ChatGPT has the potential to enhance learning outcomes and student engagement. However, it is crucial to balance its utility with the importance of interpersonal communication for developing essential skills. This study explores the moral implications and broader impacts of ChatGPT on education, focusing on learning outcomes, student engagement, and academic performance in Arabic-speaking countries through a comprehensive quantitative data collection approach. The research questions aim to gauge students' experiences with ChatGPT for educational purposes and provide insights into its value and challenges in academic settings. This investigation underscores the evolving role of technology in education and the potential for tools like ChatGPT to enrich the learning environment for students in higher education.

7. **AIoT in Education Transforming Learning Environments and Educational Technology.** This chapter discusses that within the fast-changing educational sphere, the convergence of AI and IoT gives birth to a groundbreaking paradigm called AIoT, or Artificial Intelligence of Things. Combining intelligent algorithms and connected devices, AIoT presents unparalleled possibilities to reform learning spaces and reinvent educational technology. At this crossroad of potent tech forces, education is experiencing a profound metamorphosis, departing from conventional models toward a flexible, adaptable, and customizable mode of teaching and learning.

8. **Harnessing AIoT in Transforming Education Landscape: Opportunities, Challenges and Future.** This chapter discusses that amidst the 21st century's digital transformation, AI and IoT drive significant change in education, heralding the arrival of AIoT — a fusion holding immense potential to revamp educational methods and practice. Encompassing educators, policy makers, tech developers, and students alike, interest in AIoT's implementation in education spans diverse sectors. Aiming to explore AIoT's roles, prospects, and challenges in contemporary educational settings globally, this study delves into existing applications and future advancements. With increasing internationalization and local demands for sustainable benefits, understanding AIoT's influence on education is pivotal, which highlights AI's broader purpose in fostering creativity, analytical prowess, and empathy.

9. **The Effect of Artificial Intelligence Over the Educational System (School and University).** This chapter discusses that AI represents machine intelligence, distinct from natural intelligence seen in animals, including humans. AI systems understand their surroundings and act to achieve goals, impacting various fields like self-driving vehicles, medical diagnoses, artistic creation, and more. Managing complexity is crucial in AI research across computer science branches. AI tools aid in solving intricate problems. Education complexity is being tackled with AI, as seen in software like essaygrader used by professors globally to enhance teaching processes. These programs interact with students, offering assistance and feedback. While humanoid robots are not replacing teachers yet, AI projects aim to improve education quality for both students and educators. This chapter explores AI's influence on learning across student engagement, teacher roles, monitoring and evaluation, and educational administration, highlighting AI's potential for personalized education and administrative efficiency.

10. **Machine Learning and Decision-Making.** This chapter discusses that the industrial revolution spurred new employment opportunities, markets, and companies while fueling creativity. Examples include Romain Rolland's 1921 work "The Rebellion of the Machines" and Fritz Lang's 1927 film "Metropolis," which foreshadowed AI. These concepts transitioned from fiction to reality as early as the 1930s when Konrad Zuse built the first programmable computer, the Z1. Alan Turing's Bombe computer, developed in 1939 to decrypt German army messages, further advanced computing. In 1943, Warren McCulloch and Walter Pitts proposed a neuron model based on the nervous system, laying the groundwork for AI's future development.

11. **AIoT Revolution: Transforming Networking Productivity for the Digital Age.** This chapter explores AIoT's role in enhancing networking productivity through applications like optimization, automation, enhanced security, and edge computing integration. The study examines real-world examples and discusses challenges and future research directions to maximize AIoT's benefits in networking environments.

Sajad Rezaei
University of Worcester, UK

Amin Ansary
University of the Witwatersrand, South Africa

Chapter 1

A Case Study of Big Data Analytics Capability and the Impact of Cognitive Bias in a Global Manufacturing Organisation

Helen Watts
ⓘD https://orcid.org/0000-0002-8286-8764
University of Worcester, UK

Abdulmaten Taroun
University of Worcester, UK

Richard Jones
University of Worcester, UK

ABSTRACT

With the rise of big data analytics in recent years, organisations now have more data than ever before to make decisions. Whilst AI solutions are developing to aid big data decision making, employees continue to analyse big data introducing bias and variability into the process. This chapter details a case study examining the efficacy of the big data analytic capability (BDAC) model, and how it can be augmented to account for cognitive bias to improve the model's organisational value. Qualitative semi-structured interviews were conducted within a global data-driven manufacturing organisation. Thematic analysis elucidated that cognitive bias impacts decision making when analysing big data. This case study yields recommendations for a modified big data analytics capability model to recognise an additional sub-dimension under the 'intangibles' dimension of 'objective decision-making'. Implications for manufacturing organisations and the role of AI in both removing and adding bias are discussed.

DOI: 10.4018/979-8-3693-0993-3.ch001

BACKGROUND

Industry 4.0

The 4th Industrial Revolution (Industry 4.0) is rapidly transforming the manufacturing industry, driven by advancements in data and connectivity enabling manufacturing processes to be optimised. A key component of Industry 4.0, the Industrial Internet of Things (IIoT), has enabled manufacturing organisations to increase efficiency, reduce error, improve safety, and predict maintenance through the interconnectedness of sensors and devices. This interconnectedness results in vast amounts of big data, meaning manufacturing organisations now have more data than ever before to make decisions for a range of business needs, including competitive performance gains (Mikalef et al., 2018) improved decision making (Sheng et al., 2017), agile manufacturing practices (Gunasekaran et al., 2019), and supply chain sustainability (Jeble et al., 2018). Big data can increase market competitiveness, quality management, and enhance existing manufacturing systems and it is argued as being a pre-requisite for smart, or 'intelligent' manufacturing whereby continuous monitoring, simulation, and optimization of production activities can be achieved (Tao et al., 2018) with reduced costs (Wang et al., 2022). Further, smart manufacturing cannot be realized with traditional manufacturing software and technologies, due to dependencies on multiple vendors for data, and the lack of sensory information (Cui et al., 2020). In sum, big data offers the potential for manufacturing organisations to improve their overall performance capability and competitive advantage.

However, one of the biggest limiting factors of how effective big data can be is the lack of resource to accurately interpret the data and/or for these interpretations to result in decision-making. Whilst the concept of applying Artificial Intelligence (AI) to big data is well established, recent developments in AI have radically increased the capability of businesses to make accurate decisions based on big data at scale and in a cost-effective way. For instance, AI can support three important business needs: automating business processes, gaining insight through data analysis, and engaging with customers and employees (Davenport, 2018). Further, AI has been shown to improve decision quality, efficiency, and a cultural change leading to competitive advantage (Ransbotham et al., 2021) and can facilitate the innovation of new models, means, and forms of intelligent manufacturing (Li et al., 2017). When overlaying AI with Internet of Things (IoT), Artificial Intelligence of Things (AIoT) provides the capability to analyse big data across multiple devices and sources simultaneously to make decisions and forecasts. For instance, AIoT can model potential outcomes based on different resource allocations, and proactively alert managers to potential project completions risks (El Khatib & Al Falasi, 2021). Fundamentally, AIoT arguably has the potential to offer a better return on investment due to enabling quicker, more accurate and more consistent decisions than employees.

Reluctance to Adopt AI

Whilst AI can outperform human decision making, implying significant potential for its broader application in various industrial sectors, there is still reluctance when adopting it directly at manufacturing sites, arguably due to a lack of belief that AI will offer superiority over tried-and-true practices that have been developed over time through human experience and expertise (Plathottam et al., 2023). Further, reluctance in adoption may also come from perceived compliance challenges associated with AI governance and being accountable for data handling and processing.

Also, applications of AI in manufacturing industries have been particularly challenging due to challenges in achieving accurate modelling of highly nonlinear phenomena and a high number of dimensions in datasets (Kim et al., 2022), and numerous uncertainties and interdependencies (Arinez et al., 2020). Traditionally, technologically-oriented decision tools have supported only part of an organizational or individual decision process due to the complexity and uncertainty inherent in semi-structured and unstructured decision tasks (Phillips-Wren et al., 2019).

It is also argued that AI tools, whilst aimed at adding in objectivity to the way that data is handled, are not inherently objective due to the human involvement in the provision and curation of this data, resulting in model predictions being susceptible to bias (Milaninia, 2021). Whilst AI algorithms are said to come with the "promise of neutrality,", there are issues of "algorithmic bias"; the well-documented propensity of algorithms to learn systemic bias through their reliance on historical data, as well as the potential for bias arising from human processing of AI algorithmic outputs due to selective adherence to algorithms that lead to predictions that confirm our prior assumptions (Alon-Barkat & Busuioc, 2023).

Big Data Analytics Capability

Whilst AIoT has been evolving, in the manufacturing industry big data handling and decision-making continue to be performed mostly by employees. The uncertainties surrounding the value and risk of AIoT have led to a more conservative focus on improving big data analytics (BDA). For instance, models such as the BDAC (The Big Data Analytic Capability) model (Gupta & George, 2016, Figure 1) have been developed to try and optimise use of big data by specifying the dimensions required for effective BDA in organisations. The BDAC model uses three dimensions to categorise the ability of an organisation to store, process and analyse large volumes of data. It was shown that organisations who are successfully able to utilise big data effectively

Figure 1. Classification of Big Data Resources (Gupta & George, 2016)

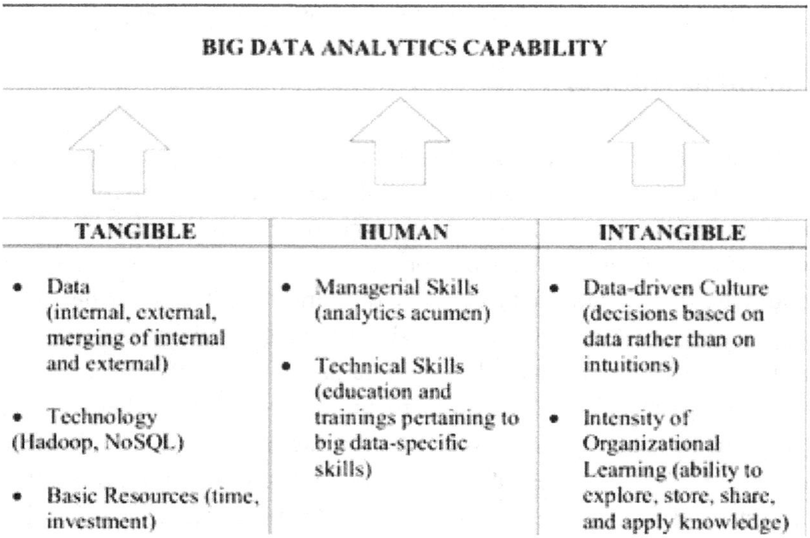

BIG DATA ANALYTICS CAPABILITY

TANGIBLE	HUMAN	INTANGIBLE
• Data (internal, external, merging of internal and external) • Technology (Hadoop, NoSQL) • Basic Resources (time, investment)	• Managerial Skills (analytics acumen) • Technical Skills (education and trainings pertaining to big data-specific skills)	• Data-driven Culture (decisions based on data rather than on intuitions) • Intensity of Organizational Learning (ability to explore, store, share, and apply knowledge)

gain a competitive advantage (Gupta & George, 2016). The BDAC model (Figure 1) conceptualises seven resources required for BDA capability, categorised into three dimensions as Tangible, Human and Intangible. Within the Tangible dimension are three resources; Data, Technology and the Basic Resources. These relate to the IT capability for storing, accessing, and connecting data. Managerial and Technical skills comprise the Human dimension, which reflects the capability of the people in an organisation to extract data and analyse it. Finally, the intangible dimension is based on the Data-driven Culture and Intensity of Organisational Learning.

The BDAC model is a well-established model, having been widely validated as a predictor of company supply chain sustainability (Shokouhyar et al., 2020), and as a predictor of organisational development (Sabharwal & Miah, 2021). More recently, the BDAC model has also been established as being a predictor of green innovation processes and environmental performance (Benzidia et al., 2023).

However, the dimensionality of the BDAC model has been questioned. For instance, organisational learning is posited by Gupta and George as being a dimension of capability, whereas Garmaki et al., (2023) consider organisational learning to be an outcome of BDA, and a mediator between BDA and firm performance. The BDAC model has also been questioned with regard to its direct link to business performance, with the argument that BDAC is more likely to impact on innovation capability during times of environmental uncertainty and turbulence; environmental hostility (Ciacci & Penco, 2023). Whilst BDAC can support decision-making, the likelihood of the decisions materialising into behavioural consequence within the organisation is low due to managers not knowing how to translate decisions into actions (Szukits, 2022). Further, the sufficiency of the BDAC model has been examined with the suggestion that employee perceptions can determine whether BDAC leads to organisational capability. For instance, it is argued that top management team capability can determine the adoption and effectiveness of BDAC (Garmaki et al., 2023).

The suggestion that BDAC is influenced by top management strengthens the 'human' dimension of BDAC; the importance of senior management buy-in to BDA. However, whilst ensuring senior management buy-in undoubtedly increases the likelihood that BDAC will yield an increase in organisational capability, seniority on its own does not ensure impartiality, objectivity and neutrality in big data decision making. The BDAC model has attracted much research in relation to the sufficiency of its dimensions and relationship with business performance, but the importance of unbiased decision making in the human dimension of BDAC has not been fully explored. It is necessary to understand how the use of the insights generated by BDAC may be maximized through articulation with individuals' intellect and other organizational processes involving the use and transformation of knowledge (Lozada et al., 2023). For instance, whilst the BDAC model describes what is required to create a successful environment, it appears that this model does not sufficiently recognise the user interactions with big data. The user is expected to interact with the system in some way to provide input or data, make choices about processing, interpret results, or come to a decision. This case study seeks to address this by explicitly focusing on the role of bias in BDAC.

Cognitive Bias in Big Data Analytics

BDA should ensure rational decision making. However, this is compromised by the often-inevitable interaction between employees and big data. Humans often make less than optional decisions from a rational viewpoint due to cognitive bias. Cognitive bias occurs when, during human decision-making, people learn and develop thinking patterns. Whilst these patterns are sometimes positive and lead to rational decision-making behaviour, other patterns lead to poor or suboptimal choices. There are many different forms of cognitive bias that affect decision making in different ways, but all cause divergence from a perceived rational decision (Haselton et al., 2015).

Even when presented with decision aids such as BDA systems, decision makers depend on personal judgement, past experience, intuition and gut feeling (Taroun & Yang, 2011). Personal judgment and instinct, whilst often feeling more comfortable for the decision maker, introduce bias into the decision-making process; a systemic deviation from rational judgement. For instance, in a manufacturing environment Big Data can be utilised to model capacity utilising measures such as Overall Equipment Effectiveness (OEE) across plants and future customer demand to drive strategic capacity planning and whether additional capital investment is required in additional production lines. A model may produce capacity information showing no investment is required. However, management may, due to pre-existing beliefs or political pressures, decide consciously, or subconsciously to invest regardless of the presented evidence. Therefore, cognitive bias may override the data leading to unnecessary investment.

The adoption of Big Data Analytics should lead to vast data sets providing unbiased evidence and therefore should lead to objective decisions. However, often effective decision making depends on the competence of the decision maker in trusting and handling the data. This introduces risk of inconsistency; a dataset may be interpreted differently in different situations based on stress (Ghasemaghaei & Turel, 2023), prior experience (Kwon et al., 2014), decision-making culture (Frisk & Bannister, 2017), and motivation (Chang et al., 2015) which inherently introduces variability into the decision making quality. For instance, subconsciously decision makers utilise cognitive schemas or heuristics to make quick judgements. Therefore, this suggests for Big Data decision making to maximise value to the organisation, bias in decision making should be mitigated.

This chapter presents a case study of how a global manufacturing organisation was used to assess and augment the BDAC model. A case study approach enabled an in-depth examination of decision making by being able to explore the decision-making processes of a diverse group of managers across one global manufacturing organisation. This enabled a higher degree of confidence that the identification of any additional variables would not be due to multiple different organisational contexts and cultures.

Research Questions

1. How applicable is the Big Data Analytics Capability (BDAC) framework to a manufacturing organisation?
2. Can a holistic framework be developed to incorporate cognitive bias into data analytics capabilities?

Research Objectives

1. Develop semi-structured interview questions, based on the BDAC framework dimensions and cognitive bias
2. Utilising the interview questions developed in objective 1, conduct qualitative interviews to:
   a. Assess the BDAC framework
   b. Explore the roles of bias in decision making

3. Conduct a thematic analysis of interviews and propose improvements to the BDAC model

METHODOLOGY

Philosophy and Approach

Decision making is a subjective and intricate process and therefore to understand decision making it was appropriate to take an interpretivist approach; understanding, exploring, and challenging the perspectives of managers and how context and personal experience can shape a decision. It has been argued that quantitative surveys dominate BDAC research, reflecting a desire to simply validate predictable responses and confirming hypothesis, rather than searching for interesting insights (Huynh et al., 2023). To address this need, a hybrid inductive/deductive approach was used to both validate the BDAC model as well as extending it further by exploring potential themes relating to bias in decision making.

Sampling

The case study organisation is global manufacturing organisation, which supports a range of market segments including the automotive and aerospace industries and has over 3000 employees across 20 locations worldwide. UK-only managers were utilised due to the data protection issues relating to storage of personal data outside of the originating country. This enabled the research to comply with the General Data Protection Regulation. Data was collected from a credible sample of 9 senior management level decision-makers, purposively selected based on seniority and experience in handling big data through working in either a financial or operational role. Whilst there were additional senior managers who could have been interviewed, these would not have been key informants as they did not have experience of directly handing big data with the purpose of decision making. A purposive sampling approach was taken (Patton, 2002) which ensured representation from both UK sites due to site-specific variances in practice, leadership style, and management structures. It is argued that given the nascent nature of BDAC research, interview studies, based on properly selected experts, are much needed (Huynh et al., 2023).

Data Collection

Procedure

Primary qualitative data was collected in this study utilising a mono-method strategy, qualitative semi-structured interviews. Interviews lasted between 30 and 45 minutes. The author carried out a mixture of face-to-face interviews and telephone interviews; it has been shown that phone interviews are as valid as face-to-face interviews (Sturges & Hanrahan, 2004). All interviews were recorded for later transcription to enable the author full engagement with the process to improve interview quality. To build dependability into the interview process, all interviews were carried out in a private meeting room environment to avoid distraction and ensure dependability and engagement (Lincoln & Guba, 1985). The author aimed to gain rapport with the interviewee and put them at ease by maintaining a professional environment and positioning the interviewee as having more knowledge on the subject area than the researcher to ensure points were not missed through assumed knowledge (Leech, 2002).

Question Selection

In relation to objective 1, for the semi structured interviews the author utilised open-ended questioning to gain content for thematic analysis. To investigate each of the 3 dimensions of the BDAC model, questions that had been proposed by Gupta & George (2016) were used as a basis to develop open-ended questions. In addition to this, questions for the investigation of bias were derived from the work of Haselton et al (2005) and Shah & Oppenheimer (2008) respectively. Grand tour questions were utilised to gain insight into typical practice, as they have been shown to be very effective at getting participants talking, in a focussed manner (Leech, 2002). An example of this was *"Could you describe a typical situation where intuition may have aided decision making?"*. In addition, the author used probing and structing questions, or redirect them back to the point of the original question and ensured that the interview stays on topic (Leech, 2002).

ANALYSIS

Primary data from semi-structured interviews was analysed through thematic analysis. Thematic analysis can be defined as the identification, extraction and categorisation of themes from descriptive data by pattern recognition, and offers a hybrid of deductive and inductive analysis (Braun & Clarke, 2006). For instance, a method of analysis was needed that could explore both the a priori themes of the BDAC model as well as allowing for additional themes to be elicited relating to bias and decision making.

Thematic Analysis

During the thematic analysis, the author utilised a six step process (Braun & Clarke, 2006) to carry out the analysis (Table 1).

Step 1 & 2

The interviews were transcribed to enable a thematic analysis of the data, and codes were generated to represent the a priori themes of the BDAC model (Gupta & George, 2016) (Table 2).

Table 1. Six Step Thematic Analysis (Braun & Clarke, 2006)

Step	
1	**Familiarise with data**
2	Generate initial codes
3	Searching for themes
4	Reviewing themes
5	Defining and naming themes
6	Producing the report

Table 2. Initial Template: A Priori Themes From BDAC Model

BDAC Dimension	Code	Theme
Tangible	DAT	Data accessibility
	TCH	Tech capability
	RES	Resources
Human	MS1	Management Skill
	TS1	Technical Skills
Intangible	DDC	Data Driven Culture
	OL	Organisation Learning

Step 3

The transcripts were examined to confirm (or disconfirm) the a priori themes and to identify 4 new themes relating to bias (Table 3).

Step 4, 5, and 6

After further review, themes were merged to combine with other themes which had been identified, to remove overlapping, indistinct themes. Following this process, 2 themes were redefined and renamed as part of the 5th step of the process. Strategy Bias, perceived to be biases which represented pursuing decisions driven to meet Key Performance Indicators, and Measurement Bias were merged into one bias; Surrogation Bias, where surrogation is defined as the measurement of a KPI replaces the KPI itself (Choi et al., 2013). For example, in manufacturing, this is when productivity performance metrics become the focus of attention rather than understanding what is truly driving productivity or questioning the accuracy and appropriateness of the metrics. After completing these two steps of analysis, Table 4 was produced.

The final template, and the implications for the BDAC model are discussed in the next section.

Table 3. Revised Template: A Priori Themes Plus Additional Themes

BDAC Dimension	Code	Theme
Tangible	DAT	Data accessibility
	TCH	Tech capability
	RES	Resources
Human	MSK	Management Skill
	TSK	Technical Skills
Intangible	DDC	Data Driven Culture
	OL	Organisation Learning
Bias	BCO	Bias - Confirmation
	BST	Bias - Strategy
	BME	Bias- Measurement
	BCH	Bias- Choice overload

Table 4. Final Template

BDAC Dimension	Code	Theme
Tangible	DAT	Data accessibility
	TCH	Tech capability
	RES	Resources
Human	MSK	Management Skill
	TSK	Technical Skills
Intangible	DDC	Data Driven Culture
	OL	Organisation Learning
Bias	BCO	Bias - Confirmation
	BSU	Bias - Surrogation
	BCH	Bias- Choice overload

DISCUSSION

Big Data Analytics Capability Model Findings

In relation to the BDAC framework (objective 2a), all 7 resources across the 3 dimensions were supported.

Tangible

Within the tangible dimension, participants indicated that they felt that data was readily available to them, *"We are very blessed with lots of the data being provided on a plate."* (Participant 1). While not involved in the technical provision of the data the participants showed some knowledge of the database technology underpinning the Big Data, *"A lot of the reports are automated either from like SQL databases"* (Participant 5) and that there are organisational resources devoted to the provision of data *"There's the professional team who do it, the business intelligence guys. Also, some people in operations who do it."* (Participant 5). Due to the nature of the organisation, this cohort of decision makers are separate to the teams and technologies involved in the provision of analytical data. Therefore, it is understandable why informants showed an appreciation of the data but a lack of detailed knowledge around the technological and basic resources involved.

Human

Looking at the human dimension participants demonstrated during the interviews that the data which was presented was done so in a meaningful way to their areas. For example, *"I must also say that the reports which are available are quite good. The people who work on that reporting side, there is a continuous drive to improve the reporting."* (Participant 9). This highlights that within the organisation, the managerial skill of the BDA teams is good, providing useful intelligence to the organisation. Therefore, the tangible BDAC dimension has been seen in this organisation. However, while participants did comment around technical skills more negatively, *"I think we haven't all got the skills to work with it properly"* (Participant 8). Whilst this is a negative reflection on this resource within the BDAC model for this organisation, because it was highlighted within the interviews it does demonstrate that decision makers are aware of its relevance to Big Data Analytics Capability. Therefore, whilst it could be argued that this organisation requires better technical skills, it provides validation for this dimension within the BDAC model.

Intangible

Considering the intangible dimension, it was clear that the use of big data was ingrained in the business, *"Part of the organisation culture, everything has to be evidenced and justified"* (Participant 2). This showed a strong link to the Data Driven Culture resource. It also emerged that Organisational Learning was seen as critical to the informants. There seemed to be a general feeling among participants that there was room for improvement in this critical area for the organisation. *"Not so much for our facility, it's really about getting the other facilities closer to this one. From a support point of view this could be a real challenge with resourcing, but it would be very valuable to the organisation."* (Participant 1).

Overall, the dimensions, represented as a priori themes were supported. However, as anticipated in the literature review, the impact of cognitive bias on BDA was also observed.

Cognitive Bias Themes

In relation to objective 2b, outside the framework of the BDAC model, there were three themes of bias identified, relating to the irrational selection (confirmation bias), avoidance (choice overload bias) and focus (surrogation bias) of big data during the decision-making process which subsequently limits the propensity for big data analytics to contribute to organisational success. After questions had been asked in relation to the BDAC dimensions, a specific question was asked to explore the general role that bias may play in BDA. Specific biases were not asked about to encourage spontaneous and unpromoted identification of specific biases (Kvale, 1996).

Confirmation Bias

Confirmation bias is one of the more widely recognised forms of cognitive bias., which was highly prevalent amongst the informants. Confirmation Bias can be defined as a conscious, or unconscious, effort to ignore data contradicting the decision maker's views (Gatlin et al., 2017). Informants provided examples such as, *"I do think it happens quite often that someone goes in with a notion, and then let's say it is like this, and then when they're analysing the data...they're going to confirm it"* (Participant 3). Informants appeared to link assumptions with experience to suggest that this improved decision making; *"with experience you can make your decisions faster and intuitively, being confident you are right"* (Participant 2). This highlights a lack of awareness of the potential negative impacts of bias based on prior experience.

There also appeared to be a situational effect of time pressure and operational crisis points that resulted in a propensity for confirmation bias; *"in normal day-to-day stuff, data has more of an impact than the non-logical, gut-feel type thing. When it gets to more of a crisis situation, that's when what the data is saying tends to take a little bit of a backseat, and you go with more of your intuition to get yourself out of this."* (Participant 7). Relating to time pressure; *"There's a lot of pressure to make sure that you have got a true record and you have got really corrective actions... think sometimes the pressure of that leads you maybe to not be taking the time to really understand what the issues are. You do kind of take a judgment based on what your gut's telling you. You might just let the data drive you just to get through, "I've got a deadline, let's just get it done," kind of thing"* (Participant 5). Another pressure that may cause confirming evidence to be sought is not wanting to deliver information that may cause negative affect; *"People don't like reporting out bad numbers, do they?"* (Participant 7).

Impression management was also found to create a need to look for confirming evidence, to help show alignment and affinity with senior management views; *"If they know that someone more senior than them, often maybe two or three levels up from them, is looking at that report, they're going to be more tuned to that data, and if someone very senior has a pet interest in something, then that's going to have a disproportionate amount of focus and effort to keep that one in check. Even if in the grand scheme of things, it's a small cog in a relatively large system"* (Participant 6).

There is often an interplay of hierarchy, which can redirect cognitive efforts; *"if that customer has called the managing director, and the managing director feels that we aren't delivering, then it doesn't matter if it's number five by value. It's going* to be number one on our to-do list. That's the way it goes sometimes" (Participant 6). Therefore, as leadership creates an environment where it is not safe to report 'bad' numbers, or where decision makers cannot manage upward to ensure focus is given where there is greatest value, impression management causes action to be taken where data would not drive activity. Where the bias is created due to an area of senior management focus, this creates 'cognitive tunnelling'; an evidentially unjustified level of focus on an area linked to negative effects on judgement and reduction in organisational success (Posavac et al., 2010).

Choice Overload Bias

Choice overload bias was also prevalent in the thematic analysis. Choice overload bias is when the presence of too much information results in attentional efforts being minimised to control cognitive capacity. For instance, the over-provision of data in reports doing similar things created confusion and data avoidance; *"These two reports that should tell me the same thing tell me different things. Which one is right?"* (Participant 6). By not knowing which data should be trusted, the informants felt that they were not able to utilise data in decision making. This could mean that data is disregarded altogether, or potentially incorrect data from an old report model is utilised rather than a corrected version, *"as the needs morph and new fields added in, sometimes you start and distrust some of the automated reports that come out."* (Participant 6). Decision makers commented that *"too much information can be overwhelming"* (Participant 2), therefore, too much data negatively affects decision making. Research has indicated that too much data can lead to cognitive and information overload (Merendino et al., 2018). Therefore, this results in moderation of organisational value because decision makers cannot process the information provided to them. Participants made several references to receiving too much information, or as with the earlier example being confused by the similar data sets providing different reports.

Paradoxically, it seems that this is driven by a perception of the decision maker's demand for more data *"I think people ask for more and more data but I'm not sure it is adding"* (Participant 3), *"it's too easy nowadays to just pull data"* (Participant 7). This over-availability of data leads to the decision makers working with different reporting tools giving similar data *"For us specifically, there are too many reports, doing very similar things."* (Participant 7). This appears to lead to a feeling that data cannot be understood properly and is ultimately not utilised in decision making, *"There are lots of reports, lots of reporting, lots of very useful data which is not necessarily used to the full extent it could be used because people are driven a lot by day-to-day actions of things they have to do quickly."* (Participant 9), "at *this organisation there's an awful lot of data, and so half the skill is cutting through what data is useful and what data might mislead" (Participant 8).* As described by Participant 6; *"the more data that you have, it can be hard to come to a consensus. Obviously, the more data you add, it's like the law of diminishing returns",* and participant 5, *"So then they can ask for a scheduled report that doesn't add value and it goes around in a bit of circle, to the point you get data overload, and stuff gets just lost. At which point, you start again, and it carries on and on and on".*

Overall, while the informants in the study acknowledged that there was too much data, it was also found that there was still a demand to gain more and more data. It is clear in this organisation that informants feel that they had ready access to data for their own analysis which is likely to amplify this effect. It was also found that too much data also led to trust issues with the data because this created confusing situations where decision makers were presented with conflicting information. This carries the risk of discarding potentially useful data and limiting decision making quality. It has been shown that in situations of high cognitive load, people exhibit lower levels of trust (Samson & Kostyszyn, 2015). Therefore, there is a compounding effect of cognitive overload and data trust issues. It has also been argued that the stress associated with big data can lead to 'premature closure'; making decisions before all alternatives have been considered, as well as 'temporal narrowing'; not allowing enough time to process data fully (Ghasemaghaei & Turel, 2023).

Surrogation Bias

Surrogation bias was observed in the thematic analysis and can be defined as the replacement of a measured construct with the measure itself (Choi et al., 2013). In a manufacturing setting this could be productivity metrics; managers become focussed on improving, or maintaining the metric number, forgetting that this is not actual productivity. Surrogation is when the use of the measure becomes the goal, rather than the strategic construct. In other words, the drive to chase a metric goal rather than drive the correct behaviour of the strategic concept behind the metric. For example, *"Metric chasing, for example. So maybe that then, you end up not doing the right thing because you want to make a metric look better, or you need to make a metric look better because that's what you've been challenged to do."* (Participant 6).

The perception seems to be that with too much focus on data and results within the organisation drives the wrong behaviour, *"Sometimes you can see it very clearly in the metrics that we run. Some business metrics drive the wrong behaviour. People can fudge the process or manipulate the data input to. With too much focus on the data people can work to fudge and improve their results. I've seen that happen a lot of times over the years."* (Participant 2). As a counterpoint, one informant explained this as a positive, *"I think it makes us very result-focused, which, personally, I like.*

It gives everyone a direction. It drives us to look at certain areas in more detail, so we can start making the right decisions, but then on the opposite side of that, it can also have the reverse effect that we discussed" (Participant 7). However, while the argument has some merit, the drive is to focus on key performance indicator which may be at risk of driving the wrong behaviour. As Participant 2 suggests, *"people are going through the motions every day on score cards or stand-ups like a morning meeting. You ramble off numbers and you look at your capacity and everything else, but no one is really doing anything to make it better."*

The very presence of a measure creates this behaviour (Black et al., 2022) and it has been shown that surrogation increases where there is a high need for consistency (Bentley, 2018). Surrogation bias negatively affects decision making because managers make measure-driven decisions rather than strategy-driven decisions. Therefore, outcomes are not always maximising improvements in organisational value. This is similar to the 'streetlight effect', whereby attention is directed phenomena where there is a plethora of data and metrics, rather than directing attention towards the problems, which may not have sufficient and appropriate data available to enable analysis (Elragal & Klischewski, 2017).

Augmented BDAC Model

After addressing objectives 1 and 2, objective 3 was to utilise the findings from the thematic analysis to improve existing models, in line with the second research question to create a holistic framework. This research set out to understand the utilisation of big data in a manufacturing organisation and to assess the efficacy of the Big Data Analytics Capability (BDAC) model. This case study has validated the BDAC model in a global manufacturing environment. However, to account for the role of bias within BDA, it is suggested that the intangible dimension be split further, to add an eighth dimension of Objective Decision-making. Data-driven culture should specifically refer to the motivation to engage with data, and the perceived value data has. However, objective decision making as a separate sub-dimension could accommodate the willingness and capability to remain objective through reducing confirmation bias, choice overload bias, and surrogation bias, by ensuring full inclusion of relevant data, sufficient capacity to process data, and keeping business goals and objectives in mind (as opposed to the metrics themselves) when analysing data.

Additional Construct Items

Whilst confirmation bias is acknowledged within the BDAC model, specifically the conceptualisation of data-driven culture through the item of DD3, "we are willing

Figure 2. The Augmented Big Data Analytics Capability Model
*Recommended 8th sub-dimension

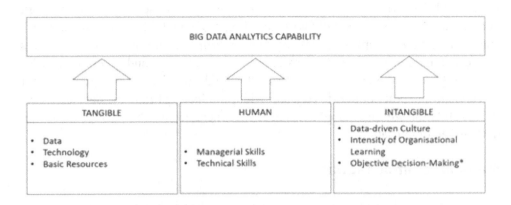

Table 5. Additional Construct Items to Measure Objective Decision-Making

Construct	Item
Objective decision making	ODM1* We are willing to override our own intuition when data contradict our viewpoints ODM2 We consider all data available and are prepared to acknowledge and reconcile any ambiguities ODM3 We continuously question the relevance of metrics and how accurately they indicate true business performance

*Taken from original BDAC model (DD3)

to override our own intuition when data contradict our viewpoints", based on this case study, it is suggested that DD3 be removed from the Data-driven Culture sub-dimension, and moved to a new construct 'Objective Decision-Making', along with 2 items as shown below (Table 5).

ODM2 measures choice overload bias, and ODM3 measures surrogation bias.

LIMITATIONS AND FUTURE DIRECTIONS

A limitation of this study was that only one cohort, decision makers, was investigated. This limitation could be overcome by comparing results to cohorts of executive leadership and the teams involved in the creation of Big Data Analytics. Also, this research was conducted within a single manufacturing organisation with 9 participants. While this was a saturated sample size allowing conclusions to be drawn from the work there are some limitations. The largest limitation of this study is that while a new model is proposed linking the BDAC model with decision making and bias, this has not been validated in the data collected. Therefore, a follow-on study from this work should be carried out. The new research would utilise the proposed framework and validate the new construct and items (Figure 2 and Table 5) in a quantitative study with a large sample size across manufacturing organisations. Most scientists are accustomed to make predictions based on consolidated and accepted theories pertaining to the domain of prediction. This study may serve as a guideline for researchers and practitioners to consider while conducting future big data analytics. After validation within a manufacturing environment, it could be broadened to other data driven organisations.

Managerial Implications

Throughout the data collection and analysis of data it has been clear that this organisation is committed to a data driven culture, using Big Data Analytics in all aspects of operations. However, several recommendations for improvements to organisational practice with a view to improving organisational value and driving continued benefit from the technology investments.

Mitigating Bias

This case study firmly establishes the role that cognitive bias plays when processing and handling big data and making decisions. It has been demonstrated that biases in decision making can negatively affect decision making quality even when senior managers, who are invested in making optimal decisions, are the decision makers. Whilst there are many variations of bias at work in the decision-making process, there are several actions which can be taken to improve the situation. Firstly, by carrying out general bias training with all decision makers the organisation will improve the situation. By simply raising awareness, it will naturally help employees challenge themselves around potential biases. Taking this further, by fostering a culture of bias free decision making, instilling a process of challenging the status quo and playing 'devil's advocate' this will moderate the impact of bias.

Data Accessibility

Through this research it became clear that decision makers were in situations where decisions were either stalling due to too much data being available or leading to data being discarded in decision making. This was identified as cognitive overload. In part, this seemed due to the availability of data for decision makers to interrogate independently. Therefore, to reduce the impact of this, accessibility of raw data needs to be carefully considered by data owners. In addition, this will improve organisational data protection and reduce organisational risk. Doing this will ensure that decision makers must utilise centrally managed Big Data Analytics resources. To help decision makers transition to the new model this will require that the data presented is in a format which is useful to decision makers, reducing the need for manual investigation. For example, participants described that summary data presented in the centrally created reports would need further investigation. By redesigning reports to enable this in a focussed fashion would support this business need. To support this, decision makers will require more training throughout the organisation to ensure that they understand the process of change and how to maximise their use of the new process. In addition, there would need to be a cultural shift away from decision makers driving data investigation. For example, where management require further investigation into a particular trend decision makers would need to partner with data analysts to build a requirement and provide the data. This may be a slower process but the investment in time would enable one-off requirements to be discussed centrally and used to augment existing reporting, ultimately reducing future needs.

To support this effort to reduce cognitive overload business analyst resources would need to be increased within the organisation. Their role would be two-fold, firstly to create and optimise analytics for the organisation, secondly, they would drive organisational learning to improve sharing of understanding across the organisation. In addition, to this the resources would be expected to curate existing reports and improve organisational trust in data.

Adoption of AI

According to Deloitte, smart production is the main area of AI deployment in manufacturing; factory automation, order management, and automated scheduling whilst over the next two years, a AI technologies will be increasingly utilised in quality monitoring and defect management, and applications that shorten design time, customize customer experiences, and enhance marketing efficiency (Deloitte, 2020). AI based production monitoring, providing operational insight on manufacturing execution, quality control and preventative maintenance have all been identified as measures to enhance productivity and provide competitive advantage to organisations. In addition to driving industry 4.0, AI technology is increasingly utilised to enhance IT security particularly to enhance Intrusion Detection Systems. By utilising machine learning to detect potential attacks, organisations reduce their cyber security risk to organisational data and productivity.

However, the adoption of AI can help manufacturing organisations reduce the risk of cognitive biases and introduce the necessary objectivity to ensure optimal use of big data. For instance, AI automatically removes the need to manually extract and consult multiple data sources, which is the root cause of choice overload. Regarding mitigating the risk of confirmation bias, AI would systematically include all relevant data sources, without risk of selectively excluding data sources or misinterpreting data due to self-serving motivations.

Finally, regarding surrogation bias, the very fact that AI goes beyond just reporting metrics, AI moves the spotlight away from the metric, towards actions and decisions. Whilst AI inherently contains human subjectivity in the way in which algorithms are designed to select and process attributes (Favaretto et al., 2019), it is well-established that AI can be used to make decisions that are on par or better than decisions made by humans (Araujo et al., 2020). There has been an increase in the interest in researching AI, i.e. automation or augmentation with the view that AI should be used for augmenting decision making (Duan et al., 2019). Taking the view that big data can introduce risk of bias coming from choice overload, it is argued that employees should be able to use analytic tools to process big data to reduce stress and protect against sub-optimal decision making (Ghasemaghaei & Turel, 2023).

However, the manufacturing industry has been relatively slow to adopt artificial intelligence, mainly due to manufacturing data being localised and too specific to a company's operations which means there is often insufficient data available for building reliable AI models (Stackpole, 2023). AI is seen as a key enabling technology for Industry 4.0, but it is important to temper this with proper expectation. On its own it will not compensate for flaws in data or solve all business challenges.

Whilst AI is meant to help us overcome our biases, automated systems might give rise to over-reliance on AI (Buçinca et al., 2021), and new and distinct biases arising from human processing of automated outputs. Similar to confirmation bias, 'automation bias' refers to undue deference to automated systems that disregard contradictory information from other sources (Alon-Barkat & Busuioc, 2023). However, providing that the business intent for AI use is clearly mapped, that there is a normal distribution of samples used for data training, and that testing should be of a full range of use cases, the effect of bias in algorithmic outcomes can be reduced, enabling AI to mitigate human bias and introduce more objectivity into decision making.

CONCLUSION

This research was undertaken to explore two research questions, firstly to explore the Big Data Analytics Capability (BDAC) model in the context of a manufacturing organisation and secondly to create a holistic framework integrating the original model with decision making theory and bias. In the literature review it was found that the BDAC model was an indicator of an organisations ability to utilise Big Data Analytics. However, applications of the BDAC model had not explored the biases of decision making. Decision making theory shows that human decision making is more subjective and affected by bias. Therefore, this research has enabled the production of an augmented Big Data Analytics Capability Model to explain the findings and pave the way for future research.

REFERENCES

Alon-Barkat, S., & Busuioc, M. (2023). Human–AI Interactions in Public Sector Decision Making: "Automation Bias" and "Selective Adherence" to Algorithmic Advice. *Journal of Public Administration: Research and Theory*, *33*(1), 153–169. doi:10.1093/jopart/muac007

Araujo, T., Helberger, N., Kruikemeier, S., & de Vreese, C. H. (2020). In AI we trust? Perceptions about automated decision-making by artificial intelligence. *AI & Society*, *35*(3), 611–623. doi:10.1007/s00146-019-00931-w

Arinez, J. F., Chang, Q., Gao, R. X., Xu, C., & Zhang, J. (2020). Artificial Intelligence in Advanced Manufacturing: Current Status and Future Outlook. *Journal of Manufacturing Science and Engineering*, *142*(11), 110804. Advance online publication. doi:10.1115/1.4047855

Bentley, J. W. (2018). *Decreasing Operational Distortion and Surrogation through Narrative Reporting* (*SSRN* Scholarly Paper 2924726). doi:10.2139/ssrn.2924726

Benzidia, S., Bentahar, O., Husson, J., & Makaoui, N. (2023). Big data analytics capability in healthcare operations and supply chain management: The role of green process innovation. *Annals of Operations Research*. Advance online publication. doi:10.1007/s10479-022-05157-6 PMID:36687515

Black, P. W., Meservy, T. O., Tayler, W. B., & Williams, J. O. (2022). Surrogation Fundamentals: Measurement and Cognition. *Journal of Management Accounting Research*, *34*(1), 9–29. doi:10.2308/JMAR-2020-071

Braun, V., & Clarke, V. (2006). Using thematic analysis in psychology. *Qualitative Research in Psychology*, *3*(2), 77–101. doi:10.1191/1478088706qp063oa

Buçinca, Z., Malaya, M. B., & Gajos, K. Z. (2021). To Trust or to Think: Cognitive Forcing Functions Can Reduce Overreliance on AI in AI-assisted Decision-making. *Proceedings of the ACM on Human-Computer Interaction*, *5*(CSCW1), 188:1-188:21. 10.1145/3449287

Chang, Y.-W., Hsu, P.-Y., & Wu, Z.-Y. (2015). Exploring managers' intention to use business intelligence: The role of motivations. *Behaviour & Information Technology, 34*(3), 273–285. doi:10.1080/0144929X.2014.968208

Choi, J., Hecht, G. W., & Tayler, W. B. (2013). Strategy Selection, Surrogation, and Strategic Performance Measurement Systems. *Journal of Accounting Research, 51*(1), 105–133. doi:10.1111/j.1475-679X.2012.00465.x

Ciacci, A., & Penco, L. (2023). Business model innovation: Harnessing big data analytics and digital transformation in hostile environments. *Journal of Small Business and Enterprise Development.* doi:10.1108/JSBED-10-2022-0424

Cui, Y., Kara, S., & Chan, K. C. (2020). Manufacturing big data ecosystem: A systematic literature review. *Robotics and Computer-integrated Manufacturing, 62,* 101861. doi:10.1016/j.rcim.2019.101861

Davenport, T. H. (2018, January 30). Artificial Intelligence for the Real World. *Harvard Business Review.* https://hbr.org/webinar/2018/02/artificial-intelligence-for-the-real-world

Deloitte. (2020). *Deloitte Survey on AI Adoption in Manufacturing.* Deloitte China. https://www2.deloitte.com/cn/en/pages/consumer-industrial-products/articles/ai-manufacturing-application-survey.html

Duan, Y., Edwards, J. S., & Dwivedi, Y. K. (2019). Artificial intelligence for decision making in the era of Big Data – evolution, challenges and research agenda. *International Journal of Information Management, 48,* 63–71. doi:10.1016/j.ijinfomgt.2019.01.021

El Khatib, M., & Al Falasi, A. (2021). Effects of Artificial Intelligence on Decision Making in Project Management. *American Journal of Industrial and Business Management, 11*(03), 251–260. doi:10.4236/ajibm.2021.113016

Elragal, A., & Klischewski, R. (2017). Theory-driven or process-driven prediction? Epistemological challenges of big data analytics. *Journal of Big Data, 4*(1), 19. doi:10.1186/s40537-017-0079-2

Favaretto, M., De Clercq, E., & Elger, B. S. (2019). Big Data and discrimination: Perils, promises and solutions. A systematic review. *Journal of Big Data, 6*(1), 12. doi:10.1186/s40537-019-0177-4

Frisk, J. E., & Bannister, F. (2017). Improving the use of analytics and big data by changing the decision-making culture: A design approach. *Management Decision, 55*(10), 2074–2088. doi:10.1108/MD-07-2016-0460

Garmaki, M., Gharib, R. K., & Boughzala, I. (2023). Big data analytics capability and contribution to firm performance: The mediating effect of organizational learning on firm performance. *Journal of Enterprise Information Management*, *36*(5), 1161–1184. doi:10.1108/JEIM-06-2021-0247

Gatlin, K., Hallock, D., & Cooley, L. (2017). Confirmation Bias among Business Students: The Impact on Decision-Making. *Review of Contemporary Business Research*, *6*(1). Advance online publication. doi:10.15640/rcbr.v6n2a2

Ghasemaghaei, M., & Turel, O. (2023). The Duality of Big Data in Explaining Decision-Making Quality. *Journal of Computer Information Systems*, *63*(5), 1093–1111. doi:10.1080/08874417.2022.2125103

Gunasekaran, A., Yusuf, Y. Y., Adeleye, E. O., Papadopoulos, T., Kovvuri, D., & Geyi, D. G. (2019). Agile manufacturing: An evolutionary review of practices. *International Journal of Production Research*, *57*(15–16), 5154–5174. doi:10.108 0/00207543.2018.1530478

Gupta, M., & George, J. F. (2016). Toward the development of a big data analytics capability. *Information & Management*, *53*(8), 1049–1064. doi:10.1016/j. im.2016.07.004

Haselton, M. G., Nettle, D., & Andrews, P. W. (2015). The Evolution of Cognitive Bias. In *The Handbook of Evolutionary Psychology* (pp. 724–746). John Wiley & Sons, Ltd. doi:10.1002/9780470939376.ch25

Huynh, M.-T., Nippa, M., & Aichner, T. (2023). Big data analytics capabilities: Patchwork or progress? A systematic review of the status quo and implications for future research. *Technological Forecasting and Social Change*, *197*, 122884. doi:10.1016/j.techfore.2023.122884

Jeble, S., Dubey, R., Childe, S. J., Papadopoulos, T., Roubaud, D., & Prakash, A. (2018). Impact of big data and predictive analytics capability on supply chain sustainability. *International Journal of Logistics Management*, *29*(2), 513–538. doi:10.1108/IJLM-05-2017-0134

Kim, S. W., Kong, J. H., Lee, S. W., & Lee, S. (2022). Recent Advances of Artificial Intelligence in Manufacturing Industrial Sectors: A Review. *International Journal of Precision Engineering and Manufacturing*, *23*(1), 111–129. doi:10.1007/s12541-021-00600-3

Kvale, S. (1996). *InterViews: An Introduction to Qualitative Research Interviewing*. SAGE Publications.

Kwon, O., Lee, N., & Shin, B. (2014). Data quality management, data usage experience and acquisition intention of big data analytics. *International Journal of Information Management*, *34*(3), 387–394. doi:10.1016/j.ijinfomgt.2014.02.002

Leech, B. L. (2002). Asking Questions: Techniques for Semistructured Interviews. *PS, Political Science & Politics*, *35*(4), 665–668. doi:10.1017/S1049096502001129

Li, B., Hou, B., Yu, W., Lu, X., & Yang, C. (2017). Applications of artificial intelligence in intelligent manufacturing: A review. *Frontiers of Information Technology & Electronic Engineering*, *18*(1), 86–96. doi:10.1631/FITEE.1601885

Lozada, N., Arias-Pérez, J., & Henao-García, E. A. (2023). Unveiling the effects of big data analytics capability on innovation capability through absorptive capacity: Why more and better insights matter. *Journal of Enterprise Information Management*, *36*(2), 680–701. doi:10.1108/JEIM-02-2021-0092

Merendino, A., Dibb, S., Meadows, M., Quinn, L., Wilson, D., Simkin, L., & Canhoto, A. (2018). Big data, big decisions: The impact of big data on board level decision-making. *Journal of Business Research*, *93*, 67–78. doi:10.1016/j.jbusres.2018.08.029

Mikalef, P., Pappas, I. O., Krogstie, J., & Giannakos, M. (2018). Big data analytics capabilities: A systematic literature review and research agenda. *Information Systems and e-Business Management*, *16*(3), 547–578. doi:10.1007/s10257-017-0362-y

Milaninia, N. (2021, March 1). *Biases in machine learning models and big data analytics: The international criminal and humanitarian law implications.* International Review of the Red Cross. https://international-review.icrc.org/articles/biases-machine-learning-big-data-analytics-ihl-implications-913

Patton, M. Q. (2002). Qualitative Research & Evaluation Methods. *Sage (Atlanta, Ga.).*

Phillips-Wren, G., Power, D. J., & Mora, M. (2019). Cognitive bias, decision styles, and risk attitudes in decision making and DSS. *Journal of Decision Systems*, *28*(2), 63–66. doi:10.1080/12460125.2019.1646509

Plathottam, S. J., Rzonca, A., Lakhnori, R., & Iloeje, C. O. (2023). A review of artificial intelligence applications in manufacturing operations. *Journal of Advanced Manufacturing and Processing*, *5*(3), e10159. doi:10.1002/amp2.10159

Posavac, S. S., Kardes, F. R., & Joško Brakus, J. (2010). Focus induced tunnel vision in managerial judgment and decision making: The peril and the antidote. *Organizational Behavior and Human Decision Processes*, *113*(2), 102–111. doi:10.1016/j.obhdp.2010.07.002

Ransbotham, S., Candelon, F., Kiron, D., LaFountain, B., & Khodabandeh, S. (2021). The Cultural Benefits of Artificial Intelligence in the Enterprise. *MIT Sloan Management Review*. https://sloanreview.mit.edu/projects/the-cultural-benefits-of-artificial-intelligence-in-the-enterprise/

Sabharwal, R., & Miah, S. J. (2021). A new theoretical understanding of big data analytics capabilities in organizations: A thematic analysis. *Journal of Big Data*, *8*(1), 159. doi:10.1186/s40537-021-00543-6

Samson, K., & Kostyszyn, P. (2015). Effects of Cognitive Load on Trusting Behavior – An Experiment Using the Trust Game. *PLoS One*, *10*(5), e0127680. doi:10.1371/journal.pone.0127680 PMID:26010489

Sheng, J., Amankwah-Amoah, J., & Wang, X. (2017). A multidisciplinary perspective of big data in management research. *International Journal of Production Economics*, *191*, 97–112. doi:10.1016/j.ijpe.2017.06.006

Shokouhyar, S., Seddigh, M. R., & Panahifar, F. (2020). Impact of big data analytics capabilities on supply chain sustainability: A case study of Iran. *World Journal of Science. Technology and Sustainable Development*, *17*(1), 33–57. doi:10.1108/WJSTSD-06-2019-0031

Stackpole, B. (2023, October 25). *For AI in manufacturing, start with data*. https://mitsloan.mit.edu/ideas-made-to-matter/ai-manufacturing-start-data

Sturges, J. E., & Hanrahan, K. J. (2004). Comparing Telephone and Face-to-Face Qualitative Interviewing: A Research Note. *Qualitative Research*, *4*(1), 107–118. doi:10.1177/1468794104041110

Szukits, Á. (2022). The illusion of data-driven decision making – The mediating effect of digital orientation and controllers' added value in explaining organizational implications of advanced analytics. *Journal of Management Control*, *33*(3), 403–446. doi:10.1007/s00187-022-00343-w

Tao, F., Qi, Q., Liu, A., & Kusiak, A. (2018). Data-driven smart manufacturing. *Journal of Manufacturing Systems*, *48*, 157–169. doi:10.1016/j.jmsy.2018.01.006

Taroun, A., & Yang, J.-B. (2011). Dempster-Shafer Theory of Evidence: Potential usage for decision making and risk analysis in construction project management. *The Built & Human Environment Review, 4*.

Wang, J., Xu, C., Zhang, J., & Zhong, R. (2022). Big data analytics for intelligent manufacturing systems: A review. *Journal of Manufacturing Systems*, *62*, 738–752. doi:10.1016/j.jmsy.2021.03.005

Chapter 2
AIoT Concepts and Integration:
Exploring Customer Interaction, Ethics, Policy, and Privacy

Ranjit Singha
ⓘD https://orcid.org/0000-0002-3541-8752
Christ University, India

Surjit Singha
ⓘD https://orcid.org/0000-0002-5730-8677
Kristu Jayanti College (Autonomous), India

ABSTRACT

Integrating AIoT technologies provide businesses with increased productivity, cost savings, data-driven insights, and enhanced consumer interactions. Nevertheless, difficulties include data privacy, ethics, regulatory compliance, and technical complexities. The recommendations include transparent practices, accountability, bias mitigation, data minimization, informed consent, and ethical design. Policymakers must develop adaptable regulations, place a premium on privacy and security, and involve stakeholders. A user-centric approach and training in data ethics are essential. AIoT offers enormous potential but requires a delicate balance between innovation and responsibility, with ethics, privacy, and policy compliance at the forefront.

DOI: 10.4018/979-8-3693-0993-3.ch002

1. INTRODUCTION

The concept of "Artificial Intelligence of Things" (AIoT) refers to the integration of Artificial Intelligence (AI) technology with the Internet of Things (IoT) ecosystem. The concept of AIoT entails the integration of artificial intelligence (AI) technologies, such as machine learning, natural language processing, and data analytics, with Internet of Things (IoT) devices and networks. The collaboration between different components enables Internet of Things (IoT) devices to independently gather, handle, evaluate data, make informed choices, and establish communication with other devices and centralized systems without human involvement. AIoT involves converting traditional IoT devices into intelligent entities that can autonomously learn and improve their functionality over time (Yan et al., 2023; Ogundokun et al., 2021). The integration of Artificial Intelligence of Things (AIoT) carries significant implications for facilitating organizational transformations, rendering it a crucial technology within the rapidly changing business environment of the present era. Integrating Artificial Intelligence of Things (AIoT) in various industries has proven advantageous in operational optimization. This is attained through the automation of jobs, accurate prediction of maintenance requirements, and improved allocation of resources. As a result, the overall efficiency of operations is boosted, leading to reduced downtime and a gain in productivity (Matin et al., 2023). In addition, AIoT systems can produce significant data that can provide vital insights for making well-informed judgments. It empowers enterprises to base their decisions on data, thus enhancing product creation, process optimization, and resource allocation.

Using AIoT confers a competitive advantage by providing enhanced client experiences, expedited response times, and the introduction of novel services. According to Quy et al. (2023), this advantage can stimulate organizational growth and establish market leadership. AIoT systems possess high scalability, rendering them well-suited for entities of varying magnitudes, encompassing both modestly sized entrepreneurial ventures and expansive corporate enterprises. Additionally, integrating Artificial Intelligence of Things (AIoT) can decrease operational expenses by implementing predictive maintenance strategies, enhancing energy efficiency, and minimizing waste generation. These measures can lead to substantial financial advantages in the long run (Bartoň et al., 2022). According to Allioui and Mourdi (2023), it assumes a crucial function in fostering innovation and facilitating digital transformation, enabling firms to embrace state-of-the-art technology, maintain flexibility, and adapt to the evolving demands of the market. According to Huang and Rust (2020), using personalization, predictive services, and responsiveness leads to increased customer satisfaction and loyalty. In the contemporary dynamic business landscape, integrating Artificial Intelligence of Things (AIoT) empowers businesses to swiftly respond to market fluctuations and emerging trends, revolutionizing organizational transitions across diverse industries. A comprehensive comprehension of the potential of AIoT and its consequential relevance is vital for firms aiming to maintain competitiveness and adaptability in the contemporary business environment (Pelet et al., 2021).

AIoT, also known as the Artificial Intelligence of Things, is characterized by its dynamic and transformative nature, as it integrates the capabilities of Artificial Intelligence (AI) with the Internet of Things (IoT). Integrating data-driven intelligence with physical devices and systems leads to a harmonious integration. This overview will examine some fundamental principles and integration features of the AIoT. The process involves incorporating sensors into Internet of Things (IoT) devices. The sensors collect data about many aspects of the physical environment, such as temperature, humidity, and geographic location. This data is the basis for developing and implementing applications within Artificial Intelligence of Things (AIoT) (Zhang et al., 2022; Qian et al., 2021). After collecting data, AIoT utilizes sophisticated data analytics methods, including machine learning and deep learning, to derive essential insights from the acquired data. These insights can be utilized to make forecasts, discern patterns, and enhance the efficiency of processes. AIoT systems can acquire knowledge and adjust their behaviour based on the data they gather. Using machine learning algorithms facilitates the ability of devices to make educated decisions and enhance their functioning iteratively as time progresses (Wang et al., 2023; Yang et al., 2021). As an illustration, a thermostat can acquire user preferences and modify temperature settings following those choices. One of the fundamental principles in Artificial Intelligence of Things (AIoT) is the capacity of interconnected objects to generate judgments autonomously.

An example of an industrial robot outfitted with AIoT capabilities can be given, enabling it to autonomously adapt its operations by utilizing real-time data. This capability eliminates human involvement and enhances efficiency and safety. Integrating Artificial Intelligence of Things (AIoT) has significantly improved customer interactions by providing tailored and proactive services (Huang & Rust, 2020b). One illustration of this phenomenon is the ability of smart home gadgets to adjust their functionality following the preferences of individual users. Similarly, e-commerce platforms can recommend things to users by analyzing their previous purchasing behaviour. The ethical implications around transparency, prejudice, and accountability are brought to the forefront with the incorporation of AIoT (Ishengoma et al., 2022). Organizations must prioritize the incorporation of ethical considerations in the design and deployment of AIoT systems. The legislative framework around Artificial Intelligence of Things (AIoT) integration is undergoing continuous development and transformation. Organizations must effectively traverse a multifaceted network of policies and regulations about data privacy, security, and standards. Adherence to these policies is paramount to mitigate legal and reputational hazards. The collection and processing of extensive data by AIoT systems have given rise to apprehensions over the privacy and security of such data (Aliahmadi & Nozari, 2022). Safeguarding the confidentiality of personal data is of utmost significance, necessitating the implementation of essential strategies such as data encryption and obtaining user consent. The field of AIoT is revolutionary as it fundamentally transforms how organizations gather, analyze, and utilize data derived from the physical realm. A comprehensive comprehension of the principles and integration facets of artificial intelligence of things (AIoT) is crucial for enterprises aiming to leverage the full capabilities of this technology while simultaneously tackling ethical, policy, and privacy concerns. The chapter explores the following topics: the concept of "Artificial Intelligence of Things" (AIoT), AIoT and its impact on customer interaction, the ethical considerations surrounding AIoT, policy and legal aspects related to AIoT, and the implications for users' privacy in the context of AIoT.

2. AIoT AND CUSTOMER INTERACTION

Businesses' primary objective in the modern digital landscape is to provide an exceptional consumer experience. (Parihar et al., 2023) AIoT systems can analyze large volumes of data, providing insights into customer preferences and behaviours. This analytical capability enables businesses to provide individualized product or service suggestions, customizing each customer's purchasing or usage experience. For instance, e-commerce platforms can recommend things by examining a customer's browsing and purchasing history. Devices and sensors equipped with the Internet of Things (IoT) are essential for capturing real-time consumer behaviour and product utilization data. This data is a rich source of information regarding how consumers interact with products and services. These insights enable organizations to make data-driven decisions and continuously improve their offerings. AIoT facilitates proactive problem resolution by anticipating potential issues before they manifest (Yang et al., 2021). Predictive maintenance in the manufacturing sector is a prime example, where AIoT is used to predict equipment failures, allowing organizations to resolve problems before they disrupt operations. This method reduces downtime and improves the customer experience. AIoT facilitates interactions across multiple consumer touchpoints, such as websites, mobile applications, smart devices, and chatbots (Hsu, 2023). Customers can begin interacting on one channel and continue on another, ensuring a seamless and practical experience regardless of the chosen channel. Moreover, chatbots and virtual assistants, propelled by AIoT technologies, provide efficient, around-the-clock customer support by responding to questions, resolving issues, and providing timely information. This effectiveness improves both customer satisfaction and response times.

Matin et al. (2023) state that AIoT's real-time data analysis enables organizations to predict customer behaviours and requirements. This predictive capability enables companies to provide proactive and anticipatory services, addressing customers' requirements before they express them. Through AIoT, products themselves can become more personalized. Smart home devices, for instance, can alter their settings based on the user's preferences, creating a comfortable and individualized environment. This level of customization enhances the user experience significantly. AIoT enables organizations to offer scaled customization, notably in the retail sector. AIoT can recommend products or services aligned with individual customer preferences, increasing sales and enhancing customer satisfaction. Organizations must deliver superior customer experiences by leveraging AIoT capabilities to prosper in today's competitive business environment. Using real-time data, predictive analytics, and personalization, businesses can develop a customer-centric strategy that fosters customer loyalty and promotes growth (Wassouf et al., 2020). However, it is essential to prioritize resolving concerns regarding privacy and security to ensure the protection and ethical, responsible use of consumer data. Incorporating Artificial Intelligence of Things (AIoT) technologies transforms consumer interactions by enabling personalization and customization (Bibri & Jagatheesperumal, 2023). AIoT systems collect information regarding consumer behaviours, preferences, and interactions; this information is then used to generate personalized recommendations. Streaming platforms, for instance, can recommend movies or music based on a user's viewing or listening history, producing a more engaging and relevant experience. AIoT can offer scalable product customization in manufacturing and retail, enabling customers to personalize products, such as configuring a car's specifications or altering a smartphone's appearance. This level of customization increases consumer engagement and satisfaction.

Real-time data analysis enabled by the IoT enables predictive personalization. AI algorithms can anticipate customer requirements and preferences, enabling businesses to provide pertinent information or services before consumers express their needs. (Zhang et al., 2023) A smart home system, for instance, can alter lighting, temperature, and security settings based on predictive insights(Parihar et al., 2023). AIoT systems analyze customer behaviour and adapt interactions accordingly. For instance, an e-commerce platform can modify its homepage layout, product recommendations, and promotions based on a customer's past website activity, resulting in a more personalized purchasing experience.AIoT considers contextual information to tailor interactions. For instance, a location-aware app may provide restaurant suggestions based on a user's current location, dietary preferences, and prior dining history, creating a personalized and relevant dining experience. AIoT systems can respond proactively to consumer requirements. Predictive maintenance, for instance, can anticipate equipment failures and schedule maintenance before their occurrence, thereby reducing downtime and ensuring fewer disruptions for consumers (Keleko et al., 2022). AIoT devices and systems can continually learn and adapt to customer preferences and varying requirements, enhancing their personalization and customization capabilities over time and generating interactions that are ever more tailored and valuable. Some AIoT systems permit the user to determine the degree of customization. For instance, the owner of a smart home can alter settings to reflect individual preferences and requirements, giving customers a sense of autonomy in their interactions.

Customer satisfaction, loyalty, and engagement are increased by the ability to provide personalized and tailored interactions. AIoT technologies enable businesses to cater to individual requirements and preferences on a large scale, creating a competitive advantage. However, organizations must manage customer data responsibly, address concerns about privacy and security, and ensure that their customization efforts adhere to ethical and regulatory standards. AIoT offers tremendous potential for enhancing customer engagement and experiences. Still, it also presents organizations with several challenges and considerations that must be addressed to maximize the benefits and maintain customer trust. AIoT systems collect and process vast consumer data, including sensitive and personally identifiable information. To prevent data intrusions and theft, it is crucial to ensure the security of this data (Lin et al., 2022; Kuzlu et al., 2021). Customers must be made aware of data collection and utilization. It is crucial to provide transparent and honest information about data use and to acquire explicit consent. AIoT systems may perpetuate bias in decision-making if they are not carefully designed and trained. Ethical considerations include ensuring impartiality and preventing discrimination (Heinrichs, 2021). Organizations must hold AIoT systems accountable for their actions. Determining who is accountable for system behaviour and outcomes is essential, particularly in technological errors or damage cases. Consumer data collection and processing is governed by data protection laws, such as GDPR (General et al.) in the European Union and other regional regulations. AIoT products and services may be subject to specific industry standards and regulations. Compliance with these requirements is required to avoid legal issues. While AIoT can handle routine duties, specific interactions necessitate empathy and complex problem-solving, which may be difficult for AI. It is essential to strike the correct balance between automation and human interaction to meet customer needs effectively (Janssen et al., 2019). It is crucial to ensure that customers control their data and interactions. It is essential to allow consumers to opt in or out of specific AIoT interactions and to provide control over personalization settings.

The precision and dependability of the data collected by AIoT systems are essential. Data quality can result in accurate predictions and decisions, negatively impacting consumer experiences. AIoT systems must integrate seamlessly with other technologies and platforms. Challenges with interoperability can result in fragmented consumer experiences. As customer interactions increase, organizations must ensure that their AIoT systems can scale to meet increasing demands without sacrificing performance or dependability and ensuring that consumers comprehend how AIoT systems function and the benefits they provide can result in more knowledgeable and contented users. Organizations should continuously monitor and evaluate AIoT system performance to identify and resolve issues. Customer feedback can also be helpful for development purposes. To develop and maintain customer trust while maximizing the potential of AIoT for customer engagement, it is crucial to address these challenges and considerations. To provide a seamless and secure customer experience, organizations must adopt a responsible and ethical approach, implement robust security measures, and comply with applicable regulations.

Multiple companies use Artificial Intelligence of Things (AIoT) to improve customer satisfaction. Amazon employs Artificial Intelligence of Things (AIoT) technology in its smart home devices, such as Echo and Ring, augmenting customer experiences. For example, the incorporation of Alexa, Amazon's AI-driven virtual assistant, into different IoT devices enables customers to have customized experiences, from voice-operated home automation to interactive entertainment. Tesla utilizes Artificial Intelligence of Things (AIoT) to do predictive maintenance and over-the-air updates, enhancing the efficiency of its electric vehicles. Periodic software upgrades for Tesla vehicles enhance operation and bring novel features, enhancing the overall customer experience. Samsung utilizes AIoT (Artificial Intelligence of Things) technology within its smart home ecosystem to provide smooth and automated connectivity. Users may manage and oversee their intelligent devices, enhancing the convenience and integration of their home environment, by utilizing Samsung's SmartThings platform. Netflix employs AIoT algorithms to monitor customer preferences, providing tailored content suggestions. Using Artificial Intelligence of Things (AIoT), Netflix employs algorithms to recommend movies and TV series to customers based on their past viewing habits. It ensures a more gratifying and personalized entertainment experience. Nest, now a subsidiary of Google, utilizes Artificial Intelligence of Things (AIoT) technology in its range of smart home products, explicitly optimizing energy use and enhancing home security. Take the Nest Learning Thermostat as an example. It adjusts based on customer preferences, leading to energy conservation and improved comfort. Microsoft has included Artificial Intelligence of Things (AIoT) into its Azure IoT services, improving client experiences in many industries. Microsoft's AIoT solutions, namely in predictive maintenance, enhance operational efficiency and minimize periods of inactivity, resulting in enhanced customer satisfaction. Starbucks employs Artificial Intelligence of Things (AIoT) in its mobile ordering system to optimize and simplify the consumer experience. By utilizing AI algorithms within the Starbucks mobile application, orders may be predicted and prepared according to consumer preferences, resulting in shorter wait times and improved overall satisfaction. These examples demonstrate the wide-ranging uses of AIoT in several industries, highlighting its ability to deliver customized, streamlined, and interconnected client experiences. It is advisable to be informed about the most recent advancements and case studies to investigate the ever-changing field of AIoT applications.

3. ETHICS AND AIoT

Integrating Artificial Intelligence of Things (AIoT) technologies gives rise to various ethical problems requiring thorough company attention to guarantee responsible and ethical implementation (Vermanen et al., 2021). Organizations must exhibit transparency regarding the data that is gathered through AIoT (Artificial Intelligence of Things) devices, how it is utilized, and the individuals or entities that possess authorization to access it. Individuals must grant informed permission for data collection (Hildt & Laas, 2022; Andreotta et al., 2021; Gesualdo et al., 2021;). It is imperative to gather the essential data for the targeted objective. Refusing excessive data-gathering practices is advisable, as they can infringe upon individuals' privacy rights. It is imperative to establish explicit protocols governing the retention and deletion of data to prevent unnecessary storage of information beyond its requisite duration. Artificial Intelligence of Things (AIoT) systems can potentially include biases within their training data. It is imperative for organizations to proactively recognize and address these biases to promote outcomes that are characterized by fairness and equity. Utilize algorithms that prioritize equity and mitigate bias, particularly regarding ethnicity, gender, age, and socioeconomic position. Establishing well-defined lines of accountability inside the organization is crucial for ensuring responsible and ethical behaviour and desired outcomes in Artificial Intelligence of Things (AIoT) systems. Assigning accountability to people or teams to resolve difficulties or errors is imperative.

It is imperative to ensure that AIoT systems are meticulously developed to incorporate the provision of explanations for their judgments, especially in situations when these decisions have a direct influence on the lives and fundamental rights of individuals. Effectively convey to consumers how their data is utilized while also furnishing them with a comprehensive understanding of the functioning of AIoT technologies. To ensure consumer data security, it is imperative to establish and enforce robust data security measures. These measures should be designed to prevent breaches, unauthorized access, and cyberattacks. One such step is the implementation of encryption techniques and secure communication protocols. Data may be protected during transmission between devices and kept in databases using these technologies. Enable users to exert authority over their engagements with

AIoT technologies. Users can modify customization preferences, exercise control over certain features by opting in or out, and exercise their right to delete their data at their discretion. It is imperative to offer consumers explicit choices that facilitate their comprehension and management of data usage, enabling them to make well-informed decisions regarding disclosing their personal information. Organizations can embrace established ethical frameworks, principles, and rules tailored to the domains of artificial intelligence (AI) and the Internet of Things (IoT). By doing so, they may effectively steer their decision-making processes and guarantee the ethical application of AIoT technologies. Implementing robust data governance protocols to guarantee the responsible utilization of data and adherence to ethical norms is imperative.

It is recommended to perform routine ethical audits on AIoT systems and procedures to detect any ethical issues or possible avenues for enhancement. Acknowledging and considering these ethical factors to establish trust among customers, guarantee the responsible implementation of AIoT, and mitigate the potential hazards associated with AIoT technologies. Implementing ethical norms in the Artificial Intelligence of Things (AIoT) not only safeguards the rights of persons but also plays a significant role in fostering an excellent societal impact of these technologies. Establishing fairness, transparency, and accountability within AIoT (Artificial Intelligence of Things) systems is paramount to cultivating trust, fostering ethical use of technology, and mitigating the potential hazards linked to biased or unverifiable decision-making. Employing diverse and representative datasets while training the AIoT model is crucial to avoid bias. Ensuring that the collected data encompasses diverse demographics, backgrounds, and use cases is imperative. Implementing bias detection algorithms is crucial to identify and effectively manage biases that may be present in AIoT (Artificial Intelligence of Things) systems. To mitigate biased outcomes, it is imperative to modify algorithms and data pre-processing techniques appropriately. It is imperative to regularly perform fairness audits on AIoT systems to detect and promptly address any potential inequalities in decision-making. Utilize artificial intelligence models and algorithms that offer justifications for their decision-making processes. Explainable AI (XAI) tools have the potential to enhance the transparency of AIoT systems for both users and stakeholders. It is imperative to ensure transparency and precision regarding the procedures involved in managing data, encompassing data collection, storage, and dissemination stages. The implementation of transparent data procedures fosters a sense of confidence among both users and regulatory bodies. This communication aims to elucidate the functioning and decision-making processes of artificial intelligence (AI) algorithms. It is advisable to refrain from employing opaque models, commonly called "black box" models, which lack transparency. Providing explicit and transparent privacy rules is imperative to inform consumers how their data will be utilized and safeguarded.

It is imperative to ensure that data utilization adheres to privacy standards. Assign specific personnel or teams to oversee the behaviour and consequences of the AIoT system. The presence of responsible entities facilitates the effective resolution of problems.

It is imperative to establish comprehensive legal frameworks that effectively address and provide appropriate compensation to those who have experienced injury due to technological advancements. It is imperative to adhere to established ethical rules and concepts, including those outlined in "AI Ethics." These recommendations establish a fundamental basis for guaranteeing accountability in implementing Artificial Intelligence of Things (AIoT). Enable users to exert autonomy and regulate their engagements with Artificial Intelligence of Things (AIoT) devices. Users can tailor customization preferences, exercise control over the inclusion or exclusion of specific functionalities, and exercise their right to delete their data at their discretion. It is imperative to offer users explicit choices that facilitate their comprehension and enable them to exercise authority over utilizing their data. It is imperative to guarantee that individuals have the necessary knowledge to make well-informed choices regarding disclosing their personal information. Establishing and enforcing comprehensive data security policies to safeguard client data against breaches, illegal access, and cyberattacks is imperative. Accountability necessitates the incorporation of security measures as a fundamental component. Utilize encryption techniques and employ secure communication protocols to protect data during its transmission across devices and storage within databases. Adopt accepted ethical frameworks, concepts, and norms related to AI and IoT to guide decision-making and assure ethical AIoT adoption. It is recommended that periodic ethical audits on AIoT systems and processes be performed to detect any issues or areas that might be enhanced. Through the conscientious implementation of these tactics, businesses can cultivate equity, openness, and responsibility within AIoT systems. Adhering to responsible and ethical principles in the Artificial Intelligence of Things (AIoT) establishes trust with users, regulators, and society, facilitating the responsible and sustainable integration of AIoT technologies. The utilization of AIoT (Artificial Intelligence of Things) technologies has promise for inducing significant and profound transformations across many industries (Nozari et al., 2022). Nevertheless, it is imperative to acknowledge that these advancements also give rise to ethical quandaries and have significant societal ramifications, necessitating meticulous deliberation.

Propose the establishment of an internal ethics committee or advisory board tasked with evaluating and resolving ethical considerations about projects involving Artificial Intelligence of Things (AIoT). To ensure a comprehensive and multifaceted approach, it is imperative to incorporate various stakeholders with varying viewpoints and interests. Perform an ethical effect evaluation for any project, including

Artificial Intelligence of Things (AIoT). It is imperative to contemplate the potential ramifications on individuals, society, and the environment. This inquiry seeks to identify potential ethical challenges and areas of concern within a given context. Adopting and strictly adhering to established ethical frameworks and guidelines specifically designed for Artificial Intelligence of Things (AIoT) is imperative. These frameworks serve as a fundamental basis for addressing ethical considerations. It is recommended that resources be allocated towards implementing public awareness and education initiatives to disseminate information regarding the functionalities and constraints of AIoT technology. It facilitates individuals to make well-informed decisions and comprehend the consequences of data sharing. To effectively manage the potential job displacement resulting from the integration of Artificial Intelligence of Things (AIoT), it is crucial to anticipate and handle this issue proactively. Offer opportunities for reskilling and upskilling to employees whose job positions may encounter potential impacts.

Collaborate with policymakers to develop solutions for supporting economic resilience and flexibility. It is imperative to pay close attention to the cultural and social dimensions associated with adopting the Artificial Intelligence of Things (AIoT). Acknowledging that other cultures may possess distinct viewpoints towards privacy, data sharing, and using artificial intelligence (AI) is essential. It is imperative to modify AIoT systems and processes following the given circumstances. It stayed updated on the dynamic legislation and standards about Artificial Intelligence of Things (AIoT is essential). It is imperative to ensure that AIoT practices follow legal obligations, particularly protecting data privacy and security. Adopting a proactive approach in adhering to current and forthcoming rules is imperative. Formulate rules and procedures that exemplify strict compliance with regulatory criteria. It is imperative to consistently do ethical audits and assessments on AIoT systems and procedures. Incorporating feedback from stakeholders and persons impacted by AIoT technology should be an integral part of this process. Developing effective channels for soliciting input from consumers, workers, and the broader community is vital. This feedback has the potential to facilitate the identification of areas that require improvement and the resolution of ethical concerns.

Establish cross-disciplinary teams comprised of specialists in artificial intelligence (AI), Internet of Things (IoT), ethics, law, sociology, and other pertinent disciplines. This collaboration facilitates the attainment of a comprehensive viewpoint about ethical quandaries and their ramifications on society. Interact with various stakeholders, encompassing customers, employees, regulatory bodies, advocacy groups, and academic institutions. Collaborating with external specialists and organizations might yield significant value insights. Establishing strict adherence to ethical rules is crucial for adopting Artificial Intelligence of Things (AIoT) systems. It is imperative to ensure that projects follow these principles starting from the planning phase. The proposed approach involves developing and disseminating ethical impact studies, which aim to enhance transparency by providing comprehensive insights into the ethical concerns, obstacles, and mitigation measures associated with initiatives involving Artificial Intelligence of Things (AIoT). Organizations can develop AIoT implementations that are technologically sophisticated, morally sound, and socially responsible by methodically resolving ethical challenges and considering their societal consequences. It promotes trust among users, regulators, and the wider society, guaranteeing the sustainable and responsible utilization of AIoT technology.

4. POLICY AND LAW IN AIoT

Incorporating Artificial Intelligence of Things (AIoT) presents many legal obstacles and shortcomings in policy and law. Significant concerns revolve around privacy since the vast data gathering in AIoT systems raises questions regarding user privacy and the possibility of misuse of sensitive information. The rapid rate of technological improvements in AIoT has rendered regulatory frameworks inadequate, leading to obsolete regulations that fail to address the complexities of this rapidly growing industry. The issue of liability ambiguity arises because of the challenge legal systems face in precisely determining accountability for decisions made by AIoT systems, especially those that involve intricate algorithms and machine learning. Inadequate safeguards within legislative frameworks exacerbate security and cybersecurity concerns, potentially leaving consumers vulnerable to privacy breaches and cyber threats. The absence of clear standards in legal frameworks highlights the ethical challenges associated with the development of AIoT systems, which can inadvertently result in biases, discrimination, or ethical issues. The worldwide scope of AIoT applications presents difficulties in establishing international standards, impacting data transmission and jurisdictional considerations. The intrinsic complexity of AIoT systems presents obstacles to achieving transparency and explainability, which challenges legal frameworks in establishing precise standards for decision-making processes. Insufficient safeguards for gaining informed consent may result in potential infringements of privacy rights. To rectify these legal deficiencies, legislators must take a proactive stance and create agile, adaptable, and all-encompassing regulations in line with the ever-changing nature of AIoT technologies. Ensuring responsible and ethical integration of AIoT requires a crucial balance between innovation and regulatory safeguards.

The regulatory environment for AIoT (Artificial Intelligence of Things) technologies is intricate and dynamic, encompassing existing legislation and emerging frameworks specifically designed to tackle the distinct difficulties posed by the integration of AI and IoT. The General Data Protection Regulation (GDPR) governs the handling of personal data in the European Union (EU) and other nations. It applies explicitly to AIoT systems that entail the gathering and processing of this data (Friedewald et al., 2022; Ivanova, 2020). AIoT initiatives must conduct a Data Protection Impact Assessment (DPIA) to evaluate data processing activities for potential risks and mitigate privacy concerns. The California Consumer Privacy Act (CCPA) is a privacy law in California, United States of America, giving inhabitants data rights. The law applies to IoT applications involving Californian residents (Rothstein & Tovino, 2019). The CCPA grants consumers the right to request access to and delete their data, which impacts the data management practices of AIoT systems. Industry regulations impact AIoT application development (Pise et al., 2023). For instance, the healthcare industry is governed by the Health Insurance Portability and Accountability Act (HIPAA), which mandates stringent data protection and privacy practices for AIoT devices related to healthcare. In the automotive industry, regulations such as ISO 26262 address functional safety for road vehicles and influence AIoT deployments in autonomous vehicles. Numerous nations are contemplating or have already enacted legislation addressing AI and IoT technologies, which inevitably include AIoT. Legislation may define AIoT device requirements, data protection, transparency, security measures, and standards for AIoT deployments in diverse industries.

Cybersecurity regulations, such as the EU's Network and Information Systems (NIS), Directivdirectly impact AIoT technologies, necessitating robust cybersecurity measures to defend against cyber threats. While lacking legal enforceability, ethical standards and principles created by governments and international organizations, such as the OECD AI Principles or the EU's Ethics standards for Trustworthy AI, offer essential direction for the ethical implementation of AIoT. Individual nations and regions can implement their AIoT-related regulations and standards. China's cybersecurity laws, for instance, impact AIoT practices in the Chinese market. AIoT technologies are governed by industry organizations and standards authorities, such as the Institute of Electrical and Electronics Engineers (IEEE) and the Industrial Internet Consortium (IIC). Regulations such as the EU-U.S. Privacy Shield and Standard Contractual Clauses address cross-border data transfer pertinent to international AIoT systems. Depending on their scope, location of operation, and the type of data and services they manage, AIoT technologies must conform to a combination of these regulatory frameworks. Organizations developing and deploying IoT solutions must remain informed of these regulations and work toward compliance. The deployment of AIoT (Artificial Intelligence of Things) technologies is accompanied by several legal challenges and repercussions that businesses must navigate. Understanding and addressing these obstacles is essential for implementing AIoT, which is responsible and compliant.

Determining who is responsible for the actions and decisions of AIoT systems can be difficult. Is it the manufacturer, the user, the AI model, or some combination of these? Legal structures must explicitly define and assign responsibility. AIoT systems may be susceptible to security incidents and data breaches. Legal difficulties include identifying the culpable parties, evaluating potential damages, and addressing regulatory requirements for data breach notifications. AIoT technologies may employ patented algorithms or data protected by intellectual property rights. Intellectual property rights, licensing agreements, and prospective infringements may give rise to legal complications. Defining possession of data generated by AIoT devices can result in legal disputes, especially when data is shared among multiple parties or systems. Manufacturers of IoT devices and service providers must establish legally enforceable terms of service that define users' rights and responsibilities, data handling practices, and dispute resolution mechanisms. Contracts should include data usage agreements detailing how user data will be processed and safeguarded and whether data may be shared with third parties. If AIoT systems perpetuate bias or discrimination, anti-discrimination laws and ethical concerns may give rise to legal challenges. Identifying and resolving bias is crucial for conformance. There may be legal repercussions if AIoT systems lack transparency in their decision-making processes. Regulations may demand explanations for AI-generated decisions, mainly when they affect the rights or opportunities of individuals. Organizations that mishandle consumer data, particularly in violation of data protection laws, are subject to legal action. Users can sue for damages resulting from data breaches or misuse. The right to access one's data is granted by regulations such as GDPR. If organizations fail to provide requested data or do not comply with such requests, they may face legal complications. Compliance with a regulatory landscape that is complex and constantly changing poses a significant legal challenge. Organizations are responsible for ensuring compliance with numerous data protection, security, and technology-specific regulations.

In some regions, the law may require organizations to appoint data protection officers (DPOs) to supervise data protection and ensure regulatory compliance. When IoT systems transfer data across borders, legal complications may arise. Some countries have data localization laws that mandate data storage on-site, which may conflict with international data transfer regulations. Legal experts, AIoT developers, and data protection specialists must collaborate to address these legal challenges and ramifications. Organizations must remain vigilant and proactive to mitigate legal risks associated with IoT deployment to ensure compliance with applicable laws and ethical guidelines. Privacy and security are the top priorities in AIoT (Artificial Intelligence of Things) applications. Integrating AI and IoT technologies presents organizations with unique challenges that must be addressed to safeguard user data and guarantee the secure operation of AIoT systems. Often, AIoT devices accumulate vast quantities of data, including sensitive and private information. Unauthorized or excessive data collection may violate users' privacy. Applications utilizing IoT may share information with third parties or cloud platforms. Users require assurances that their data is only used for its intended purpose and is not shared without their consent. Collecting only the data required for the intended functionality and minimizing data storage can reduce privacy concerns. Data transmitted between IoT devices, cloud platforms, and other endpoints must be encrypted to prevent unauthorized access and surveillance. Data security requires only authorized devices and users to access AIoT systems. There should be robust device authentication mechanisms in effect.

Implement stringent access controls to restrict who can access, modify, or delete data in AIoT systems; role-based access control can be advantageous. Unsecured IoT devices may be susceptible to infiltration and exploitation. To prevent unauthorized access and attacks, manufacturers must address device security issues. It is critical to update device firmware to patch security vulnerabilities regularly. Unsecured devices may serve as entry points for intrusions. Implement secure launch processes and hardware security measures to prevent tampering or compromise of devices. To protect individual privacy, anonymize and de-identify personal data when handling it. It should not be possible to revert anonymized data to their original state. Users should be able to give informed consent for acquiring and processing their data. Critical are consent management mechanisms. Users should be informed of how their data is used and be able to assess and comprehend data management practices. For national security or law enforcement purposes, government agencies in some regions may request access to AIoT data. Creating a balance between government access and individual privacy rights is complicated. Regulations may impose data retention requirements on organizations, dictating the length of time data must be kept. Compliance is required with these regulations.

AIoT systems are vulnerable to data intrusions, which can have grave privacy repercussions. Implementing robust security measures and responding quickly to security breaches (Abdullahi et al., 2022; Sarker et al., 2021; Wheelus & Zhu, 2020). AIoT devices are susceptible to malware and ransomware attacks, compromising user privacy and security. Exploiting IoT devices has led to the creation of botnets used in cyberattacks. Protecting AIoT devices from being incorporated into botnets is a significant concern. To address these privacy and security concerns in AIoT applications, a comprehensive strategy combining technical, legal, and policy measures is required. Organizations must prioritize the protection of user data and the security of their AIoT systems to instil confidence and promote the responsible application of these technologies.

5. USERS' PRIVACY IN AIoT

Ensuring data privacy and protection in AIoT (Artificial Intelligence of Things) ecosystems is essential for establishing user trust, complying with regulations, and upholding ethical data practices (Vermanen et al., 2022; Ren et al., 2021). Collect only the data required to accomplish the intended goal. Reduce the quantity of data collected to reduce the likelihood of privacy breaches. Explicitly establish the objectives for gathering data and guarantee that it is not utilized for unrelated reasons without obtaining the necessary consent. Protect data during transmission between IoT devices, cloud platforms, and other endpoints using encryption protocols. It prevents surveillance and data interception—Encrypt data for protection against unauthorized access in the event of data breaches or physical theft of devices. Establish role-based access control mechanisms to limit data access, modification, and deletion in AIoT systems. Only authorized personnel should be permitted entry. Ensure that solid authentication and authorization processes are in place to verify the identity of users and devices before granting access to data or system features. Before beginning data collection, inform users about the data and how it will be used and obtain their informed consent. Permit users to provide granular consent, enabling them to select the specific data elements or services to which they consent. Define and communicate data retention policies that specify the duration of data storage and its eventual deletion. Ensure compliance with applicable data protection laws.

Reduce the risk of retaining extraneous data by regularly reviewing and removing data no longer required for its defined purposes. Consider anonymizing data to remove personally identifying information, making it unidentifiable and protecting user privacy. Replace personally identifiable information with pseudonyms to make it more difficult to identify individuals while preserving the ability to analyze data. Communicate how user data is collected, stored, processed, and shared, including data collection, storage, and processing practices. Transparency fosters confidence. If AIoT systems make decisions that impact users, provide justifications for these decisions to increase openness. Implement intrusion detection systems and firewalls to protect against cyberattacks and unauthorized access. Patch and update IoT devices regularly to eliminate vulnerabilities and prevent potential security breaches. Organizations can safeguard user data, uphold privacy rights, and ensure ethical data practices in AIoT ecosystems by adhering to these principles and implementing appropriate technical and organizational measures. It helps organizations comply with data protection laws, develop user trust, and protect individual privacy. Consent, data ownership, and control are essential AIoT (Artificial Intelligence of Things) interactions for ensuring privacy and user agency. In interactions involving AIoT, informed consent is essential. Users must know and consent to collecting, using, and sharing their data with AIoT devices and systems. Users should be provided with concise, readily digestible information regarding data practices. It includes the types of data collected, their purposes, and any sharing with third parties. Provide users with granular consent options. Allow them to selectively consent to specific data collection or utilization instead of a blanket agreement, thereby enhancing user control. Clarify the data ownership concept in AIoT interactions. Users should comprehend that their data belongs to them and that organizations are the data's custodians. When data is generated or shared by multiple parties (e.g., device manufacturers and cloud service providers), explicitly define ownership arrangements for shared data in contracts.

Facilitate data access and transfer for users. Data portability ensures users retain control over their data even if they transfer AIoT products or services. Enable consumers to tailor their interactions with AIoT systems. Permit them to set preferences, customize settings, and opt-in or opt-out of certain features. Provide consumers with the option to delete their data. This control ensures that users have the option to withdraw their consent and have their data deleted. Implement mechanisms that allow users to easily opt out of data collection or specific functionalities without impacting the AIoT system's overall usability. Enhance the data literacy of users. Inform them of the data collected by AIoT systems, how it is utilized, and how they can exercise control and consent. Inform users on how to access and modify their privacy settings. Clearly instruct them on how to customize their AIoT interactions. Promote interoperability between AIoT devices and systems. Users should be able to transition between devices and services without losing access to their data.

Ensure that users can easily access and export their data to other services or platforms while retaining control and ownership. Integrate privacy considerations into AIoT system design and development. It requires early consideration of data control, consent, and ownership. Address ethical concerns by reducing bias in AIoT systems. Prioritize impartiality and transparency to ensure equitable results. Enable users to comprehend the rationale behind AIoT decisions and actions to increase trust and transparency. Organizations can cultivate confidence, protect user privacy, and ensure responsible and ethical AIoT interactions by adhering to informed consent, data ownership, and user control principles. Within AIoT ecosystems, these practices enable users to make informed decisions and retain control over their data. Protecting user privacy in AIoT (Artificial Intelligence of Things) systems requires privacy-enhancing techniques and methods. Anonymization eradicates personally identifiable information (PII) from data to render it unidentifiable. It ensures that data cannot be explicitly linked to specific users. By replacing PII with pseudonyms, pseudonymization makes it more difficult to identify individuals. It permits a certain degree of traceability while maintaining privacy. This technique adds noise to data to safeguard individual privacy while still permitting the extraction of valuable insights. It ensures no particular data point can be associated with a specific individual. In this method, artificial intelligence models are trained on data disseminated across multiple devices without transferring user data to a central server. It improves privacy by keeping information on the periphery.

Homomorphic encryption allows for the execution of computations on encrypted material without the need to decode it. It protects sensitive data while enabling AI to process the data. Secure Multiparty Computation (SMC) enables many parties to collectively perform a computation on their respective inputs while ensuring the privacy of those inputs. It is valuable for AIoT computations that protect privacy. Blockchain technology can provide secure and decentralized identity management, enabling users to control their data and access services without exposing data unnecessarily. These cryptographic techniques enable one party to demonstrate to another that it possesses specific information without revealing it. They are utilized for identity verification without revealing sensitive information. Collect only the data required for a particular purpose. Limiting the amount of data collected diminishes the likelihood of privacy breaches. Specify the objectives for data collection and verify that information is not utilized for unrelated reasons without obtaining the appropriate consent. Incorporate privacy considerations into the design and development of IoT systems, ensuring privacy is essential to the system's architecture. Inform consumers about data privacy, how their data is used, and their level of control. Informed consumers are more likely to make decisions with privacy in mind. Maintain openness in AIoT data practices and operations. Users should have access to information regarding the handling of their data. Enable users to comprehend the rationale behind AIoT decisions and actions to increase trust and transparency. Ensure AIoT systems and practices comply with applicable data protection and privacy regulations, such as the GDPR and CCPA. These techniques and approaches for enhancing privacy should be implemented following legal and ethical requirements. They aid organizations in protecting user privacy, gaining user trust, and ensuring AIoT applications are responsible.

CASE STUDY 1: AIoT-BASED PERSONALIZED MARKETING IN THE RETAIL INDUSTRY

AIoT has revolutionized the marketing strategies of the retail industry. With the help of AIoT technologies, retailers are enhancing the customer experience and increasing sales through personalized marketing campaigns. A well-known retail chain installed IoT systems in its physical stores and online platforms. These systems collect data from in-store sensors, consumer mobile devices, and online activities. AI algorithms analyze this data to generate personalized marketing campaigns. For example, a consumer entering a store could receive real-time, personalized offers and recommendations based on their past purchases and preferences.

Customers receive personalized product suggestions, discounts, and special offers, which enhances their shopping experience—personalized marketing results in higher conversion rates and increased sales for customers and retailers. The retailer places a premium on data privacy, obtaining users' express consent and implementing stringent data security measures. It is challenging to strike a balance between personalized marketing and data privacy. It is essential to strike the correct balance to avoid invading customers' privacy. A persistent concern is ensuring impartiality in personalized recommendations and preventing discriminatory results. The retailer must navigate complex data protection regulations, such as the General Data Protection Regulation (GDPR), and ensure AIoT operations are compliant.

CASE STUDY 2: ETHICAL CONSIDERATIONS IN AIoT-ENABLED HEALTHCARE SYSTEMS

Healthcare providers are increasingly incorporating AIoT technologies to improve patient care. However, ethical considerations predominate. A hospital implemented IoT devices to monitor patients' health, collect data, and predict health trends. These devices provide healthcare professionals with real-time data to assist them in making more accurate diagnoses and treatment decisions. By facilitating faster, more accurate diagnoses and individualized treatment plans, AIoT devices enhance patient care. Patients can be remotely monitored, which reduces hospital visits and improves the quality of life for those with chronic illnesses. The hospital prioritizes ethical AI by ensuring its IoT systems are open, accountable, and fair. Priority number one is protecting patient data from unauthorized access or intrusions. Patients must provide informed consent for artificial intelligence-driven data collection and diagnosis. Healthcare providers in the United States must adhere to the Health Insurance Portability and Accountability Act (HIPAA) and similar legislation.

CASE STUDY 3: POLICY AND LEGAL CHALLENGES IN AUTONOMOUS VEHICLES POWERED BY AIoT

Connected autonomous vehicles to the IoT promise safer and more efficient transportation. However, they pose complex policy and legal issues. A notable automaker has introduced a fleet of autonomous, IoT-connected ride-sharing vehicles. These autonomous vehicles acquire real-time traffic data, communicate with infrastructure, and make decisions to ensure passenger safety and efficiency.

Self-driving vehicles hold the potential to mitigate accidents stemming from human errors. Connected IoT automobiles enhance traffic flow and reduce congestion. This technology can revolutionize the transportation industry, creating employment and decreasing costs. Determining liability in a catastrophe involving autonomous vehicles is problematic from a legal standpoint. It is challenging to balance safety data collection and passenger privacy concerns. Policymakers are responsible for developing exhaustive regulations to govern the operation and deployment of autonomous vehicles propelled by IoT. These case studies highlight the diversified applications of AIoT technologies and their associated benefits and challenges in a variety of industries, with an emphasis on ethical considerations, data privacy, and regulatory compliance.

6. DISCUSSION

Striking an equilibrium between personalization and user privacy is a crucial challenge. Implementing AIoT necessitates deliberation over ethical principles. In addition to being an ethical necessity, guaranteeing fairness, transparency, and accountability in AIoT systems is intrinsically linked to legal and regulatory compliance. Policies must address ethical concerns and provide guidelines for the responsible application of AIoT. The accumulation and analysis of massive data in IoT systems can pose significant privacy and security hazards. Strong privacy safeguards and cybersecurity measures are necessary for protecting user data and ensuring privacy. To remain competitive and offer cutting-edge products and services, businesses must embrace AIoT innovation. However, innovation must be accompanied by responsible deployment of AIoT, with user privacy and ethical concerns taking precedence. Organizations should adopt a proactive approach to the responsible implementation of AIoT. This requires conducting thorough impact assessments, establishing ethical guidelines, and engaging stakeholders to ensure AIoT solutions are developed and deployed responsibly.

Continuous monitoring and accountability are necessary for deploying AIoT responsibly. Organizations must be capable of resolving issues, addressing ethical dilemmas, and adjusting to shifting privacy and policy environments. Multiple stakeholders, including businesses, governments, regulatory bodies, and technology providers, must collaborate to implement the AIoT sustainably. These stakeholders should work together to develop comprehensive and ethical AIoT ecosystems. Transparent communication with stakeholders is required. Organizations should engage in an open dialogue with users, data subjects, and relevant authorities to establish trust and ensure that AIoT practices are well understood. Policymakers play a crucial role in shaping the AIoT landscape. They should develop futuristic, adaptable regulations that balance innovation, user protection, and ethical standards. By incorporating end-users into the AIoT design and deployment processes, organizations can better understand user needs, concerns, and expectations. It can result in more user-centric and accountable IoT solutions. Public-private partnerships can facilitate collaboration in addressing the AIoT's ethical, policy, and privacy issues. These alliances can aid in the creation of exhaustive AIoT implementation frameworks. A multidimensional strategy is required to incorporate AIoT across multiple industries successfully. Organizations must navigate a complex web of consumer interaction, ethics, policy, and privacy to deploy AIoT responsibly. It requires a balance between innovation and responsible practices, with a strong emphasis on transparency, accountability, and stakeholder collaboration for the implementation of AIoT that is ethical and sustainable.

7. CONCLUSION

By automating duties and processes, optimizing resource allocation, and improving decision-making, AIoT integration can increase productivity. AIoT can reduce operational costs through predictive maintenance in manufacturing and energy-efficient building management in smart cities, for example. AIoT generates valuable data that gives businesses insights into customer behaviour, operational performance, and market trends. In AIoT-driven interactions, personalization and customization can increase consumer satisfaction and loyalty. Managing the enormous data in AIoT systems while preserving data privacy and security is challenging. To avoid bias and discrimination, ensuring impartiality, transparency, and accountability in AIoT systems is a significant ethical challenge. Complying with evolving data protection and AI regulations in various regions presents obstacles. Combining AI and IoT technologies can be technically challenging and costly, requiring specialized knowledge and resources. Ensure the transparency of data practices and AI decision-making to develop user trust and avoid potential ethical pitfalls. Implement mechanisms to hold organizations accountable for the outcomes of AIoT systems, significantly when individual decisions are affected. Ensure that AIoT systems do not reinforce discrimination or injustice by prioritizing bias mitigation. To safeguard user privacy, collect only the data required for the intended purpose and minimize data storage. Prioritize obtaining informed consent for data collection and processing and allow users to opt in or out of particular data usage. Integrate ethical considerations from the onset of AIoT system development.

Policymakers must devise flexible and exhaustive regulations to address the ever-changing AIoT landscape. These regulations should guarantee privacy protection, data security, and ethical artificial intelligence. Organizations should adopt robust data protection methods, including encryption, access limits, and anonymization, to safeguard user data. Prioritize AIoT applications that enhance user experiences, respect privacy, and offer transparent and user-controlled personalized interactions. To cultivate a culture of responsible AIoT practices, organizations should provide training in data ethics to employees and stakeholders. Engage stakeholders, such as users, regulatory authorities, and data subjects, to develop AIoT applications that meet their requirements and address their concerns. The prospective benefits of AIoT integration in organizational transitions are substantial, but they come with complex ethical, privacy, and policy compliance challenges. Organizations should prioritize transparency, accountability, and bias mitigation to promote ethical AIoT practices. Policymakers play a crucial role in the creation of regulations that strike a balance between innovation and privacy protection. AIoT integration should prioritize customer-centric applications that enhance user experiences while respecting privacy and data ethics.

REFERENCES

Abdullahi, M., Baashar, Y., Alhussian, H., Alwadain, A., Aziz, N., Capretz, L. F., & Abdulkadir, S. J. (2022). Detecting cybersecurity attacks in Internet of Things Using Artificial intelligence Methods: A Systematic Literature review. *Electronics (Basel)*, *11*(2), 198. doi:10.3390/electronics11020198

Aliahmadi, A., & Nozari, H. (2022). The Neutrosophic decision-making method evaluates security metrics in AIoT and blockchain-based supply chains. *Supply Chain Forum, 24*(1), 31–42. 10.1080/16258312.2022.2101898

Allioui, H., & Mourdi, Y. (2023). Exploring the full potentials of IoT for better financial growth and stability: A comprehensive survey. *Sensors (Basel)*, *23*(19), 8015. doi:10.3390/s23198015 PMID:37836845

Andreotta, A. J., Kirkham, N., & Rizzi, M. (2021). AI, big data, and the future of consent. *AI & Society*, *37*(4), 1715–1728. doi:10.1007/s00146-021-01262-5 PMID:34483498

Bartoň, M., Budjač, R., Tanuška, P., Gašpar, G., & Schreiber, P. (2022). Identification overview of industry 4.0 essential attributes and resource-limited embedded artificial-intelligence-of-things devices for small and medium-sized enterprises. *Applied Sciences (Basel, Switzerland)*, *12*(11), 5672. doi:10.3390/app12115672

Bibri, S. E., & Jagatheesaperumal, S. K. (2023). Harnessing the potential of the metaverse and artificial intelligence for the internet of city things: Cost-Effective XReality and synergistic AIOT technologies. *Smart Cities*, 6(5), 2397–2429. doi:10.3390/smartcities6050109

Gesualdo, F., Daverio, M., Palazzani, L., Dimitriou, D., Díez-Domingo, J., Fons-Martínez, J., Jackson, S., Vignally, P., Rizzo, C., & Tozzi, A. E. (2021). Digital tools in the informed consent process: A systematic review. *BMC Medical Ethics*, 22(1), 18. Advance online publication. doi:10.1186/s12910-021-00585-8 PMID:33639926

Heinrichs, B. (2021). Discrimination in the age of artificial intelligence. *AI & Society*, 37(1), 143–154. doi:10.1007/s00146-021-01192-2

Hildt, E., & Laas, K. (2022). Informed consent in digital data management. The International library of ethics, law and technology (pp. 55–81). doi:10.1007/978-3-030-86201-5_4

Hsu, H. (2023). Facing the era of smartness – delivering excellent smart hospitality experiences through cloud computing. *Journal of Hospitality Marketing & Management*, 1–27. doi:10.1080/19368623.2023.2251144

Huang, M., & Rust, R. T. (2020). Engaged with a robot? The role of AI in service. *Journal of Service Research*, 24(1), 30–41. doi:10.1177/1094670520902266

Ishengoma, F., Shao, D., Alexopoulos, C., Saxena, S., & Nikiforova, A. (2022). Integration of artificial intelligence of things (AIoT) in the public sector: Drivers, barriers and future research agenda. *Digital Policy, Regulation & Governance*, 24(5), 449–462. doi:10.1108/DPRG-06-2022-0067

Ivanova, Y. (2020). The data protection impact assessment is a tool to enforce non-discriminatory AI. In Lecture Notes in Computer Science (pp. 3–24). doi:10.1007/978-3-030-55196-4_1

Janssen, C. P., Donker, S. F., Brumby, D. P., & Kun, A. L. (2019). History and future of human-automation interaction. *International Journal of Human-Computer Studies*, 131, 99–107. doi:10.1016/j.ijhcs.2019.05.006

Keleko, A. T., Kamsu-Foguem, B., Ngouna, R. H., & Tongne, A. (2022). Artificial intelligence and real-time predictive maintenance in industry 4.0: A bibliometric analysis. *AI and Ethics*, 2(4), 553–577. doi:10.1007/s43681-021-00132-6

Kuzlu, M., Fair, C., & Güler, Ö. (2021). Role of Artificial Intelligence in the Internet of Things (IoT) cybersecurity. *Discover the Internet of Things*, *1*(1), 7. Advance online publication. doi:10.1007/s43926-020-00001-4

Lin, S., Chang, T., Jorswieck, E. A., & Lin, P. (2022). Applications in AIoT: IoT security and secrecy. In Information Theory, Mathematical Optimization, and Their Crossroads in 6G System Design (pp. 331–373). doi:10.1007/978-981-19-2016-5_9

Matin, A., Islam, R., Wang, X., Huo, H., & Xu, G. (2023). AIoT for sustainable manufacturing: Overview, challenges, and opportunities. *Internet of Things : Engineering Cyber Physical Human Systems*, *100901*. Advance online publication. doi:10.1016/j.iot.2023.100901

Nozari, H., Szmelter-Jarosz, A., & Ghahremani-Nahr, J. (2022). Analysis of artificial intelligence of things (AIOT) challenges for the smart supply chain (Case Study: FMCG Industries). *Sensors (Basel)*, *22*(8), 2931. doi:10.3390/s22082931 PMID:35458916

Ogundokun, R. O., Awotunde, J. B., Misra, S., Abikoye, O. C., & Folarin, O. (2021). Application of machine learning for ransomware detection in IoT devices. In Springer eBooks (pp. 393–420). doi:10.1007/978-3-030-72236-4_16

Parihar, V., Malik, A., Bhawna, B. B., & Chaganti, R. (2023). from smart devices to smarter systems: The evolution of artificial intelligence of things (AIoT) with characteristics, architecture, use cases and challenges. In Springer eBooks (pp. 1–28). doi:10.1007/978-3-031-31952-5_1

Pelet, J., Lick, E., & Taieb, B. (2021). The internet of things in upscale hotels: Its impact on guests' sensory experiences and behaviour. *International Journal of Contemporary Hospitality Management*, *33*(11), 4035–4056. doi:10.1108/IJCHM-02-2021-0226

Pise, A. A., Yoon, B., & Singh, S. (2023). Enabling ambient intelligence of things (AIoT) healthcare system architectures. *Computer Communications*, *198*, 186–194. doi:10.1016/j.comcom.2022.10.029

Qian, K., Zhang, Z., Yamamoto, Y., & Schuller, B. (2021). Artificial intelligence internet of things for the elderly: From assisted living to Healthcare Monitoring. *IEEE Signal Processing Magazine*, *38*(4), 78–88. doi:10.1109/MSP.2021.3057298

Quy, V. K., Thành, B. T., Chehri, A., Linh, D. M., & Tuan, D. A. (2023). AI and digital transformation in higher education: Vision and approach of a specific university in Vietnam. *Sustainability (Basel)*, *15*(14), 11093. doi:10.3390/su151411093

Ren, W., Xin, T., Du, J., Wang, N., Li, S., Min, G., & Zhao, Z. (2021). Privacy enhancing techniques in the internet of things using data anonymization. *Information Systems Frontiers*. Advance online publication. doi:10.1007/s10796-021-10116-w

Rothstein, M. A., & Tovino, S. A. (2019). California takes the lead on data privacy law. *The Hastings Center Report*, *49*(5), 4–5. doi:10.1002/hast.1042 PMID:31581323

Sarker, I. H., Furhad, H., & Nowrozy, R. (2021). AI-Driven Cybersecurity: An overview, security intelligence modelling and research directions. *SN Computer Science*, *2*(3), 173. Advance online publication. doi:10.1007/s42979-021-00557-0 PMID:33778771

Vermanen, M., Rantanen, M. M., & Koskinen, J. (2022). Privacy in the internet of things ecosystems – a prerequisite for companies' ethical data collection and use. *IFIP Advances in Information and Communication Technology*, 18–26. doi:10.1007/978-3-031-15688-5_2

Wang, C., He, T., Zhou, H., Zhang, Z., & Lee, C. (2023). Artificial intelligence enhanced sensors - enabling technologies to next-generation health-care and biomedical platforms. *Bioelectronic Medicine*, *9*(1), 17. Advance online publication. doi:10.1186/s42234-023-00118-1 PMID:37528436

Wassouf, W. N., Alkhatib, R., Salloum, K., & Balloul, S. (2020). Predictive analytics using big data for increased customer loyalty: Syriatel Telecom Company case study. *Journal of Big Data*, *7*(1), 29. Advance online publication. doi:10.1186/s40537-020-00290-0

Yan, B., Yang, Y., & Huang, R. (2023). Memristive dynamics enabled neuromorphic computing systems. *Science China. Information Sciences*, *66*(10), 200401. Advance online publication. doi:10.1007/s11432-023-3739-0

Yang, C., Chen, H. W., Chang, E. J., Kristiani, E., Nguyen, K. L. P., & Chang, J. S. (2021). Current advances and future challenges of AIoT applications in particulate matter (PM) monitoring and control. *Journal of Hazardous Materials*, *419*, 126442. doi:10.1016/j.jhazmat.2021.126442

Zhang, Z., Wen, F., Sun, Z., Guo, X., He, T., & Lee, C. (2022). Artificial intelligence-enabled sensing technologies in the 5G/internet of things era: From virtual reality/augmented reality to the digital twin. *Advanced Intelligent Systems*, *4*(7), 2100228. Advance online publication. doi:10.1002/aisy.202100228

KEY TERMS AND DEFINITIONS

Artificial Intelligence of Things (AIoT): AIoT refers to integrating artificial intelligence technologies, including machine learning and data analytics, with IoT devices. This integration allows the devices to independently acquire knowledge, make decisions, and enhance performance as time progresses.

Customer Interaction: Customer Interaction refers to all interactions and transactions between a company and its customers, focusing on communication, service, and tailored experiences to establish and sustain favourable connections.

Ethics: Ethics encompasses the fundamental principles and norms that govern moral judgments and actions, specifically in domains like technology and commerce, to ensure accountable and equitable conduct.

Integration: Integration is merging various components or systems to achieve smooth and efficient operation, intending to improve efficiency, communication, and overall performance.

IoT (Internet of Things): IoT, short for Internet of Things, denotes a network including interconnected objects that possess embedded sensors and communication capabilities. It enables them to gather, exchange, and utilize data to improve efficiency and performance.

Policy: Policy encompasses a collection of regulations, directives, or doctrines put forward by an institution, governing body, or authoritative entity to govern behaviours, choices, and protocols within a defined framework.

Privacy: It refers to safeguarding individuals' personal information from unwanted access or exposure, focusing on controlling one's data and guaranteeing confidentiality in different settings, such as technology and business.

Chapter 3
AIoT–Education:
The Effects of ChatGPT Use Among Postgraduate Students in Arab Countries

Qasim Hamakhurshid Hamamurad
Management Information System, UTM-Skudai, Malaysia

Normal Mat Jusoh
Azman Hashim International Business School, Malaysia

Farahakmar Anorsalim
Azman Hashim International Business School, Malaysia

Sathiswaran Uthamaputhran
Azman Hashim International Business School, Malaysia

Muslim Amin
Azman Hashim International Business School, Malaysia

ABSTRACT

Artificial intelligence and technological advancements impact human life and activities. Through structured surveys to find the effects of ChatGPT use among postgraduate students in Arab countries, we collected data on students' experiences and thoughts about using ChatGPT in their academic pursuits. In this study, one hundred postgraduate students from various academic subjects and universities in Arab countries participated. We used inferential statistics (including regression analysis and t-tests) to investigate the relationships between ChatGPT usage and academic achievement and descriptive statistics to summarise demographic data. We assessed the collected data using the Statistical Package for the Social Sciences (SPSS). The findings, which show that ChatGPT usage positively affects learning outcomes, support past research outlining the potential benefits of AI-assisted learning technology.
DOI: 10.4018/979-8-3693-0993-3.ch003

INTRODUCTION

The rapid advancement of Artificial Intelligence (AI) technology impacts various facets of modern culture, including education. Among these technological developments, OpenAI's ChatGPT, a highly developed language model, stands out as a tool that might transform college students' learning (Motlagh et al., 2023). AI is a vital part of modern technology that operates covertly to imitate the human mind and help humans in various ways. AI is expanding and developing at an unprecedented rate right now. ChatGPT illustrates how far AI has come with its innovations (Gupta et al., 2023). The traditional educational paradigm is changing because of integrating AI-driven tools. ChatGPT, with its natural language processing capabilities, offers a unique potential to improve the learning experience by providing students instant access to information, stimulating ideas, and helping with complex problem-solving. Even though ChatGPT's capabilities have exceeded expectations, its uses and misuse must be contemplated. Its versatility and ability to understand context make it an invaluable ally for students across various academic subjects.

Using ChatGPT, students can learn and get acquainted with new material. Furthermore, ChatGPT provides students instant access to the most recent technological advancements and instruction. How ChatGPT is implemented and applied to a particular use case will determine how it impacts student performance. For example, ChatGPT could assist with language and learning tasks like summarising, writing assignments, translating, and paraphrasing by giving students personalised, context-sensitive feedback on their work (Gupta et al., 2023). Although, it's important to remember that ChatGPT cannot replace having honest conversations with people. Students must regularly communicate with their teachers and peers to develop interpersonal skills and emotional regulation.

This study aimed to discover how college students felt about the moral implications of the tool for academic work and investigate the broad impacts of ChatGPT on essential aspects of education, including learning outcomes, student engagement, and academic performance. Under these circumstances, it becomes imperative to investigate the effects of ChatGPT usage on college students. Furthermore, this study contributes to the expanding body of research on AI and education, allowing students to make informed decisions in the dynamic world of higher education. This study is novel because it synthesises the impact of AI, specifically ChatGPT, on higher education students in Arabic-speaking countries through a thorough scoping of quantitative data collection. The main research questions (RQ) for the interviews that form the basis of this study are:

RQ1 Have you previously used ChatGPT for educational purposes?
RQ2 What is the frequency of your use of ChatGPT for learning?
RQ3 What is the duration of your usage of ChatGPT for educational purposes?

RQ4 Do you find ChatGPT a valuable tool for your academic work?

RQ5 Are the instructors happy with ChatGPT data and its usage?

RQ6 What types of academic tasks do you use ChatGPT for?

RQ7 How much do you rely on ChatGPT for your academic work?

RQ8 Have you faced any challenges or difficulties with ChatGPT?

RQ9 Would you recommend ChatGPT to other postgraduate students for academic purposes?

As we begin this investigation, it is essential to consider the changing role of technology in education and how tools like ChatGPT can improve the educational experience for students, giving them the knowledge and skills needed for success in the twenty-first century. This study is evidence of the ongoing search for novel approaches to promote a rich learning environment in higher education.

The rest of this study is structured: The strategy used for the literature review is described in Section 2. The method of the study is explored in Section 3; the highlighted results are discussed in depth in Section 4. The article is concluded in Section 5.

LITERATURE REVIEW

Artificial Intelligence (AI)

Artificial Intelligence encompasses hardware and software capable of reasoning, absorbing, gathering data, interacting, controlling, and recognising objects. John McCarthy used the term Artificial Intelligence (AI) in 1956 to describe a relatively new area of computer science that aims to mimic human behaviour in machines. The topic "mid-twentieth century" is relatively new to the field of study (Alafnan et al., 2023).

Advances in AI have produced numerous applications in education and wellness in recent years. With various data, AI systems can be trained to mimic the human brain and perform repetitive tasks. AI has also been applied to education to enhance academic support and administrative functions. Among the best examples are Intelligent Instructional Systems (ITS), which can mimic private tutoring sessions one-on-one. A meta-findings analysis indicated that IT has had a somewhat favorable effect on college students' academic performance. However, the development of ITS can be challenging since it calls for the improvement of feedback phrasing and dialogue strategies besides the creation of content and design (Lo, 2023).

ChatGPT

Eaton et al. state that Open AI's ChatGPT is a language model that can understand and converse in the standard human language Eaton *et al.* (2023). It can generate numerous texts and codes with practice and other text input. A computer programme called ChatGPT mimics human speech and comprehension (Haluza & Jungwirth, 2023).

ChatGPT can respond to context and have naturalistic dialogues. OpenAI released a new version of ChatGPT at the end of 2022. Better than expected, ChatGPT's features have sparked much conversation about misuse scenarios (Malinka *et al.*, (2023). ChatGPT is an intelligent chatbot developed by OpenAI, an American AI company. Within a week of its launch on November 30, 2022, ChatGPT had over a million users. A LLM that can converse naturally and adapt to the situation is called ChatGPT. There are three versions of ChatGPT available: GPT-4 (upgraded), ChatGPT Plus (paid), and ChatGPT (free) (Sok & Heng, 2023). Users can use ChatGPT, homework, essays, and even the creation of contracts to answer exam-style questions. It is acknowledged that one extensive study and "creative intelligence." ChatGPT and other AIs are facilitating automation and promoting novel forms of creativity (Zhai, 2023).

The increased accessibility of innovative AI technology in recent years has significantly impacted various fields, including research and education. One such technique is OpenAI's ChatGPT, a mighty big language model. With the help of this technology, teachers and students can take advantage of exciting opportunities like customised feedback, improved accessibility, interactive discussions, lesson planning, assessment, and innovative ways to teach challenging subjects. However, ChatGPT presents several risks to the established research and education system, such as the potential for online exam cheating, the creation of human-like text, a decline in critical thinking abilities, and challenges assessing information generated by ChatGPT (Kasneci et al., 2023).

ChatGPT in Educational Contexts

In educational settings, ChatGPT facilitates the creation of exams, the writing of essays, and language translation. It also allows users to ask and respond to various questions, summarise readings, and work with classmates. Gupta *et al.*, (2023), claim that this model can also show creativity in writing on nearly any subject, from an argumentative or almost persuasive research study to a single paragraph. According to Atlas, (2023), implementing ChatGPT and other language models could significantly improve postsecondary education. These models can be used in various contexts, including writing support, language acquisition, research, and administration. Therefore, it is possible to argue that ChatGPT has the potential to be an important tool for analysis and education. Many academics have indicated ChatGPT's serious disadvantages, although it might have several educational benefits. Based on a preliminary review, (Baidoo-Anu & Owusu Ansah, 2023; Creeger, 2006) found that the current version of ChatGPT did not consistently provide accurate answers to questions about anatomical facts. Therefore, there are advantages and disadvantages to this well-known AI technology (Sok & Heng, 2023).

Opportunities and Challenges

A prominent advantage of ChatGPT in education is its capacity to increase student engagement. Atlas, (2023) to foster student creativity, provide individualised coaching, and effectively prepare students for working with AI systems in the workplace, educational institutions should use ChatGPT with caution. Additionally, ChatGPT can enhance learning outcomes by encouraging student motivation and engagement during asynchronous online learning sessions or activities (Alves de Castro, 2023). The personalisation of learning provided by ChatGPT is another significant advantage for education. Every student may receive responses from ChatGPT that are suitable for their learning style and performance level. Personalised feedback from ChatGPT can aid students in learning more efficiently than through more conventional means. ChatGPT can help pupils gain new languages (Cano et al., 2023). Students can improve their language abilities by imitating real-life conversations in a secure, judgement-free setting. Students are free to experiment with new idioms, grammar rules, and phrases without worrying about making mistakes. Students can also receive tailored language feedback and suggestions for improvement from ChatGPT.

However, privacy is one of the biggest issues with using ChatGPT in the classroom. Student data collected and stored by ChatGPT may be misused, giving rise to privacy concerns (Tlili et al., 2023). Additionally, students might feel ashamed to provide private information to an AI-powered system. Academic integrity presents another obstacle to the use of ChatGPT in the classroom. Students may use ChatGPT to cheat on tests or assignments because it might offer responses like a human's. Instructors who use ChatGPT in the school must create strategies to prevent cheating (Cano et al., 2023; Tlili et al., 2023). Another drawback of ChatGPT in education is prejudice (Cano et al., 2023). ChatGPT learns from the data it is taught; if biased, the system may reinforce the bias (Ray, 2023), for example, the system may produce limited results if the data used to train ChatGPT is biased towards a specific gender or race. Highlights how important it is for educators to ensure that ChatGPT's training data is representative and diverse (Alves de Castro, 2023).

Postgraduate Performance in Arab Countries

Arab postgraduate performance refers to the academic and professional achievements of Arab postgraduates; this includes science, technology, engineering, arts, and humanities. Research output, publication contributions, conference participation, and master's and doctorate completion are used to evaluate postgraduate performance (Al-Zubaidi, 2012). It examines how postgraduate education affects professional advancement and social help. Postgraduate performance is emphasised to promote higher education and research excellence in Arab countries, boosting knowledge economies and fostering innovation and expertise in many domains (Al-Zubaidi, 2012; Lightfoot, 2011).

Arab postgraduate performance suffers several academic issues that hinder intellectual potential. Research funding and resources are few, preventing postgraduate students from undertaking rigorous and impactful research (Abdulkareem, 2013) [21]. Many Arab educational institutions lack experienced faculty; therefore, mentorship and direction are lacking. Postgraduate students rarely see groundbreaking research or curricula because they often fall below global norms (Hadidi & Al Khateeb, 2015; Muassomah et al., 2023). Postgraduate students struggle to participate in international academic debates because of language difficulties and a lack of high-quality academic journals and resources (Al-Zubaidi, 2012; Rechards & Al-Zubaidi, 2015). A complete educational infrastructure reform, more significant research and development funding, and effective mentorship programmes for Arab scholars are needed to address these issues (Elgamri et al., 2023).

ChatGPT is essential to promote academic quality and innovation in Arab postgraduate performance (Mohammed et al., 2023). Advanced students benefit from ChatGPT's dynamic conversation and knowledge exchange platform. Students can have meaningful discussions, clarify complex subjects, and access a wealth of knowledge using its natural language processing. ChatGPT offers postgraduate students a personalised and interactive learning experience beyond standard methods (Abouammoh et al., 2023).

METHOD

Structured Questionnaire

Develop a questionnaire with a clear structure consisting of closed-ended questions. Closed-ended questions facilitate quantitative data collection by providing predetermined response options. Include a mix of Likert scale items to measure attitudes and perceptions, multiple-choice questions for demographic information, and ranking questions for prioritisation (Crewswell, W, 2009). Data for the study was gathered through a questionnaire. To collect data, each participant answers the same questions on a questionnaire in precisely the same order. There are no open-ended questions. Obtaining primary data from large samples through questionnaires is a productive method.

Cross-Sectional Design

This study was conducted among academic students, and interviews and data collection were conducted from July 1, 2023, to July 15, 2023. Data were provided through university mailing lists, online platforms, and class announcements. Communication of the research aim, voluntary nature of participation, and confidentiality of responses were the hallmarks of this study. Initially, participants were given a consent form outlining the study's aims, procedures, and rights. Only participants who agreed to participate and provide information in the study were included.

Data Collection Process

The structured questionnaires and Google Forms will be distributed electronically through email and online survey platforms. Participants also accessed the questionnaires via Q.R. codes in the print announcements. Reminder emails were sent to participants during data collection to encourage timely responses. One hundred seventy questionnaires were distributed to Middle Eastern Countries (MEC) students. Online surveys were used as a platform to distribute the questionnaire. Respondents are students who are studying at several universities in this region. A student is requested to fill out the online survey. The students were asked to ensure they had used ChatGPT for the last three months as a filter question. Students who did not use ChatGPT were ignored from the survey. Close the data collection on November 15, 2023. Ensure that all responses are collected within the specified timeframe to maintain the study's cross-sectional nature. One hundred questionnaires were returned and used for further analysis.

Data Analysis

This study used quantitative analysis. Once data collection is complete, perform quantitative analysis using statistical software. The research used SPSS and conducted the comments. Analyse demographic information, frequency of ChatGPT use, perceived impact, and other relevant variables.

Reporting and Limitation

As a result, we generate a comprehensive report presenting key findings, including the prevalence of ChatGPT use, participants' perceptions, and any significant correlations between variables. Acknowledge that the results represent a snapshot of the participants' experiences at a specific time. Recognise that changes in technology or educational practices may not be captured after the data collection period.

RESULTS

Arab postgraduate students were asked to complete questionnaires to provide demographic and academic ChatGPT usage data. Participants gave their informed consent for the ethical collection of their data. For postgraduate students in the area, the study looked at ChatGPT's effects on academic performance and its advantages, disadvantages, and ethical concerns.

Section A: Respondents' Profile

The demographic information of the students who participated in this study is displayed in Table 1, Respondents' Profile. Eighty per cent of the 100 respondents are male students, making them most respondents. There are 20% more female students than male students overall. This gender disparity within the respondent pool prompts further exploration into the factors influencing participation rates, potentially shedding light on the study's generalizability and the broader context of gender dynamics within the educational environment under examination.

In the demographic distribution of the surveyed student population, a notable 61% fell within the 25–30 age range, indicating a predominant concentration of individuals in the early stages of adulthood. Following closely, 18% of the students were aged between 30 and 35, showcasing a diverse age spectrum. The study further revealed that 12% of participants were in the 35-40 age bracket, underlining the continued presence of mature students. Notably, 9% of the surveyed individuals were 45 years old or older, emphasising the inclusion of a more seasoned demographic within the study. Regarding marital status, a clear majority of 66% of the students identified as single, while the remaining 34% reported being married, providing insights into the varied life stages and relationship statuses within the academic community under examination.

Participants in the study come from various Arabic-speaking countries, highlighting the area's diversity. Remarkably, 46 percent of the students were from MEC, including those with rich histories and cultures. That 33% of the participants

Table 1. Respondents Demographic Information

	Description	Frequency	Percent
Gender	Male	80	80.0
	Female	20	20.0
	Total	**100**	**100.0**
Age	25 - 30	61	61.0
	30 - 35	18	18.0
	35 - 40	12	12.0
	45+ above	9	9.0
	Total	**100**	**100.0**
Marital Status	Single	66	66.0
	Married	34	34.0
	Total	**100**	**100.0**
Countries: *Based on the regions of the Arab World*	Middle East countries	46	46.0
	North African countries.	33	33.0
	Horn of Africa countries.	17	17.0
	Indian Ocean islands countries.	4	4.0
	Total	**100**	**100.0**
Education	Master Degree	77	77.0
	Doctoral Degree	23	23.0
	Total	**100**	**100.0**
Field of study	Business and Management	38	39.0
	Engineering and Technology	40	39.0
	Education and Health	13	13.0
	Natural and Applied sciences	9	9.0
	Total	**100**	**100.0**

were from North Africa illustrates how wide the study's geographic scope was. With 17% of the student body coming from the Horn of Africa, the cultural mosaic under study became distinctive. Four percent of the participants were from the islands in the Indian Ocean, highlighting the study's dedication to include different viewpoints from the Arabic-speaking community. This composition's diversity guarantees a thorough comprehension of these regions' subtle cultural, linguistic, and educational aspects.

The research project revealed a significant variation in the educational background of the participating students, with a discernible majority holding graduate degrees.

A noteworthy 77% of the participants indicated they had earned a master's degree, underscoring the large percentage of postgraduate-qualified individuals in the study population. Notably, 23 percent of research participants had earned a doctorate, the highest degree possible. The distribution highlights the diversity and validity of the research findings by showcasing a highly educated and varied pool of respondents. It also alludes to the study's focus on individuals who strongly commit to their academic pursuits. The significant involvement of those holding master's and doctoral degrees shows the high caliber of academic work and breadth of expertise in the field covered by the study.

According to a recent study that looked at the students' academic interests who took part, 40% of the respondents said they were interested in studying engineering and technology, making it the most sought-after field. Business and management came in second, with 38% of students opting to learn in this field. According to the survey, students were more interested in practical and professional areas; education and health combined attracted 13% of their attention. Despite being fundamental and essential to many industrial sectors, just 9% of respondents preferred the natural and applied sciences. This distribution emphasises the application-oriented and business-driven parts of education among the student population questioned.

Section B: How Students Utilise ChatGPT

Table 2 provides an explanation of the research question's answer and shows how students use ChatGPT. It also provides an example of the research questions' results:

RQ1: Have you previously used ChatGPT for educational purposes?

The study's participants showed high usage of ChatGPT for learning; 87% said they had previously used the language model for teaching. This indicates that numerous students who answered the survey were aware of ChatGPT and had encountered it in an educational context. Different academic settings may have used the language model differently, as indicated by the 13% of participants who said they had never used ChatGPT for educational purposes. The findings highlight ChatGPT's value as a teaching aid by highlighting how students find information and support for their academic endeavors.

RQ2: What is the frequency of your use of ChatGPT for learning?

A recent study found that students are incorporating ChatGPT into their academic routines more and more. According to the findings, 47 percent of participants use ChatGPT in their regular educational activities. Twenty-four percent of students who participated in the survey said they rely on the language model for academic support every week. Eighteen per cent of students use ChatGPT as a useful study tool each month. Interestingly, only 11% of respondents claimed they had never used ChatGPT for academic work, suggesting that most students embrace technology

Table 2. Detail About how Students Utilize ChatGPT

Questions	Status	Frequency	Percent
Have you previously used ChatGPT for educational purposes?	Yes	87	87.0
	No	13	13.0
	Total	**100**	**100.0**
What is the frequency of your use of ChatGPT for learning?	Daily	47	47.0
	Monthly	18	18.0
	Weekly	24	24.0
	Never	11	11.0
	Total	**100**	**100.0**
How long have you been using ChatGPT for academic purposes?	Less than three months	57	57.0
	3 - 6 months	28	28.0
	Over 12 months	15	15.0
	Total	**100**	**100.0**
Do you find ChatGPT to be a useful tool for your academic work?	Yes, very useful	58	58.0
	Yes, somewhat useful	34	34.0
	No, not very helpful	6	6.0
	No, pointless	2	2.0
	Total	**100**	**100.0**
Are the instructors happy with ChatGPT data and its usage?	No, they claim that it "encourages academic dishonesty"	67	67.0
	Indeed, they say it "improves knowledge and data discovery in a short amount of time"	31	31.0
	Total	**98**	**98.0**
What types of academic tasks do you use ChatGPT for?	Writing assignments	28	28.0
	Research	19	19.0
	Exam preparation	9	9.0
	Summarisation	2	2.0
	Interpretation and rewording	14	14.0
	All of Them	28	28.0
	Total	**100**	**100.0**
How much do you rely on ChatGPT for your academic work?	Not at all	34	34.0
	A little	41	41.0
	Moderately	25	25.0
	Total	**100**	**100.0**

Table 2. Continued

Questions	Status	Frequency	Percent
Have you faced any challenges or difficulties with ChatGPT?	Sometimes, I lose points because of the ChatGPT context	17	17.0
	It is necessary to confirm the ChatGPT context twice	31	31.0
	Occasionally, I get incorrect context	18	18.0
	Every One of Them	34	34.0
	Total	**100**	**100.0**
Would you recommend ChatGPT to other postgraduate students for academic purposes?	Yes, without a doubt	58	58.0
	Neutral	30	30.0
	No, most likely not	12	12.0
	Total	**100**	**100.0**

and frequently use it to aid their studies. This pattern highlights the platform's significance as a useful and approachable resource for educational carrier.

RQ3: What is the duration of your usage of ChatGPT for educational purposes?

Students' participation with ChatGPT has varied throughout time when used for instructional purposes. A recent poll found that 57% of participants said they had used ChatGPT for educational work for less than three months, suggesting that most users had only used the platform for a short while. In contrast, 28% of students said they had used ChatGPT for three to six months in a learning environment, showing a slightly extended period of incorporation into their daily study routines. Remarkably, just 15% of the students who responded to the poll stated they had used ChatGPT for learning for over a year. This shows that a comparatively lesser percentage of users have continued to depend on the programme for their academic work over an extended period. These results highlight how ChatGPT's function in education is dynamic since numerous users integrate it into their learning processes over shorter periods.

RQ4: Do you find ChatGPT a valuable tool for your academic work?

It has become clear that ChatGPT is a valuable tool for academic work; in fact, 58% of students firmly believe that it offers significant advantages. An additional 34% of respondents think it is somewhat helpful, underscoring its beneficial effects on the academic environment. This broad endorsement implies that ChatGPT is a valuable tool for students to use when conducting research, writing, and engaging in other educational activities. Remarkably, just 6% believe it to be useless, and even fewer, at 2%, believe it to be completely meaningless. These figures reflect how well-received ChatGPT has been by students and show how it can improve and simplify several academic tasks, as shown in Figure (1).

Figure 1. Regarding the Utility of ChatGPT as a Tool for Academics

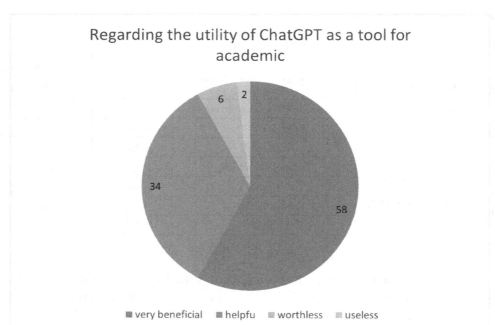

RQ5: Are the instructors happy with ChatGPT data and its usage?

While 67% of teachers are concerned that ChatGPT encourages academic dishonesty, students' perspectives on the technology are divided. They are worried that students may abuse the platform instead of participating in independent learning to get pre-made answers. Only 31% of teachers think ChatGPT makes a beneficial difference by improving data discovery and speeding up knowledge acquisition. This difference in viewpoints draws attention to the current discussion about the potential effects of AI-driven language models, such as ChatGPT, on education. While some teachers value the possibility of instantaneous access to knowledge, others are still apprehensive about how it may affect student autonomy and academic integrity.

RQ6: What types of academic tasks do you use ChatGPT for?

According to a recent study, ChatGPT is essential for several areas of students' academic endeavours. Approximately 28% of students use ChatGPT to create their written assignments because they value its help in producing well-organised and compelling text. ChatGPT's features also benefit research work; 19% of students use the model to make their research procedures more efficient. 14% of students use ChatGPT for paraphrasing and translation assignments, demonstrating the platform's adaptability in language applications. Remarkably, just 9% of respondents use ChatGPT to study for exams, showing that the model may help the review process.

Figure 2. ChatGPT is Essential for Several Areas of Students' Academic Endeavours

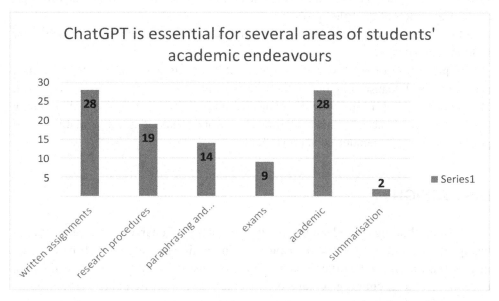

Merely 2% of pupils use ChatGPT for information summarisation. Notably, 28% of students use ChatGPT only for their academic needs, demonstrating the platform's broad acceptance and use among students, as shown in Figure (2).

RQ7: How much do you rely on ChatGPT for your academic work?

Regarding academic efforts, students' use of ChatGPT differs, showing various ways to incorporate this sophisticated language model into their work. A significant percentage of students, 34%, don't use ChatGPT and instead use more conventional techniques for their research and study. Conversely, 41% of students use ChatGPT only infrequently and only sometimes, adding its features to their toolset of academic resources. A handful of students, 25% of the total, fit the description of moderate users using ChatGPT as an additional tool for their coursework. In the quest for academic success, this distribution highlights the complex and dynamic relationship between students and AI.I. technologies.

RQ8: Have you faced any challenges or difficulties with ChatGPT?

Students face various obstacles and hurdles when using ChatGPT, as evidenced by a user experience survey. A noteworthy 31% of students report continuously verifying the context of ChatGPT, highlighting worries regarding the precision and consistency of responses. 18% of users describe situations in which they are given incorrect context, which could cause misinterpretations or false information. Notably, 17% of students have had points deducted from their assignments because of content generated by ChatGPT that does not fit the desired context. That 34% of students experience a mix of all these difficulties is perhaps the most startling sign of the frequency of complex problems.

RQ9: Would you recommend ChatGPT to other postgraduate students for academic purposes?

Despite these obstacles, most students 58% of students suggest their friends use ChatGPT because they understand its usefulness in particular situations. Still, 30% express no opinion, which is cautious, and 12% advise against others using ChatGPT, presumably because of lingering worries about its dependability and potential effects on academic achievement.

DISCUSSION

The findings thoroughly understand the study participants' characteristics, educational backgrounds, and geographic distribution. The main conclusions will be covered in this talk, along with other noteworthy patterns and trends that were noticed. Foremost, one fascinating feature of the study is the participants' demographic makeup. The study's inclusivity and representation of varied perspectives are called into doubt by the preponderance of male respondents, which may indicate a gender gap in the sample. This result calls for more investigation into the causes of gender distribution and how it affects the study's conclusions. Another exciting feature of the participants' age distribution is that most were in the 25 to 30 age group. This focus on a specific age range may affect how broadly applicable the study's conclusions are. The analysis would gain more depth if the causes of this age distribution were investigated and its effects on the study's findings were considered.

Regarding academic levels, the large number of respondents with doctorates (23%) and master's degrees (77%) suggests that the sample is well-educated. This may improve the calibre of answers, but it also raises questions about whether the results can apply to groups of people with lesser levels of education. Evaluating how educational levels might influence the study's conclusions and addressing any biases from this educational skew would strengthen the research. As indicated by the split of study areas, the respondents' top choices were engineering and technology, followed by business and management. This conclusion highlights a specific intellectual bent within the sample and calls into question the study's focus and applicability. It would be instructive to investigate the reasons behind these preferences and consider how they impact the R.Q. and findings.

The geographical dispersion of respondents among the regions gives the study an extra depth. The results may have regional biases because of the majority's (46%) and (33%) Middle Eastern and North African origins. Conversely, the Indian Ocean islands and the Horn of Africa account for smaller percentages (17%). (4 percent).

If the reasons for this regional concentration were identified and evaluated for any impact on the study's conclusions, a more nuanced interpretation might be conceivable.

Table 2's results shed light on how widely student participants in the field use ChatGPT for academic purposes. That 87% of students said they had used the tool at least once suggests a noteworthy occurrence, according to the data. This diagram illustrates how ChatGPT and other AI.I. language models are increasingly used in academic settings. The results show an interesting trend: nearly half of the students surveyed said they used ChatGPT daily for schooling. The frequency of communication indicates that ChatGPT has developed into an essential part of the academic toolkits for many postgraduate students. Moreover, 58% of students believe ChatGPT is extremely beneficial, which indicates that they view its impact on their academic endeavours favourably.

However, the study also emphasises how challenging it can be to use ChatGPT in a schoolroom. Two main issues are the need for users to get double context and the potential for receiving inaccurate information. This concern casts doubt on the tool's dependability and raises the possibility of unintentional academic dishonesty, a crucial consideration in learning environments that respect integrity and authenticity. Despite these challenges, the most common educational uses of ChatGPT are for research and writing assignments. Interestingly, only 25% of students use the tool moderately, suggesting that academic assignments should only occasionally use AI-generated content. Despite the challenges, most students (58%) recommend ChatGPT to their peers. This may indicate that users still find ChatGPT to have more benefits than drawbacks despite the acknowledged restrictions, at least for those who have integrated it into their regular study routines.

The findings of the study suggest that the student participants in the Arab geographic areas that were included in the sample represent the overall student population. Interestingly, the results indicate that PhD students frequently use ChatGPT as a helpful academic resource. However, the study also shows how incomplete ChatGPT integration into these students' academic schedules is. Notwithstanding the tool's acknowledged value, application-related challenges cause a nuanced perspective. The study emphasises the importance of considering the moral quandaries and difficulties of implementing ChatGPT in educational environments. Despite these challenges, a few participants indicated they knew the tool's potential by saying it would help other postgraduate students. This puts out a call to action for educators and students to thoroughly weigh the moral and practical implications of ChatGPT before fully incorporating it into Arab academic procedures.

CONCLUSION

The study found that postgraduate students in the area, primarily those studying business, management, engineering, and technology, frequently use ChatGPT for academic purposes. Concerns have been raised regarding how it might promote academic dishonesty and the challenges of the correctness and context of its responses. However, most students find it very beneficial for their academic work. These findings imply that universities and educators should provide moral guidelines for the ethical use of AI-based tools in academic research and that students should notice the risks associated with using these services. Furthermore, it is critical to support the creation of AI-based tools specially made for academic use with strong anti-bias and plagiarism protection. Arab nations can enhance their postgraduate performance by using ChatGPT to link students to educational resources as they advance in their studies.

Future studies could look at how well ChatGPT improves learning outcomes, how it affects writing and critical thinking abilities, and whether it's workable to create AI-based tools to help with summarisation tasks. Furthermore, future studies may investigate academic institutions' and teachers' attitudes and guidelines about using AI-based teaching aids. Finally, even though ChatGPT has become a popular and widely used tool in higher education, it is crucial to deliberate the ethical consequences and limitations. The study's findings provide valuable information about the intricate nature of students' conversations with ChatGPT and emphasise the significance of thoroughly understanding the platform's function in academic settings. Government organisations, researchers, and educators must address the ethical implications of integrating AI.I. tools into educational processes as their use expands. This entails creating guidelines for moral usage, raising awareness of potential risks, and cultivating an intellectual honesty culture in the AI.I. era.

REFERENCES

Abdulkareem, M. N. (2013). An investigation study of academic writing problems faced by arab postgraduate students at Universiti Teknologi Malaysia (UTM). *Theory and Practice in Language Studies, 3*(9), 1552–1557. doi:10.4304/tpls.3.9.1552-1557

Abouammoh N. Alhasan K. Raina R. Malki K. A. Aljamaan F. Tamimi I. Muaygil R. Wahabi H. Jamal A. Al-Tawfiq J. A. Al-Eyadhy A. Soliman M. Temsah M.-H. (2023). *Exploring Perceptions and Experiences of ChatGPT in Medical Education: A Qualitative Study Among Medical College Faculty and Students in Saudi Arabia.* doi:10.1101/2023.07.13.23292624

Al-Zubaidi, K. O. (2012). The Academic Writing of Arab Postgraduate Students: Discussing the Main Language Issues. *Procedia: Social and Behavioral Sciences*, *66*, 46–52. doi:10.1016/j.sbspro.2012.11.246

Alafnan, M. A., Dishari, S., Jovic, M., & Lomidze, K. (2023). ChatGPT as an Educational Tool: Opportunities, Challenges, and Recommendations for Communication, Business Writing, and Composition Courses. *Journal of Artificial Intelligence and Technology*, *3*(2), 60–68. doi:10.37965/jait.2023.0184

Alves de Castro, C. (2023). A Discussion about the Impact of ChatGPT in Education: Benefits and Concerns. *Journal of Business Theory and Practice*, *11*(2), 28. doi:10.22158/jbtp.v11n2p28

Atlas, S. (2023). ChatGPT for Higher Education and Professional Development: A ChatGPT for Higher Education and Professional Development: A Guide to Conversational AI Guide to Conversational AI Terms of Use. *DigitalCommons@ URI*. https://digitalcommons.uri.edu/cba_facpubs

Baidoo-Anu, D., & Owusu Ansah, L. (2023). Education in the Era of Generative Artificial Intelligence (AI): Understanding the Potential Benefits of ChatGPT in Promoting Teaching and Learning. SSRN *Electronic Journal, 7*(December), 52–62. doi:10.2139/ssrn.4337484

Cano, Y. M., Venuti, F., & Martinez, R. H. (2023). *ChatGPT and AI Text Generators: Should Academia Adapt or Resist?* https://hbsp.harvard.edu/inspiring-minds/chatgpt-and-ai-text-generators-should-academia-adapt-or-resist

Creeger, M. (2006). Evolution or Revolution? *ACM Queue; Tomorrow's Computing Today*, *4*(10), 56. doi:10.1145/1127854.1127873

Crewswell, W. J. (2009). *Research Design Qualitative*. Quantitave and Mixed Methods Approaches.

Eaton, S. E., Brennan, R., Wiens, J., & Mcdermott, B. (2023). *Artificial Intelligence and Academic Integrity: The Ethics of Teaching and Learning with Algorithmic Writing Technologies*. Academic Press.

Elgamri, A., Mohammed, Z., El-Rhazi, K., Shahrouri, M., Ahram, M., Al-Abbas, A. M., & Silverman, H. (2023). Challenges facing Arab researchers in conducting and publishing scientific research: A qualitative interview study. *Research Ethics*. Advance online publication. doi:10.1177/17470161231214636

Gupta, B., Mufti, T., Sohail, S. S., & Madsen, D. Ø. (2023). ChatGPT : A Brief Narrative Review. Preprints.*Org,* 1–12. doi:10.20944/preprints202304.0158.v2

Hadidi, M. S., & Al Khateeb, J. M. (2015). Special Education in Arab Countries: Current challenges. *International Journal of Disability Development and Education*, *62*(5), 518–530. doi:10.1080/1034912X.2015.1049127

Haluza, D., & Jungwirth, D. (2023). Artificial Intelligence and Ten Societal Megatrends: An Exploratory Study Using GPT-3. *Systems*, *11*(3), 120. Advance online publication. doi:10.3390/systems11030120

Kasneci, E., Sessler, K., Küchemann, S., Bannert, M., Dementieva, D., Fischer, F., Gasser, U., Groh, G., Günnemann, S., Hüllermeier, E., Krusche, S., Kutyniok, G., Michaeli, T., Nerdel, C., Pfeffer, J., Poquet, O., Sailer, M., Schmidt, A., Seidel, T., ... Kasneci, G. (2023). ChatGPT for good? On opportunities and challenges of large language models for education. *Learning and Individual Differences*, *103*(February), 102274. Advance online publication. doi:10.1016/j.lindif.2023.102274

Lightfoot, M. (2011). Promoting the knowledge economy in the Arab world. *SAGE Open*, *1*(2), 1–8. doi:10.1177/2158244011417457

Lo, C. K. (2023). What Is the Impact of ChatGPT on Education? A Rapid Review of the Literature. *Education Sciences*, *13*(4), 410. Advance online publication. doi:10.3390/educsci13040410

Malinka, K., Peresíni, M., Firc, A., Hujnák, O., & Janus, F. (2023). On the Educational Impact of ChatGPT: Is Artificial Intelligence Ready to Obtain a University Degree? *Annual Conference on Innovation and Technology in Computer Science Education, ITiCSE, 1*, 47–53. 10.1145/3587102.3588827

Mohammed, A. A. Q., Al-ghazali, A., & Khalid, A. S. (2023). Exploring ChatGPT Uses in Higher Studies. *Journal of English Studies in Arabia Felix*, *2*(2), 9–17. doi:10.56540/jesaf.v2i2.55

MotlaghN. Y.KhajaviM.SharifiA.AhmadiM. (2023). The Impact of Artificial Intelligence on the Evolution of Digital Education: A Comparative Study of OpenAI Text Generation Tools including ChatGPT, Bing Chat, Bard, and Ernie. http://arxiv.org/abs/2309.02029

Muassomah, M., Hakim, A. R., & Salsabila, E. L. (2023). *Arabic Learning Challenges*. Atlantis Press SARL. doi:10.2991/978-2-38476-002-2_45

Ray, P. P. (2023). ChatGPT: A comprehensive review on background, applications, key challenges, bias, ethics, limitations and future scope. *Internet of Things and Cyber-Physical Systems*, *3*(March), 121–154. doi:10.1016/j.iotcps.2023.04.003

Rechards, C., & Al-Zubaidi, K. O. (2015). Arab Postgraduate Students in Malaysia : Identifying and overcoming the cultural and language barriers Dr. Khairi Obaid Al-Zubaidi Language Academy, Technology Malaysia (UTM) Arab Postgraduate Students in Malaysia : Identifying and overcoming the cult. *English Journal, 1,* 107–129.

Sok, S., & Heng, K. (2023). ChatGPT for Education and Research: A Review of Benefits and Risks. SSRN *Electronic Journal, January.* doi:10.2139/ssrn.4378735

Tlili, A., Shehata, B., Adarkwah, M. A., Bozkurt, A., Hickey, D. T., Huang, R., & Agyemang, B. (2023). What if the devil is my guardian angel: ChatGPT as a case study of using chatbots in education. *Smart Learning Environments, 10*(1), 15. Advance online publication. doi:10.1186/s40561-023-00237-x

Zhai, X. (2023). ChatGPT for Next Generation Science Learning. SSRN *Electronic Journal*, 1–8. doi:10.2139/ssrn.4331313

Chapter 4
AIoT in Education Transforming Learning Environments and Educational Technology

Tarun Kumar Vashishth
iD https://orcid.org/0000-0001-9916-9575
IIMT University, India

Bhupendra Kumar
iD https://orcid.org/0000-0001-9281-3655
IIMT University, India

Vikas Sharma
iD https://orcid.org/0000-0001-8173-4548
IIMT University, India

Sachin Chaudhary
iD https://orcid.org/0000-0002-8415-0043
IIMT University, India

Kewal Krishan Sharma
iD https://orcid.org/0009-0001-2504-9607
IIMT University, India

Rajneesh Panwar
iD https://orcid.org/0009-0000-5974-191X
IIMT University, India

ABSTRACT

The integration of artificial intelligence and the internet of things (AIoT) in education has ushered in a paradigm shift, transforming traditional learning environments and revolutionizing educational technology. This chapter explores the multifaceted impact of AIoT on education, encompassing personalized learning experiences, intelligent tutoring systems, and the creation of smart educational spaces. By leveraging AI algorithms and IoT-connected devices, educators can tailor instructional content to individual student needs, fostering a dynamic and adaptive learning environment. The interconnectivity of devices enables real-time data collection, facilitating

DOI: 10.4018/979-8-3693-0993-3.ch004

insights into student performance, engagement, and preferences. Moreover, AIoT contributes to the development of smart classrooms equipped with interactive technologies, enhancing collaboration and active participation. As we navigate this transformative era, the potential of AIoT in education holds promise for optimizing educational outcomes and shaping a future where technology seamlessly integrates with the learning experience.

1. INTRODUCTION

In the rapidly evolving landscape of education, the confluence of Artificial Intelligence (AI) and the Internet of Things (IoT) has given rise to a transformative paradigm known as AIoT, or the Artificial Intelligence of Things. This amalgamation of intelligent algorithms and interconnected devices holds unprecedented potential to revolutionize learning environments and redefine educational technology. As we stand at the intersection of these two powerful technological forces, the educational landscape is undergoing a profound shift, moving away from traditional models towards a dynamic, adaptive, and personalized approach to learning. Hasko et al. (2020) propose a mixed fog/edge/AIoT/robotics education approach based on tripled learning.

1.1 Overview of AIoT (Artificial Intelligence of Things)

In the ever-evolving landscape of education, the convergence of Artificial Intelligence (AI) and the Internet of Things (IoT) has given rise to a transformative paradigm known as AIoT, or the Artificial Intelligence of Things. This amalgamation represents a watershed moment in the educational sector, where traditional pedagogical models are being reshaped and redefined by the infusion of intelligent algorithms and connected devices. AIoT stands at the intersection of AI-driven analytics and the vast network of IoT-enabled devices, offering a synergistic approach to education that transcends conventional boundaries. At its core, AIoT in education embodies a commitment to personalized and adaptive learning experiences. The integration of AI algorithms with IoT devices allows educational platforms to discern and respond to individual student needs in real time. This level of adaptability transforms the educational landscape into a dynamic, responsive environment, where content is based on each student's unique learning profile. This personalization extends beyond the curriculum, encompassing the entire educational journey and fostering a level of engagement and comprehension that traditional methods struggle to achieve. Stewart, Davis, & Igoche (2020) provide a impacts of AI, IoT, and AIoT on the artificial intelligence curriculum. Intelligent Tutoring Systems (ITS) form a pivotal component of AIoT in education. These systems leverage AI algorithms to provide personalized guidance, assistance, and feedback to students. Virtual assistants, powered by AI, offer real-time support, creating an interactive and adaptive learning environment. The integration of IoT ensures that data from students' interactions with educational content and devices is collected seamlessly, enabling ITS to refine and tailor their responses continually. Lin et al. (2021) conducted a study that aimed to explore the potential of augmented reality and AIoT learning for developing computational thinking skills.

AIoT further manifests in the creation of smart educational spaces, notably smart classrooms. Srivastava & Pathak (2023) provide AIoT-based smart education and online teaching. These environments are equipped with an array of interconnected devices, including interactive whiteboards, sensors, and IoT-enabled educational tools. The result is an immersive learning experience where students actively engage with content, collaborate in real time, and benefit from the insights derived from AI-driven analytics. Learning analytics, driven by the wealth of data generated by AIoT, plays a central role in shaping educational strategies. The continuous collection and analysis of data provide educators with invaluable insights into student performance, preferences, and areas of improvement. This data-driven approach enables a more proactive and responsive educational system, where interventions can be precisely targeted to address individual needs. While the promises of AIoT in education are substantial, challenges such as security, privacy, and ethical considerations must be navigated. The integration of these technologies requires a delicate balance between innovation and safeguarding sensitive information, emphasizing the need for robust cybersecurity measures and ethical frameworks. Ming et al. (2022) discusses the use of AIoT integration technology to facilitate multicultural knowledge and information literacy learning.

In subsequent sections, this chapter will delve into specific facets of AIoT in education, exploring personalized learning experiences, intelligent tutoring systems, smart educational spaces, learning analytics, and the challenges and considerations inherent in this transformative integration. Through an in-depth examination of case studies, success stories, and future trends, we aim to provide a comprehensive understanding of how AIoT is reshaping the educational landscape and propelling it into a future where technology seamlessly integrates with the learning experience.

1.2 Significance of AIoT in the Education Landscape

The significance of Artificial Intelligence of Things (AIoT) in the education landscape is underscored by its transformative potential to redefine traditional educational paradigms and chart a course towards more dynamic, adaptive, and personalized learning environments. In an era where educational technology plays an increasingly

Figure 1. AIoT in Education

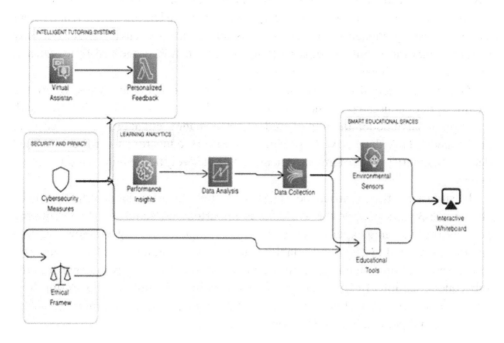

pivotal role, the integration of AI and IoT heralds a new chapter in how knowledge is imparted and acquired. At the heart of the significance lies the ability of AIoT to revolutionize personalized learning experiences. Traditional education systems often grapple with the challenge of catering to diverse learning styles and individualized needs. AIoT addresses this challenge by harnessing the power of AI algorithms and the connectivity of IoT devices to tailor educational content in real-time. The result is a learning experience that is not only more engaging but also highly adaptive, ensuring that each student's unique learning profile is taken into account.

Furthermore, the significance of AIoT in education is illuminated by its role in the creation of Intelligent Tutoring Systems (ITS). These systems act as virtual mentors, leveraging AI to provide personalized guidance, feedback, and support to students. The amalgamation of AI-driven analytics with IoT-connected devices enables these tutoring systems to comprehend the nuances of student interactions, refining their responses and recommendations continuously. This dynamic and responsive support mechanism transcends the limitations of traditional teaching methods, offering students a personalized educational journey that aligns with their pace and comprehension levels. Smart educational spaces, another crucial facet of AIoT, augment the significance of this integration. The concept of smart classrooms, enriched by an ecosystem of IoT-enabled devices, transforms traditional teaching environments into hubs of innovation and interaction. Interactive whiteboards, sensors, and other connected devices create an immersive learning experience where students actively engage with content. This interconnected environment not only fosters collaboration but also ensures that educators can leverage real-time data and insights to tailor their teaching strategies.

The significance of AIoT extends to the realm of learning analytics, where the wealth of data generated by these interconnected systems provides educators with a comprehensive understanding of student performance, engagement, and learning preferences. This data-driven approach empowers educators to make informed decisions, identify areas of improvement, and proactively address individual student needs.

In summary, the significance of AIoT in the education landscape lies in its ability to revolutionize how knowledge is imparted and acquired. By facilitating personalized learning experiences, powering Intelligent Tutoring Systems, creating smart educational spaces, and enabling data-driven insights, AIoT emerges as a transformative force shaping the future of education. As we delve deeper into the subsequent sections of this chapter, the exploration will unfold, illustrating how AIoT is not merely a technological integration but a catalyst for a more effective, engaging, and personalized educational journey.

1.3 Objectives of the Chapter

The objectives of this chapter are multifaceted, aiming to provide a comprehensive understanding of the transformative impact of Artificial Intelligence of Things (AIoT) in the realm of education. As we embark on this exploration, the chapter seeks to achieve the following key objectives:

Clarify the Fundamentals of AIoT in Education: The chapter endeavors to elucidate the fundamental concepts of AIoT, unraveling the intricate interplay between Artificial Intelligence and the Internet of Things specifically within the educational context. This includes defining the core components, mechanisms, and the overarching synergy that propels AIoT as a disruptive force in education.

Illuminate the Significance of Personalized Learning: One primary objective is to highlight the pivotal role of AIoT in facilitating personalized learning experiences. By leveraging AI algorithms and IoT connectivity, educational content can be dynamically tailored to the individual needs, preferences, and learning styles of each student, fostering a more engaging and effective learning journey.

Explore Intelligent Tutoring Systems (ITS) in Depth: Delve into the intricacies of Intelligent Tutoring Systems powered by AIoT. This involves an examination of how AI algorithms guide and support students in real-time, creating virtual mentors that adapt to individual learning curves. The exploration encompasses the capabilities of virtual assistants, adaptive feedback mechanisms, and the continuous refinement of tutoring strategies.

Examine the Evolution of Smart Educational Spaces: Explore the concept of smart classrooms and educational environments enhanced by IoT-connected devices. The objective is to analyze how these interconnected technologies create immersive learning experiences, fostering collaboration, interactivity, and real-time adaptability. The chapter will delve into the impact of smart spaces on traditional teaching methodologies.

Uncover the Role of Learning Analytics and Data-Driven Insights: Investigate the significance of data generated by AIoT in education. The objective is to showcase how learning analytics, powered by the wealth of data from interconnected systems, provides educators with actionable insights into student performance, engagement patterns, and areas for improvement. This includes a focus on the ethical considerations surrounding data collection and utilization.

In pursuit of these objectives, this chapter endeavors to contribute a holistic and nuanced perspective on the transformative role of AIoT in education, equipping readers with a comprehensive understanding of its applications, challenges, and the future trajectory of this dynamic integration in learning environments.

2. FUNDAMENTALS OF AIOT IN EDUCATION

2.1 Definition and Components of AIoT

Artificial Intelligence of Things (AIoT) in education represents a groundbreaking fusion of two transformative technologies: Artificial Intelligence (AI) and the Internet of Things (IoT). The definition of AIoT, in the educational context, encapsulates a dynamic synergy between intelligent algorithms and a network of interconnected devices. At its essence, AIoT harnesses the analytical power of AI to interpret data generated by IoT-connected devices, creating an ecosystem that adapts and responds intelligently to the ever-evolving needs of students and educators.

The components of AIoT in education constitute the building blocks of this innovative integration. The AI aspect involves sophisticated machine learning algorithms capable of processing large datasets, making predictions, and discerning patterns in student behavior and learning styles. These algorithms drive the intelligence behind adaptive learning systems, virtual tutors, and personalized content delivery. On the other hand, the IoT components comprise a diverse array of devices seamlessly connected within the educational environment. Wu, Chin and Lai (2022) present an interesting concept of utilizing environmental IoT detection tools for environmental education in college students. Sensors, smart boards, wearables, and other IoT-

Figure 2. AIoT in Education Architecture

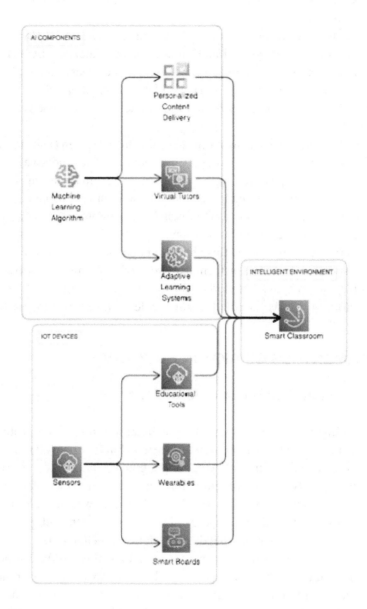

enabled educational tools form the intricate web that collects real-time data on student interactions, engagement levels, and environmental factors.

The definition and components of AIoT in education go beyond the conventional understanding of technological integration. It signifies a paradigm shift from traditional, one-size-fits-all educational models to a dynamic and adaptive approach that tailors the learning experience to individual students. In this ecosystem, AI-driven analytics constantly analyze the data flowing from IoT devices, enabling a nuanced understanding of students' strengths, weaknesses, and learning preferences.

2.2 Integration of Artificial Intelligence and IoT in Education

The integration of Artificial Intelligence (AI) and the Internet of Things (IoT) in education marks a revolutionary stride towards a more dynamic and responsive learning ecosystem. At its essence, this integration goes beyond a mere juxtaposition of technologies; it represents a symbiotic relationship where the analytical prowess of AI converges with the connectivity of IoT to reimaging the traditional educational landscape. AI, with its capacity to analyze vast datasets and derive meaningful insights, seamlessly intertwines with IoT, a network of interconnected devices, creating an educational synergy that adapts to the individual needs and preferences of learners.

In the context of education, this integration unfolds across multiple dimensions. Adaptive Learning Systems, a prime example, leverage the amalgamation of AI and IoT to tailor educational content based on individual student progress. These systems continuously analyze data generated by IoT-connected devices, discerning patterns in student interactions and learning behaviors. As a result, the curriculum becomes a fluid entity, dynamically adjusting to cater to the unique pace and comprehension levels of each student. This adaptability is not only transformative but also represents a departure from traditional static teaching methods, offering a personalized and engaging learning experience.

Virtual Intelligent Tutoring Systems (ITS) emerge as a focal point in the integration of AI and IoT. These systems harness AI algorithms to provide real-time guidance and support to students, functioning as virtual mentors. The interconnectedness with IoT allows these tutoring systems to gather data on student interactions, learning preferences, and areas of struggle. This continuous feedback loop refines the tutoring process, creating a personalized and adaptive learning journey. Virtual assistants, powered by AI, offer immediate responses to queries, creating an interactive environment that transcends the limitations of traditional classroom settings.

The integration of AI and IoT further extends into the creation of smart educational spaces. Classrooms equipped with IoT-enabled interactive whiteboards, sensors, and collaborative tools form intelligent environments where AI algorithms facilitate real-time adaptation of teaching strategies. For instance, the system can identify when a particular concept is challenging for a group of students and dynamically adjust the instructional approach, creating a collaborative and data-driven learning experience.

2.3 Synergies and Interactions between AI and IoT in Educational Settings

The dynamic interplay between Artificial Intelligence (AI) and the Internet of Things (IoT) within educational settings establishes a transformative synergy, redefining traditional learning environments and educational technology. At the core of this synergy lies the ability of AI to harness the rich tapestry of data woven by IoT-connected devices, creating an interconnected ecosystem that adapts and responds intelligently to the needs of both educators and students. This collaboration is not merely technological; it is a holistic approach that envisions a future where data-driven insights and intelligent algorithms converge to create a personalized and adaptive educational experience. One of the primary manifestations of these synergies is evident in the realm of Adaptive Learning Systems. AI algorithms, powered by data from IoT devices, dynamically tailor educational content to match the unique learning profiles of individual students. The continuous flow of real-time data allows AI to discern patterns in student interactions, understand their learning preferences, and adjust the curriculum on the fly. Consequently, the traditional one-size-fits-all approach to education is replaced with a nuanced and adaptive model, where each student's journey is personalized for optimal engagement and comprehension. Kuka, Hörmann & Sabitzer (2022) provide a study on Teaching and Learning with AI in Higher Education.

Intelligent Tutoring Systems (ITS) exemplify the collaborative interactions between AI and IoT. These systems leverage AI algorithms to provide personalized guidance and support to students in real time. The IoT component facilitates the collection of data on student interactions, areas of difficulty, and learning styles. This constant feedback loop refines the tutoring process, ensuring that interventions are precisely tailored to individual needs. Virtual assistants, an integral part of ITS, engage with students, offering adaptive feedback and creating an interactive learning environment that transcends the constraints of traditional classrooms. The creation of smart educational spaces is another tangible outcome of the synergies between AI and IoT. Classrooms equipped with IoT-enabled devices, such as interactive whiteboards and sensors, become intelligent environments where AI algorithms optimize teaching strategies based on real-time data. For instance, the system can recognize when a specific concept is challenging for a group of students and dynamically adjust the instructional approach to enhance comprehension and engagement. This collaborative approach fosters an environment where technology becomes an enabler, seamlessly integrating with the teaching process to elevate the overall learning experience.

Figure 3. AI and IoT Synergies in Education

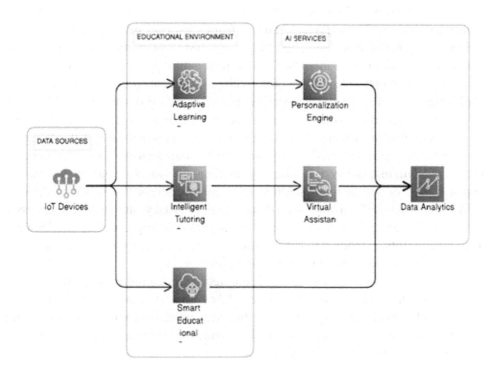

3. PERSONALIZED LEARNING EXPERIENCES

3.1 Adaptive Learning Systems

Adaptive Learning Systems stand as a hallmark of innovation within the realm of Artificial Intelligence of Things (AIoT) in education, representing a paradigm shift from conventional pedagogical models to a dynamic and personalized approach. At the heart of these systems lies the synergy between Artificial Intelligence (AI) and the Internet of Things (IoT), creating an ecosystem that tailors educational content in real-time based on individual student needs. The adaptive nature of these systems transcends the limitations of traditional one-size-fits-all approaches, offering a responsive and engaging educational experience. Tsai et al. (2021) suggest that an AIOT implementation course can improve the learning outcomes of senior high school students.

The defining characteristic of Adaptive Learning Systems is their ability to dynamically adjust the curriculum to suit the unique pace, learning styles, and comprehension levels of each student. This adaptation is made possible by the sophisticated AI algorithms that analyze and interpret vast datasets generated by IoT-connected devices. These devices, ranging from sensors to smart educational tools, create a network that continuously collects data on student interactions, engagement patterns, and performance metrics. The real-time nature of this data empowers AI to discern nuanced patterns, allowing for precise adjustments to the learning journey. Vashishth et al., 2023 provides a comprehensive overview of the potential benefits of IT for the education sector, as well as the challenges of its effective integration.

In practical terms, when a student engages with educational content, the Adaptive Learning System assesses their progress and comprehensiveness. If a concept proves challenging, the system adapts, providing additional resources, alternative explanations, or interactive exercises tailored to address the specific areas of difficulty. Conversely, if a student demonstrates mastery, the system seamlessly advances the curriculum, ensuring that learning remains engaging and challenging. This adaptability extends beyond traditional classroom settings, accommodating various learning environments, styles, and preferences.

As educational institutions worldwide embrace the potential of Adaptive Learning Systems, the transformative impact of AIoT becomes increasingly evident. These systems not only elevate the quality of education but also pave the way for a future where personalized, adaptive, and technology-enhanced learning experiences become the cornerstone of educational excellence.

3.2 AI-driven Content Customization

In the landscape of Artificial Intelligence of Things (AIoT) in education, AI-driven Content Customization emerges as a pivotal force, reshaping how educational material is presented and consumed. This facet of AIoT harnesses the power of Artificial Intelligence (AI) to dynamically tailor content based on individual student profiles and interactions within the Internet of Things (IoT) ecosystem. Unlike traditional static curriculum approaches, AI-driven Content Customization adapts to the diverse learning needs, preferences, and progress rates of each student, offering a truly personalized learning experience.

At its core, AI-driven Content Customization employs sophisticated algorithms to analyze the vast datasets generated by IoT-connected devices. These devices, ranging from smart educational tools to sensors capturing real-time student interactions, contribute to a wealth of information. AI processes this data, discerning patterns and understanding the nuances of how each student engages with educational content. This analysis facilitates the creation of a personalized curriculum that caters to individual strengths, weaknesses, and learning styles.

The adaptive nature of AI-driven Content Customization becomes particularly evident when students encounter challenging concepts or excel in certain areas. If a student grapples with a particular topic, the system can intuitively provide additional resources, alternative explanations, or interactive exercises tailored to address the specific areas of difficulty. Conversely, if a student demonstrates mastery, the system seamlessly progresses to more advanced material, ensuring that learning remains engaging and aligned with the student's pace.

As AIoT continues to shape the educational landscape, AI-driven Content Customization stands as a testament to the transformative potential of combining AI with IoT. The result is an educational experience where technology becomes an intuitive and responsive partner in the learning journey, fostering engagement, adaptability, and a deeper understanding of subject matter for each student.

3.3 Individualized Educational Paths

Individualized Educational Paths stand as a cornerstone in the revolution brought about by Artificial Intelligence of Things (AIoT) in education. In this paradigm, the integration of Artificial Intelligence (AI) with the Internet of Things (IoT) culminates in the creation of personalized and adaptive learning trajectories for each student. Unlike traditional educational models that follow a uniform curriculum, Individualized Educational Paths leverage the insights derived from AI analysis of IoT-generated data to tailor the educational journey to the specific needs, strengths, and learning styles of individual learners. Xie, H., Hwang, G. J., & Wong, T. L. (2021) on the transition from conventional to modern AI in education is an interesting read.

At the heart of Individualized Educational Paths is the capability of AI algorithms to discern intricate patterns in student interactions and performance. IoT-connected devices, such as smart tools and sensors, continuously collect data on how students engage with educational content. This real-time data serves as a dynamic canvas upon which AI paints personalized educational paths. When a student encounters a challenging concept, the system can intelligently adapt, offering supplementary resources, targeted exercises, or alternative explanations tailored to address the specific areas of difficulty. Conversely, if a student demonstrates mastery, the system seamlessly advances, ensuring that learning remain both challenging and engaging.

The concept of Individualized Educational Paths extends beyond the adaptation of content. It embraces a holistic approach, considering diverse factors that influence learning. This includes the student's preferred learning environment, the time of day when they are most receptive, and even their individual preferences for certain types of instructional media. By accounting for these nuances, Individualized Educational Paths create an immersive and personalized learning experience that resonates with the individual characteristics and preferences of each learner. Yadava et al. (2022)

Figure 4. Individualized Educational Paths in AIoT System

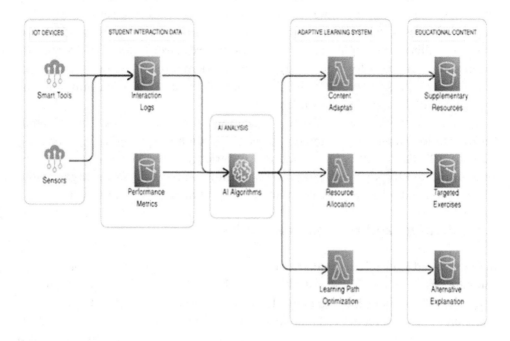

provides an analysis of the current state of AIoT, it fails to propose strategies to overcome the challenges faced by AIoT

Moreover, these paths are not static but evolve with the student's progress. The continuous feedback loop between AI and IoT ensures that the educational trajectory is refined and optimized over time. This iterative process enables educators to offer targeted interventions, measure progress accurately, and provide timely support to students.

4. INTELLIGENT TUTORING SYSTEMS

Intelligent Tutoring Systems (ITS) represent a pivotal advancement in education, and their integration with Artificial Intelligence of Things (AIoT) brings forth a new era of personalized and adaptive learning experiences. This section explores the multifaceted role of AIoT within Intelligent Tutoring Systems, encompassing the deployment of virtual assistants, AI-guided feedback and assessment mechanisms, and real-time monitoring to provide proactive support for students.

4.1 A. Role of AIoT in Tutoring and Mentorship

The fusion of AI and IoT within Intelligent Tutoring Systems redefines the role of tutoring and mentorship in education. AIoT empowers these systems to go beyond traditional tutoring by creating virtual mentors that leverage the connectivity of IoT devices. These mentors analyze real-time data from educational tools and sensors to understand students' learning patterns, strengths, and weaknesses. By adapting to individualized needs, AIoT-infused tutoring becomes a dynamic and responsive process, providing personalized guidance that goes beyond static, one-size-fits-all approaches.

4.2 Virtual Assistants in Education

Virtual assistants, as integral components of Intelligent Tutoring Systems, showcase the transformative impact of AIoT in education. These assistants leverage AI algorithms to interact with students, offering immediate responses to queries and providing contextual guidance. In the AIoT ecosystem, virtual assistants are connected to a network of devices, including smart boards and educational tools. This connectivity enables them to gather real-time data on student interactions, facilitating adaptive responses and creating an interactive learning environment. The role of virtual assistants extends beyond answering questions; they become AI-driven companions, aiding in problem-solving and enhancing the overall learning journey.

4.3 AI-guided Feedback and Assessment

AIoT plays a crucial role in revolutionizing feedback and assessment mechanisms within Intelligent Tutoring Systems. By analyzing data from IoT-connected devices, such as online quizzes, interactive exercises, and collaborative platforms, AI algorithms gain insights into students' performance and engagement. This data-driven approach enables personalized feedback that addresses individual strengths and areas for improvement. AIoT-infused assessment tools not only streamline the evaluation process but also offer a nuanced understanding of students' progress, allowing educators to tailor interventions effectively.

4.4 Real-time Monitoring and Support for Students

Real-time monitoring and support for students become a reality with the integration of AIoT in Intelligent Tutoring Systems. IoT devices, ranging from wearables to online learning platforms, continuously capture data on students' interactions, attention spans, and emotional states. AI algorithms analyze this data to identify patterns

Figure 5. Intelligent Tutoring System With AIoT

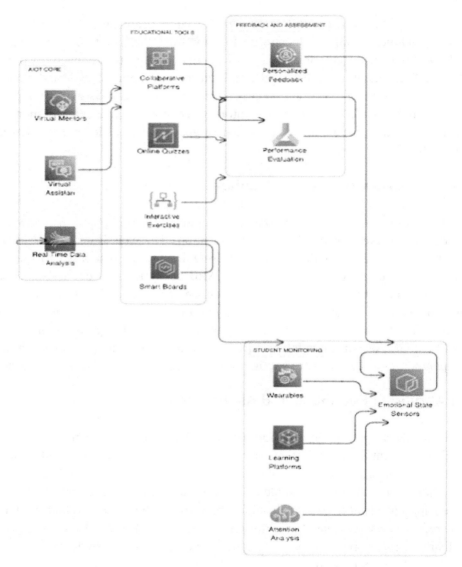

and potential challenges, enabling proactive support mechanisms. Educators can intervene promptly, offering targeted assistance and creating a supportive learning environment that adapts in real-time to students' evolving needs.

5. SMART EDUCATIONAL SPACES

Smart Educational Spaces represent a transformative approach to learning environments, leveraging the integration of Internet of Things (IoT) technologies to create dynamic and adaptive classrooms. This section explores the conceptual framework of Smart Classrooms, the integration of IoT-enabled interactive technologies, the role of connected devices and infrastructure, and the overall enhancement of collaboration and active learning within educational spaces.

5.1 Concept of Smart Classrooms

The concept of Smart Classrooms revolves around creating technologically enriched learning environments that foster interactive and engaging educational experiences. In these spaces, traditional teaching methods are augmented with cutting-edge technologies to create an immersive and dynamic atmosphere. Smart Classrooms leverage IoT to connect devices, capture real-time data, and provide adaptive learning experiences tailored to individual student needs. This shift from a static to a dynamic learning environment enhances overall student engagement and comprehension.

5.2 IoT-Enabled Interactive Technologies

The integration of IoT-enabled interactive technologies is a cornerstone of Smart Educational Spaces. These technologies encompass interactive whiteboards, smart projectors, and other collaborative tools that facilitate seamless communication between students, educators, and digital content. Quy et al (2023) discuss the potential application of AI and digital transformation in higher education. IoT connectivity allows these devices to interact with each other and gather data on student interactions. For example, an interactive whiteboard can capture collaborative exercises, enabling educators to gauge group dynamics and individual contributions. This integration transforms traditional lectures into interactive sessions, promoting student participation and comprehension.

5.3 Connected Devices and Infrastructure

Smart Educational Spaces rely on a robust network of connected devices and infrastructure. IoT devices such as sensors, cameras, and smart devices are strategically deployed to capture data on various facets of the learning environment. Connected infrastructure ensures the seamless flow of information, enabling real-time analysis and adaptation. For instance, a network of sensors can monitor classroom temperature, lighting, and noise levels, creating an optimal environment for learning.

This connectivity not only enhances the efficiency of educational processes but also contributes to the creation of adaptive and personalized learning experiences. Zhang et al. (2021) presents an interesting approach to talent education for AI-based IoT system development and implementation using the CDIO concept.

Smart Educational Spaces redefine the traditional classroom by infusing it with IoT technologies. This evolution goes beyond the integration of devices; it envisions an environment where connectivity, adaptability, and collaboration converge to create a holistic and enriching educational experience for both educators and students.

6. LEARNING ANALYTICS AND DATA-DRIVEN INSIGHTS

In the realm of education, Learning Analytics and Data-Driven Insights play a pivotal role in shaping effective teaching strategies, personalized learning experiences, and overall educational outcomes. This section delves into the importance of data in educational environments, the integration of Artificial Intelligence of Things (AIoT) for Learning Analytics, ethical considerations in data collection, and the utilization of insights for continuous improvement.

6.1 Importance of Data in Educational Environments

The importance of data in educational environments cannot be overstated. Data serves as the foundation for informed decision-making, allowing educators and administrators to gain insights into student performance, engagement levels, and overall learning outcomes. In the digital age, the vast amount of data generated in educational settings, including assessments, online interactions, and attendance records, provides a comprehensive view of student progress. Analyzing this data enables educators to identify areas of strength and weakness, tailor instructional approaches, and implement interventions that cater to the unique needs of each student. Singh et al. (2023) aimed to evaluate the effectiveness of online learning in comparison to traditional classrooms in the digital age.

6.2 AIoT for Learning Analytics

AIoT emerges as a transformative force in Learning Analytics, enhancing the depth and efficiency of data analysis in educational environments. The integration of AI and IoT technologies enables the collection of real-time, granular data from various sources, such as smart devices, online platforms, and interactive tools. AI algorithms process this data, uncovering patterns, trends, and correlations that may not be immediately apparent through traditional methods. AIoT-driven Learning

Analytics empowers educators to create adaptive learning paths, predict student performance, and personalize educational experiences, ultimately optimizing the teaching and learning process.

6.3 Ethical Considerations in Data Collection

Ethical considerations in data collection are paramount in educational settings. As institutions gather and analyze vast amounts of student data, ensuring privacy, security, and transparency becomes imperative. Ethical guidelines must address issues such as informed consent, data anonymization, and the responsible use of AI algorithms. Striking a balance between harnessing the potential of AIoT for educational insights and safeguarding individual privacy is crucial to maintaining trust and ethical integrity within the educational community.

In conclusion, the intersection of Learning Analytics, Data-Driven Insights, and AIoT represents a powerful catalyst for positive change in educational environments. By recognizing the importance of data, navigating ethical considerations, and embracing continuous improvement through insights, educational institutions can harness the full potential of these technologies to enhance teaching and learning outcomes.

7. CHALLENGES AND CONSIDERATIONS

Navigating the landscape of AIoT in education presents various challenges and considerations that demand careful attention. This section explores the critical aspects of security and privacy concerns, the complexities of integration and potential solutions, the technological infrastructure requirements, and the imperative of addressing ethical dilemmas associated with AIoT in education.

7.1 Security and Privacy Concerns

Security and privacy concerns stand out as paramount challenges in the implementation of AIoT in education. The vast amount of sensitive student data collected through IoT devices and AI algorithms raises the stakes for cybersecurity. Safeguarding this data against unauthorized access, data breaches, and malicious activities becomes a critical imperative. Institutions must establish robust security protocols, encryption mechanisms, and access controls to protect the privacy and confidentiality of student information. Additionally, educating stakeholders about data security measures is essential to foster a culture of awareness and responsibility.

7.2 Integration Challenges and Solutions

Integrating AIoT technologies into existing educational systems poses significant challenges. Compatibility issues, interoperability challenges, and resistance to change from traditional teaching methods are hurdles that institutions must overcome. Solutions involve investing in comprehensive training programs for educators, ensuring seamless integration with existing technologies, and adopting flexible systems that can adapt to evolving educational needs. Collaborative partnerships between educational technology providers and institutions can facilitate smoother integrations and enhance the overall effectiveness of AIoT in education.

7.3 Technological Infrastructure Requirements

The successful deployment of AIoT in education necessitates robust technological infrastructure. High-speed and reliable internet connectivity, the deployment of IoT devices, and the implementation of AI algorithms require substantial investments in infrastructure. Educational institutions must assess their existing technological capabilities and make strategic investments to upgrade and expand their infrastructure. This includes considerations for network capacity, device compatibility, and scalability to accommodate the increasing demands of AIoT applications in educational settings.

7.4 Addressing Ethical Dilemmas in AIoT Education

The infusion of AIoT in education raises ethical dilemmas that require careful consideration. Issues such as data ownership, algorithmic bias, and the responsible use of AI technologies demand ethical frameworks and guidelines. Institutions must establish ethical guidelines for data collection, storage, and utilization, ensuring transparency and accountability. Moreover, addressing algorithmic biases and ensuring fairness in AI-driven decisions becomes crucial to avoid perpetuating existing inequalities. Ethical education and ongoing dialogue within the educational community are essential components of navigating these complex ethical considerations.

Addressing the challenges and considerations in the implementation of AIoT in education requires a holistic and proactive approach. By prioritizing security and privacy, finding effective solutions to integration challenges, investing in technological infrastructure, and establishing ethical guidelines, educational institutions can unlock the transformative potential of AIoT while safeguarding the well-being and privacy of students and educators.

Figure 6. Challenges in AIoT With Education

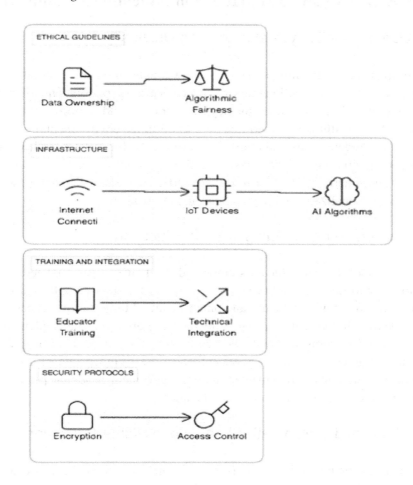

8. CASE STUDIES, INDUSTRIAL REPORT, AND SUCCESS STORIES

The implementation of AIoT in educational institutions has yielded transformative outcomes, enhancing teaching methodologies, personalizing learning experiences, and improving overall educational outcomes. Here are a few compelling case studies and success stories that showcase the successful integration of AIoT technologies in diverse educational settings:

8.1 Showcase of Implementations in Educational Institutions

Smart Campus Initiatives at a Leading University

A prestigious university embarked on a Smart Campus initiative, integrating AIoT technologies to create an intelligent and connected learning environment. IoT-enabled sensors monitored classroom occupancy, temperature, and lighting conditions in real time. AI algorithms analyzed this data to optimize environmental conditions for enhanced student focus and comfort. Additionally, the system tracked student movements, facilitating efficient resource allocation and security measures. The success of this initiative led to improved energy efficiency, enhanced campus security, and a more personalized and adaptive learning atmosphere.

Personalized Learning Paths in K-12 Education

In a K-12 school district, educators embraced AIoT-driven personalized learning paths to cater to diverse student needs. IoT-connected devices collected data on student interactions with educational content, while AI algorithms processed this data to identify individual learning styles and preferences. The implementation of personalized learning paths allowed educators to tailor instruction, providing additional resources for challenging concepts and fostering independent learning. The result was a significant improvement in student engagement, academic performance, and a more student-centric approach to education.

Virtual Labs and Collaborative Learning in Higher Education

A higher education institution revolutionized its approach to practical education by implementing AIoT-enabled virtual labs. Through IoT devices and AI algorithms, students gained access to virtual experiments and simulations, replicating real-world scenarios in a controlled digital environment. The system facilitated collaborative learning, enabling students to interact with virtual experiments in real time and share insights. This initiative not only enhanced access to practical education but also promoted collaborative problem-solving, critical thinking, and a deeper understanding of complex concepts.

8.2 Demonstrated Improvements in Learning Outcomes

In a pioneering educational institution, the implementation of AIoT technologies resulted in significant and demonstrable improvements in learning outcomes. This case study focuses on how the strategic integration of Artificial Intelligence and the Internet of Things positively impacted the educational experience and academic achievements of students.

Background

The educational institution, committed to leveraging cutting-edge technologies, embarked on an initiative to enhance the overall learning environment and outcomes for its diverse student body. The focus was on harnessing AIoT to create a dynamic and adaptive ecosystem that responded to the individual needs and learning styles of students across various disciplines.

Implementation

Personalized Learning Paths:

- IoT devices were deployed in classrooms and learning spaces to capture real-time data on student interactions with educational content.
- AI algorithms processed this data to create personalized learning paths for each student, identifying areas of strength and weakness.

Adaptive Assessments:

- AI-driven assessment tools were introduced, tailoring quizzes and exams based on individual progress and comprehension levels.
- The system dynamically adjusted the difficulty of questions, ensuring an optimal level of challenge for each student.

Real-time Feedback Mechanisms:

- Students received instant, AI-generated feedback on their assignments and assessments, providing guidance on areas that needed improvement.
- Educators accessed comprehensive analytics dashboards to track individual and collective performance trends.

RESULTS

The implementation of AIoT technologies demonstrated remarkable improvements in learning outcomes:

Individualized Academic Progress:

- Students experienced a more personalized learning journey, with adaptive content and assessments aligning with their unique learning styles.
- The system identified and addressed knowledge gaps, resulting in a more comprehensive understanding of subjects.

Increased Engagement and Motivation:

- The interactive nature of AIoT-infused learning environments increased student engagement and motivation.
- Real-time feedback and personalized challenges fostered a sense of accomplishment and intrinsic motivation.

Higher Retention Rates:

- The institution observed an increase in retention rates, as students benefited from tailored support and a more engaging educational experience.

CONCLUSION

This case study exemplifies how the strategic integration of AIoT technologies can lead to demonstrated improvements in learning outcomes. By personalizing learning paths, introducing adaptive assessments, and implementing real-time feedback mechanisms, the educational institution created an environment where students thrived academically and educators were empowered to deliver impactful instruction. The success of this initiative underscores the transformative potential of AIoT in shaping the future of education.

8.3 Lessons Learned From Real-World Applications

In the realm of AIoT in education, real-world applications have provided valuable insights and lessons that shape the future of technological integration in educational settings. This section highlights lessons learned from specific case studies and success stories, showcasing the practical wisdom gained through hands-on experiences.

Seamless Integration Enhances User Experience

Case Study: A university successfully integrated AIoT into its learning management system, creating a seamless experience for both educators and students.

Lesson Learned: The importance of ensuring that AIoT technologies seamlessly integrate into existing educational systems cannot be overstated. A user-friendly experience fosters acceptance and adoption among educators and students.

Ethical Considerations Are Paramount

Case Study: A school district implemented AI-driven tutoring systems but faced ethical concerns regarding data privacy and algorithmic bias.

Lesson Learned: Ethical considerations, including data privacy and fairness in algorithms, must be at the forefront of AIoT implementations. Establishing clear ethical guidelines is essential to build trust within the educational community.

Scalability and Flexibility Are Key

Case Study: A community college implemented IoT-connected devices in classrooms, but challenges arose when trying to scale the solution to accommodate a growing student population.

Lesson Learned: Planning for scalability and flexibility is crucial. Educational institutions should choose AIoT solutions that can adapt to changing needs and accommodate a growing number of users without compromising performance.

Educator Training Is a Critical Success Factor

Case Study: A high school introduced AI-driven content customization, but some educators struggled to effectively utilize the new technology.

Lesson Learned: Comprehensive training programs for educators are vital. Investing in professional development ensures that educators can leverage AIoT technologies to their full potential, maximizing the impact on student learning.

Continuous Monitoring and Evaluation Are Essential

Case Study: A college implemented real-time monitoring systems to track student engagement but faced challenges in interpreting and acting on the generated data.

Lesson Learned: Continuous monitoring and evaluation of AIoT systems are essential. Regular assessments and feedback loops help institutions refine their strategies, address challenges, and make data-driven improvements over time.

These lessons learned from real-world applications emphasize the multidimensional nature of AIoT in education. As institutions continue to innovate, understanding and applying these lessons will be instrumental in navigating challenges and maximizing the positive impact of AIoT technologies on teaching and learning experiences.

Industrial Report

The intersection of Artificial Intelligence (AI) and the Internet of Things (IoT) is revolutionizing education, redefining learning environments, and reshaping educational technology. This industry report provides a comprehensive analysis of the transformative impact of AIoT in the education sector. It delves into key trends, applications, challenges, and future prospects, offering insights for educators, policymakers, and technology providers.

Key Findings:

a.   Personalized Learning: AIoT facilitates personalized learning experiences by analyzing individual student data from IoT devices, adapting content, and providing tailored educational materials.

b.   Smart Classrooms: Connected devices and sensors powered by AI optimize classroom management, automate administrative tasks, and enhance the overall teaching and learning experience.

c.   Predictive Analytics: AIoT leverages predictive analytics to identify at-risk students, allowing early intervention strategies for improved academic outcomes.

d.   Virtual and Augmented Reality (VR/AR): The integration of AI and IoT enhances immersive learning experiences through VR/AR technologies, creating dynamic and interactive educational content.

e.   Security and Privacy Challenges: The report addresses concerns related to data security, privacy, and ethical considerations, emphasizing the need for robust frameworks to safeguard student information.

f.   Professional Development: AIoT supports educators through intelligent tools for professional development, offering insights, resources, and adaptive learning modules.

g.   Emerging Technologies: The report explores emerging technologies like blockchain for secure credentialing and decentralized learning platforms, showcasing their potential impact on the education landscape.

Challenges and Recommendations: The industry report outlines challenges such as infrastructure limitations, data governance, and ethical use of AIoT in education. Recommendations include investing in teacher training programs, establishing clear data protection policies, and fostering collaboration between educators and technology developers.

Future Prospects: As AIoT continues to evolve, the report envisions a future where smart educational ecosystems seamlessly integrate with emerging technologies, fostering innovation, inclusivity, and accessibility in education. The industry is poised for continuous growth, with potential applications ranging from adaptive learning systems to revolutionizing assessment methodologies.

Conclusion: The AIoT revolution in education marks a paradigm shift, offering unprecedented opportunities to enhance learning outcomes and revolutionize the educational landscape. Stakeholders must collaborate to address challenges, ensuring responsible deployment and maximizing the potential of AIoT to create dynamic, adaptive, and equitable learning environments.

9. FUTURE TRENDS AND INNOVATIONS

9.1 Emerging Technologies in AIoT for Education

The future of education is intricately woven with the rapid evolution of emerging technologies within the realm of Artificial Intelligence of Things (AIoT). One prominent trend poised to reshape learning environments is the integration of Edge Computing. This technology brings computational power closer to the data source, enabling real-time processing of AIoT-generated data. In educational settings, this translates to instantaneous feedback, adaptive learning experiences, and swift decision-making based on real-time insights, fostering a dynamic and responsive learning environment. Another pivotal advancement on the horizon is the widespread implementation of 5G connectivity. As the education landscape becomes increasingly reliant on Internet of Things (IoT) devices, 5G networks promise ultra-fast and reliable connectivity, facilitating seamless communication between devices and AI systems. This not only ensures smoother data transmission but also opens doors to more advanced applications, such as augmented reality (AR) and virtual reality (VR), ushering in a new era of immersive and interactive learning experiences. Lai et al. (2021) explores the potential of VR (Virtual Reality) in enhancing AIoT computational thinking skills.

Blockchain technology is emerging as a cornerstone for securing and transparently managing data transactions in AIoT applications. In the educational context, blockchain has the potential to fortify data security, protect student privacy, and streamline administrative processes like secure sharing of educational records and credential verification. Lioupa et al. (2023) provide the Integration of 6G and Blockchain into an Efficient AIoT-Based Smart Education Model.

Figure 7. Future Aspects of AIoT in Education

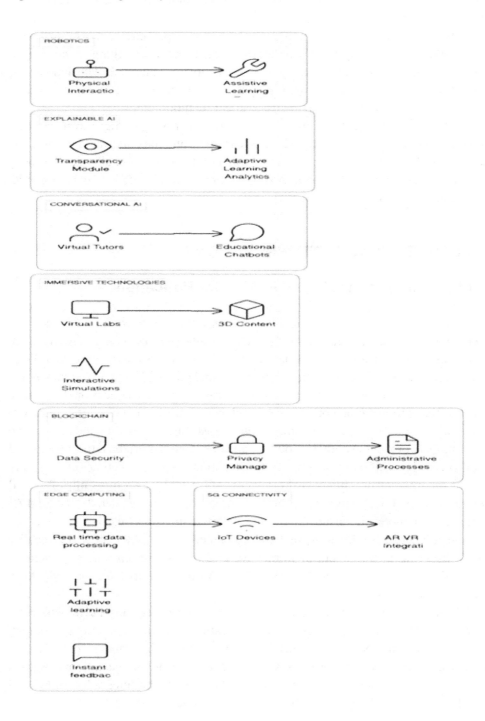

The integration of Augmented Reality (AR) and Virtual Reality (VR) stands as a transformative force, promising to revolutionize learning experiences. By providing virtual labs, simulations, and interactive 3D content, AR and VR technologies bring abstract concepts to life, making learning more engaging, dynamic, and tailored to individual needs.

Natural Language Processing (NLP) is set to play a pivotal role in creating conversational interfaces within AIoT applications for education. From virtual tutors to educational chatbots, NLP enables machines to understand and generate human-like language, fostering more accessible and interactive educational resources.

Finally, the concept of Swarm Intelligence is gaining traction for collaborative problem-solving. Inspired by decentralized and self-organized systems in nature, this approach optimizes resource allocation, improves scheduling, and enhances group activities, fostering a collaborative and synergistic learning environment. As these emerging technologies continue to mature, the future of AIoT in education promises a paradigm shift towards more personalized, interactive, and adaptive learning ecosystems, redefining the boundaries of what is possible in the realm of education technology.

9.2 Anticipated Developments in Adaptive Learning

The future landscape of education is marked by a profound shift towards increasingly adaptive learning environments, driven by the symbiotic integration of Artificial Intelligence of Things (AIoT) technologies. Anticipated developments in adaptive learning are poised to revolutionize how students engage with educational content, tailoring experiences to individual needs and optimizing learning outcomes. One key development is the refinement of personalized learning paths. AIoT, leveraging data from IoT devices, will enable the creation of more granular and responsive learning paths for students. These paths will not only consider academic strengths and weaknesses but also factor in real-time behavioral and environmental data, ensuring a holistic understanding of each learner's preferences and learning style. Furthermore, the incorporation of Predictive Analytics is set to play a pivotal role in adaptive learning. AI algorithms will analyze historical data, user interactions, and performance metrics to predict future learning needs and potential areas of difficulty for individual students. This proactive approach allows educators to intervene early, providing targeted support and resources, ultimately preventing learning gaps from widening.

Adaptive Assessments represent another frontier in the evolution of learning evaluation. Future developments in this realm will witness the integration of AI-driven assessments that dynamically adjust based on a student's real-time performance. These assessments will go beyond traditional methods, adapting difficulty levels, question types, and content delivery in response to the student's evolving comprehension and skill levels.

Moreover, the evolution of Intelligent Tutoring Systems (ITS) is anticipated to elevate the one-on-one learning experience. AIoT will enable ITS to utilize real-time data from IoT devices, offering instant and personalized feedback to students. These systems will not only address academic challenges but also consider emotional and cognitive factors, providing a comprehensive support system for learners.

In summary, the anticipated developments in adaptive learning within the AIoT paradigm herald a future where education is truly personalized, responsive, and attuned to the unique needs of each learner. By harnessing the power of data from IoT devices and leveraging advanced AI algorithms, these developments pave the way for a transformative era in education, where adaptive learning is not merely a feature but an integral and dynamic aspect of the educational journey.

9.3 Shaping the Future of Smart Educational Environments

The convergence of Artificial Intelligence of Things (AIoT) in education is poised to shape the future of smart educational environments, ushering in an era where learning spaces become dynamic, responsive, and interconnected ecosystems. A pivotal element in this transformation is the integration of Smart Campus Initiatives. AIoT technologies will imbue educational institutions with the capability to create intelligent campuses, where IoT sensors and devices collect real-time data on various facets of campus life. From monitoring energy usage to optimizing classroom environments, these initiatives will enhance operational efficiency, sustainability, and overall campus well-being. Moreover, the future holds the promise of Interactive Technologies within Smart Classrooms. AIoT will facilitate the integration of interactive technologies such as IoT-enabled whiteboards, collaborative platforms, and augmented reality tools. These technologies will redefine traditional classroom dynamics, fostering greater student engagement, collaboration, and interaction with educational content. Real-time data analytics will enable educators to gauge student responses and adapt teaching strategies on the fly, creating a more personalized and responsive learning environment.

Furthermore, the evolution of Connected Devices and Infrastructure will be instrumental in shaping the smart educational environments of the future. AIoT will catalyze the proliferation of connected devices, from smart desks to wearable technology. These devices will not only gather data on student interactions but also seamlessly communicate with each other, creating an integrated network that supports a variety of educational activities. This interconnected infrastructure will facilitate the seamless flow of information, enabling educators to tailor their approaches based on a holistic understanding of each student's learning journey.

The future of smart educational environments, shaped by the integration of AIoT, envisions a paradigm shift in the way education is delivered and experienced. From intelligent campuses to interactive classrooms and collaborative learning spaces, AIoT technologies are poised to redefine the educational landscape, creating environments that are not only technologically advanced but also human-centric, responsive, and conducive to the diverse needs of 21st-century learners.

10. CONCLUSION

10.1 Recapitulation of Key Findings

In the culmination of our exploration into the transformative intersection of Artificial Intelligence of Things (AIoT) in education, a comprehensive recapitulation of key findings unveils a landscape marked by innovation, adaptability, and a profound impact on learning environments. Throughout this journey, it has become evident that the integration of AIoT technologies holds the potential to revolutionize traditional paradigms of education, ushering in a new era characterized by personalized learning, intelligent campuses, and dynamic, interconnected classrooms.

One of the salient findings underscores the pivotal role of Adaptive Learning. The marriage of AI and IoT facilitates the creation of adaptive learning paths that respond to individual student needs, leveraging real-time data to tailor educational experiences. This adaptability not only addresses diverse learning styles but also ensures that each student's journey is unique, optimizing the assimilation of knowledge and skills.

Predictive Analytics emerges as a powerful tool in the educational arsenal, foreseeing potential challenges and learning needs. By analyzing historical data and user interactions, AIoT systems can predict future academic hurdles, enabling proactive interventions and personalized support. This anticipatory approach not only prevents learning gaps but also fosters a more proactive and responsive educational ecosystem.

The integration of Intelligent Tutoring Systems (ITS) stands out as a cornerstone in the transformative potential of AIoT. The utilization of real-time data from IoT devices allows for instant and personalized feedback, creating a symbiotic relationship between technology and education. This personalized support system extends beyond academic challenges to consider emotional and cognitive factors, addressing the holistic development of learners.

In conclusion, the recapitulation of key findings affirms that the fusion of AI and IoT in education is a catalyst for positive change. The transformative potential of AIoT lies not just in the technology itself, but in its capacity to create a more adaptive, personalized, and interconnected educational experience. As we navigate the future of education, the findings underscore the importance of embracing AIoT as an enabler, fostering environments where learners are empowered, educators are supported, and the boundaries of educational possibilities are continually expanded.

10.2 Envisioning the Future Landscape of AIoT in Education

As we conclude this exploration into the transformative fusion of Artificial Intelligence of Things (AIoT) in education, the vision for the future landscape of learning environments is one that transcends current boundaries and redefines the very essence of education. Envisioning this landscape entails a profound shift towards a dynamic, interconnected, and highly adaptive educational ecosystem. The amalgamation of AI and IoT technologies propels us into an era where education becomes a personalized, anticipatory, and collaborative journey, tailored to the unique needs of each learner.

In this future landscape, Adaptive Learning emerges as a cornerstone, reshaping the traditional one-size-fits-all approach. The AIoT synergy allows for the creation of adaptive learning paths that evolve in real-time, responding to the nuances of individual learning styles and preferences. This adaptive nature ensures that education becomes not just instructive but inherently responsive, fostering an environment where learners can thrive at their own pace.

Predictive Analytics, another facet of this envisioned landscape, instills a sense of foresight into the educational journey. AIoT systems, through the analysis of vast datasets, anticipate potential stumbling blocks and future learning needs. This predictive capability transforms education from a reactive process to a proactive one, allowing educators to intervene early and provide targeted support, ultimately paving the way for more successful learning outcomes.

Intelligent Tutoring Systems (ITS) further shape the future educational landscape by fostering a personalized and emotionally intelligent support system. Real-time data from IoT devices enables these systems to offer instant, tailored feedback that extends beyond academic realms to encompass emotional and cognitive aspects. This not only enhances academic achievement but also nurtures holistic student development.

The future, as envisioned through the lens of AIoT in education, is characterized by Smart Educational Environments. Intelligent campuses, IoT-enabled classrooms, and collaborative technologies redefine physical spaces and interactions. These environments are not static; they adapt to the diverse needs of students and educators, fostering a dynamic ecosystem where technology augments the human experience rather than replacing it.

10.3 Call to Action for Educators, Technologists, and Policymakers

As we conclude our journey through the transformative realm of Artificial Intelligence of Things (AIoT) in education, a resounding call to action echoes for educators, technologists, and policymakers alike. The profound impact of AIoT on learning environments demands a collective and intentional effort to harness its potential responsibly and ethically.

Educators, positioned at the frontline of this educational revolution, are urged to embrace a paradigm shift in pedagogy. The call is to foster a mindset that views technology not as a replacement but as an enabler. By integrating AIoT into teaching practices, educators can tailor learning experiences, address individual needs, and cultivate the essential skills students need for the future. Continuous professional development becomes crucial, empowering educators to navigate the evolving landscape of educational technology effectively.

Technologists, the architects of the AIoT landscape, bear the responsibility of developing technologies that prioritize inclusivity, transparency, and ethical considerations. The call to action is to design AIoT systems that enhance accessibility, address algorithmic bias, and prioritize data privacy. Collaborative endeavors between technologists and educators are essential to ensure that technological advancements align with the pedagogical goals of education, fostering an ecosystem where technology seamlessly integrates with the human aspect of learning.

Policymakers, entrusted with shaping the regulatory framework of education technology, are called upon to enact policies that foster innovation while safeguarding ethical standards. The call is to create an environment where AIoT in education is governed by principles of equity, accessibility, and data protection. Policymakers play a pivotal role in fostering collaboration between educational institutions, technology developers, and communities to ensure that the benefits of AIoT are equitably distributed and that no learner is left behind.

Collectively, this triumvirate of educators, technologists, and policymakers is essential for realizing the full potential of AIoT in education. The call to action extends beyond individual efforts to a collaborative commitment to creating an educational landscape that prepares learners for the challenges and opportunities of the future. By heeding this call, we can collectively shape an educational paradigm where the fusion of artificial intelligence and the internet of things becomes a force for positive transformation, empowering learners and society as a whole.

REFERENCES

Chou, C. M., Shen, T. C., Shen, T. C., & Shen, C. H. (2022). Influencing factors on students' learning effectiveness of AI-based technology application: Mediation variable of the human-computer interaction experience. *Education and Information Technologies*, *27*(6), 8723–8750. doi:10.1007/s10639-021-10866-9

Hasko, R., Shakhovska, N., Vovk, O., & Holoshchuk, R. (2020). A Mixed Fog/Edge/AIoT/Robotics Education Approach based on Tripled Learning. In COAPSN (pp. 227-236). Academic Press.

Kim, J., Lee, H., & Cho, Y. H. (2022). Learning design to support student-AI collaboration: Perspectives of leading teachers for AI in education. *Education and Information Technologies*, *27*(5), 6069–6104. doi:10.1007/s10639-021-10831-6

Kuka, L., Hörmann, C., & Sabitzer, B. (2022). Teaching and Learning with AI in Higher Education: A Scoping Review. *Learning with Technologies and Technologies in Learning: Experience, Trends and Challenges in Higher Education*, 551-571.

Lai, Y. H., Chen, S. Y., Lai, C. F., Chang, Y. C., & Su, Y. S. (2021). Study on enhancing AIoT computational thinking skills by plot image-based VR. *Interactive Learning Environments*, *29*(3), 482–495. doi:10.1080/10494820.2019.1580750

Lin, Y. S., Chen, S. Y., Tsai, C. W., & Lai, Y. H. (2021). Exploring computational thinking skills training through augmented reality and AIoT learning. *Frontiers in Psychology*, *12*, 640115. doi:10.3389/fpsyg.2021.640115 PMID:33708166

Lioupa, A., Memos, V. A., Stergiou, C. L., Ishibashi, Y., & Psannis, K. E. (2023, September). The Integration of 6G and Blockchain into an Efficient AIoT-Based Smart Education Model. In *2023 6th World Symposium on Communication Engineering (WSCE)* (pp. 1-5). IEEE.

Ming, H. X., Ti, L. R., Chen, H. C., Yu, T. C., Ming, H. Y., & Lai, Y. H. (2022, October). Multicultural Knowledge and Information Literacy Learning Using AIoT Integration Technology. In *2022 IET International Conference on Engineering Technologies and Applications (IET-ICETA)* (pp. 1-2). IEEE. 10.1109/IET-ICETA56553.2022.9971523

Omonayajo, B., Al-Turjman, F., & Cavus, N. (2022). Interactive and innovative technologies for smart education. *Computer Science and Information Systems*, *19*(3), 1549–1564. doi:10.2298/CSIS210817027O

Quy, V. K., Thanh, B. T., Chehri, A., Linh, D. M., & Tuan, D. A. (2023). AI and Digital Transformation in Higher Education: Vision and Approach of a Specific University in Vietnam. *Sustainability (Basel)*, *15*(14), 11093. doi:10.3390/su151411093

Saadé, R. G., Zhang, J., Wang, X., Liu, H., & Guan, H. (2023). Challenges and Opportunities in the Internet of Intelligence of Things in Higher Education—Towards Bridging Theory and Practice. *IoT*, *4*(3), 430–465. doi:10.3390/iot4030019

Sharma, V., Sharma, K. K., Vashishth, T. K., Panwar, R., Kumar, B., & Chaudhary, S. (2023, November). Brain-Computer Interface: Bridging the Gap Between Human Brain and Computing Systems. In *2023 International Conference on Research Methodologies in Knowledge Management, Artificial Intelligence and Telecommunication Engineering (RMKMATE)* (pp. 1-5). IEEE. 10.1109/RMKMATE59243.2023.10369702

Sibanda, M., Khumalo, N. Z., & Fon, F. N. (2023, November). A review of the implications of artificial intelligence tools in higher education. Should we panic? In *The 10th Focus Conference (TFC 2023)* (pp. 128-145). Atlantis Press.

Singh, B., Gupta, V. K., Jain, A. K., Vashishth, T. K., & Sharma, S. (2023). *Transforming education in the digital age: A comprehensive study on the effectiveness of online learning*. Academic Press.

. Srivastava, S., & Pathak, D. M. (2023). AIoT-based smart education and online teaching. AIoT Technologies and Applications for Smart Environments, 229.

Stewart, J. C., Davis, G. A., & Igoche, D. A. (2020). AI, IoT, and AIoT: Definitions and impacts on the artificial intelligence curriculum. *Issues in Information Systems*, *21*(4).

Tsai, C. C., Cheng, Y. M., Tsai, Y. S., & Lou, S. J. (2021). Impacts of AIOT implementation course on the learning outcomes of senior high school students. *Education Sciences*, *11*(2), 82. doi:10.3390/educsci11020082

Vashishth, T. K., Sharma, V., Sharma, K. K., Panwar, R., & Chaudhary, S. (2023). The Impact of Information Technology on the Education Sector: An Analysis of the Advantages, Challenges, and Strategies for Effective Integration. *International Journal of Research and Analytical Review*, *10*(2), 265–274.

Wu, T. N., Chin, K. Y., & Lai, Y. C. (2022, July). Applications of Intelligent Environmental IoT Detection Tools in Environmental Education of College Students. In *2022 12th International Congress on Advanced Applied Informatics (IIAI-AAI)* (pp. 240-243). IEEE. 10.1109/IIAIAAI55812.2022.00055

Xie, H., Hwang, G. J., & Wong, T. L. (2021). Editorial note: from conventional AI to modern AI in education: reexamining AI and analytic techniques for teaching and learning. *Journal of Educational Technology & Society, 24*(3).

Yadava, A. K., Chouhan, V., Uke, N., Kumar, S., Banerjee, J., & Benjeed, A. O. S. (2022). A study of the progress, challenges, and opportunities in artificial intelligence of things (AIoT). *International Journal of Health Sciences.* doi:10.53730/ijhs.v6nS1.7224

Zhang, Y., Ning, Y., Li, B., & Jun, Y. (2021, June). The research on talent education for AI-based IoT system development and implementation by the CDIO concept. In *2021 2nd International Conference on Artificial Intelligence and Education (ICAIE)* (pp. 110-114). IEEE. 10.1109/ICAIE53562.2021.00031

KEY TERMS AND DEFINITIONS

Artificial Intelligence (AI): Refers to the development of computer systems capable of performing tasks that typically require human intelligence. These tasks include learning from experience (machine learning), understanding natural language, recognizing patterns, solving problems, and making decisions. AI aims to create machines that can mimic human cognitive functions and improve their performance over time without explicit programming.

Artificial Intelligence and the Internet of Things (AIoT): Is a synergistic integration of artificial intelligence (AI) technologies with the vast network of interconnected physical devices enabled by the Internet of Things (IoT). It combines AI's ability to analyze and make decisions with the data-gathering capabilities of IoT devices, creating a powerful ecosystem where devices communicate, adapt, and respond intelligently to enhance efficiency, automation, and user experiences across various domains.

Intelligent Tutoring Systems (ITS): Are computer-based educational platforms that utilize artificial intelligence to provide personalized and adaptive learning experiences. These systems assess individual student progress, tailor instructional content to meet specific needs, and offer real-time feedback, creating an interactive and customized learning environment. ITS aims to enhance the effectiveness of education by providing dynamic and responsive support to learners.

Internet of Things (IoT): Is a network of interconnected physical devices embedded with sensors, software, and connectivity, enabling them to collect, exchange, and act upon data. IoT enables devices to communicate and interact with each other, facilitating seamless data sharing and automation to enhance efficiency, convenience, and functionality in various domains such as home automation, industrial processes, and healthcare.

Virtual Reality (VR): Is a technology that creates a computer-generated immersive environment, allowing users to interact with and experience a three-dimensional digital world as if it were real.

Chapter 5
AIoT Revolution:
Transforming Networking Productivity for the Digital Age

C. V. Suresh Babu

https://orcid.org/0000-0002-8474-2882
Hindustan Institute of Technolgy and Science, India

M. Sowmi Saltonya

https://orcid.org/0009-0007-5160-9130
Hindustan Institute of Technology and Science, India

Suresh Ganapathi
Hindustan Institute of Technology and Science, India

A. Gunasekar
Hindustan Institute of Technology and Science, India

ABSTRACT

This chapter provides a comprehensive exploration of networking challenges in today's evolving landscape. It introduces the concept of artificial intelligence of things (AIoT) and demonstrates the synergy between AI and IoT in addressing these challenges. Through a holistic AIoT strategy, the chapter guides readers in designing and implementing proactive network monitoring, predictive maintenance, enhanced security measures, quality of service management, automation, and data-driven decision-making. It emphasizes the tangible benefits of AIoT, including optimized networking performance, cost reduction, and improved user experiences. Real-world case studies highlight successful AIoT implementations, offering valuable lessons. The chapter also delves into emerging trends, ethical considerations, and concludes by emphasizing the implications and the need for a forward-looking approach in networking productivity.

DOI: 10.4018/979-8-3693-0993-3.ch005

1. INTRODUCTION

The Raise of Networking Productivity Challenges

The landscape of networking has undergone significant transformations in recent years, driven by technological advancements and changing business requirements. Cisco's Visual Networking Index (VNI) Forecast predicts that global IP traffic will reach 4.8 zettabytes per year by 2022, highlighting the substantial growth in data transmission (Cisco, 2019). Key factors contributing to this evolution include:

Evolving Technologies: Staying updated with advancements in AI, IoT, and networking technologies ensures that networking strategies remain relevant and effective in the face of evolving challenges and opportunities.

Real-Time Data Processing: AI-driven analytics facilitate real-time data processing, enabling swift and data-informed decision-making.

Enhanced Security Measures: AI-powered threat detection strengthens network defenses against evolving cyber threats, allowing devices to collaborate in the learning process while keeping sensitive information localized.

AI-Powered Cybersecurity Solutions: Security concerns remain paramount, and advancements in AI-powered cybersecurity solutions are set to revolutionize threat detection and response, autonomously identifying and mitigating potential risks (Suresh Babu, C. V., Dhanusha, T., et al., 2023).

Moreover, AIoT (Alahi, E. E & Sukkuea, 2023) offers dynamic scalability and resource allocation, ensuring networks can adapt to changing demands. As data-driven insights become integral, organizations can personalize user experiences and make strategic decisions.

Aim

This chapter paper aims to explore the potential of integrating Artificial Intelligence of Things (AIoT) in networking environments to enhance productivity. It seeks to investigate how AIoT technologies can optimize networking operations, improve efficiency, and drive organizational productivity in both traditional and emerging networking paradigms.

Scope of the Study

The scope of this chapter paper encompasses exploring the integration of Artificial Intelligence of Things (AIoT) in networking to enhance productivity. It will examine AIoT applications such as network performance optimization, automation, security enhancement, and edge computing integration. Real-world case studies and use cases will be presented to illustrate AIoT's impact. Additionaly, challenges and future research directions will be discussed to provide insights into maximizing AIoT's potential for productivity improvement in networking environments. By leveraging AIoT, businesses can optimize operations, reduce costs, and stay competitive in the digital era.

Proliferation of Devices

The proliferation of internet-connected devices, ranging from smartphones and smart home appliances to IoT sensors, has been remarkable. These devices continuously generate and consume data, significantly increasing the complexity of network management (Botta et al., 2016).

The primary objective is to create a more livable, sustainable, and technologically advanced urban environment. Through the utilization of AI and IoT, cities can optimize their operations, reduce resource consumption, and enhance the quality of life for their residents. This innovative approach to urban management is at the core of the digital age, propelling cities towards a smarter, more interconnected future (Suresh Babu et al., 2023).

Security and safety are pivotal aspects of IoT. IoT devices communicate with each other to enhance security in their environment. IoT sensors monitor the surroundings and provide real-time information about the environmental conditions, thus preventing unwanted events. For example, in an autonomous car, sensors monitor tire pressure to prevent tire bursts if the pressure exceeds a certain threshold. In heavy industrial settings, IoT can prevent machine damage by monitoring their temperature. Similarly, smart wearables can monitor hazardous environments, and wireless body area networks (WBAN) can oversee physiological parameters to ensure safety and security (Abid MA, Afaqui N, et al., 2022).

According to Zhang, Y. (2018), reducing complexity by integrating AI into IoT in networking and mobile edge computing advances technologies in mobile sensing and wireless communication. The concept of the Internet of Things (IoT) has evolved significantly, envisioning IoT as a global network of pervasive smart mobile devices that measure and understand environmental characteristics and interact with one another. IoT networks have introduced numerous novel mobile applications and have the potential to enable many more. However, the inherent limitations of IoT devices, such as low computational power, storage, and battery capacity, pose a significant challenge in supporting computation-intensive applications on resource-constrained devices.

Theories

Addressing Networking Challenges

The evolving networking landscape has presented several significant challenges:

Network Congestion

The constant growth in data flow and the increasing number of devices have made network congestion a common issue, resulting in performance bottlenecks and a deteriorated user experience (Buj et al., 2018).

Productivity challenges in networking environments encompass a range of obstacles that can hinder the seamless flow of data and communication. These obstacles often include limited bandwidth, leading to slow data transfers and delays in accessing critical resources. Latency problems also arise, causing delays in real-time interactions and negatively impacting user experience. Network congestion can result in packet loss and reduced performance during high-traffic periods (Tie Qiu et al., 2020).

Security Risks

The expanding attack surface of networks renders them susceptible to cyber threats. Cyberattacks can lead to data breaches, financial losses, and damage to an organization's reputation (Chen et al., 2018).

Security concerns pose a significant threat, as striking a balance between stringent security measures and usability is a delicate task. Additionally, managing a diverse array of devices with varying configurations and operating systems can be complex and time-consuming. Ensuring scalability, reliability, and redundancy in the network infrastructure is crucial for uninterrupted operations. As the work landscape evolves, facilitating secure remote connectivity becomes paramount. Proactive monitoring, troubleshooting, and compliance with regulatory standards are equally vital in effectively addressing these productivity challenges (Pereira, A., 2016). Addressing these issues requires a combination of advanced technology solutions, robust best practices, and vigilant monitoring and management practices.

Scalability

As businesses expand, networks must scale to accommodate additional users and devices. Conventional network architectures may struggle to efficiently expand, leading to network performance issues (Zhang et al., 2014).

AIoT enables dynamic scalability and resource allocation, ensuring that networks can adapt to changing demands. As data-driven insights become integral, organizations can personalize user experiences and make strategic decisions (Sukkuea et al., 2023).

Reliability

Downtime and network outages can result in substantial financial losses. A study by the Ponemon Institute found that the average cost of a data center outage was $9,000 per minute in 2016 (Ponemon Institute, 2016).

According to Gray et al. (2003), Quality of Service (QoS) encompasses a range of strategies and techniques designed to maintain a high standard of performance, responsiveness, and efficiency across the network. Central to this effort is the establishment of clear Service Level Agreements (SLAs) that outline precise performance expectations, including metrics like latency, reliability, availability, and throughput. By defining these benchmarks, organizations can effectively measure and uphold the quality of services provided.

BACKGROUND

The Need for Innovative Solutions

To address these challenges, it is essential to consider the following innovative solutions:

AI and IoT Integration

Artificial Intelligence of Things (AIoT) combines the analytical capabilities of AI with real-time data collection in IoT, enabling networks to become intelligent and adaptive. For instance, AI-driven predictive maintenance can reduce network downtime and enhance reliability (Perera et al., 2017).

Companies like Siemens, IBM, and Cisco have actively incorporated AIoT technologies into their solutions:

Siemens utilizes AIoT to optimize manufacturing processes, predict maintenance needs, and enhance overall operational efficiency.

IBM's Watson IoT platform combines AI and IoT to help businesses make sense of the vast amount of data generated by connected devices. It provides predictive maintenance solutions, supply chain optimization, and more.

Cisco, a global leader in networking technology, explores AIoT applications, particularly in the context of IoT security and network optimization, with the aim of using AI to enhance network performance and security (Alahi, E et al., 2023).

Proactive Monitoring

Proactive network monitoring and management should be embraced. Machine learning algorithms can detect network anomalies and address them before they disrupt operations (Zhang et al., 2019).

A proactive approach to network monitoring involves the detection of abnormal states in monitored channels to protect computing infrastructures. This approach utilizes cutting-edge technologies, such as Deep Learning (DL), large-scale data processing, and modern network Intrusion Detection Systems (IDS). These technologies improve monitoring and security protection by modeling multiple monitoring channels using DL techniques (Nguyen & Dlugolinsky, 2020).

Automation

Automating routine network tasks can optimize network management, reduce human errors, and enhance productivity. This can be achieved through the implementation of Software-Defined Networking (SDN) and network orchestration (McKeown et al., 2008).

Automation also enables self-healing capabilities, allowing the network to autonomously detect and rectify issues, minimizing downtime and disruptions, which is crucial for mission-critical applications.

Security Enhancements

Implementing advanced security measures, such as AI-driven threat detection and response, is vital for network protection. AI can analyze network traffic patterns to detect unusual behavior and prevent potential threats (Kim et al., 2020).

Embracing an AIoT strategy for networking productivity yields various tangible benefits, including improved automation, efficiency, security, and user experiences, which can significantly enhance operational performance and competitiveness.

Cybersecurity measures play a pivotal role in safeguarding critical infrastructure. Robust encryption protocols and authentication mechanisms are imperative to protect against unauthorized access or tampering of systems (Suresh Babu, C. V., Dhanusha, T, et al., 2023).

2.0 THE ROLE OF AIoT IN NETWORKING OPTIMIZATION

2.1 Exploring the Concept of Artificial Intelligence of Things (AIoT)

Artificial Intelligence of Things (AIoT) is an emerging paradigm that combines the capabilities of Artificial Intelligence (AI) and the Internet of Things (IoT) to revolutionize various domains, including networking. AIoT represents a powerful fusion of AI's analytical and decision-making abilities with IoT's real-time data collection and connectivity.

Triantafyllou et al. (2018) explore AIoT applications, particularly in the context of IoT security and network optimization, aiming to enhance network performance and security. AIoT empowers devices to collaborate in the learning process while keeping sensitive information localized. It addresses security concerns and promises to revolutionize threat detection and response.

AIoT aims to imbue "things" or connected devices with intelligence, enabling them to make data-driven decisions and communicate with each other, humans, and the broader network ecosystem (Perera, Liu, & Jayawardena, 2017). For example, in a smart factory, AIoT may enable sensors to detect machinery anomalies, predict maintenance needs, and autonomously schedule maintenance, reducing downtime and optimizing operations.

2.2 The Synergy of AI and IoT in Networking

The combination of AI and IoT within the AIoT framework creates a synergy that enhances networking in several ways:

2.2.1 Real-Time Data Analysis

AI processes vast amounts of real-time data from IoT devices, providing immediate insights and actionable information. For instance, in healthcare, IoT devices can monitor patients' vital signs, and AI algorithms can detect irregularities and send alerts to medical professionals (Rajasegarar et al., 2018).

Integrating AI for analysis and optimization is a pivotal advancement in IoT deployments. This integration incorporates sophisticated algorithms and machine learning models into the IoT ecosystem, enabling autonomous data processing, pattern recognition, anomaly detection, and informed decision-making without human intervention (Souag-Gamane D. et al., 2020).

2.2.2 Predictive Maintenance

AIoT enables predictive maintenance by analyzing data from sensors on industrial machinery or vehicles, preventing costly breakdowns and disruptions (Pan et al., 2018).

Sverko et al. (2022) note that predictive maintenance in AIoT represents a significant advancement, shifting from reactive, schedule-based approaches to proactive, data-driven methods. This not only improves operational efficiency but also leads to substantial cost savings and increased competitiveness.

2.2.3 Network Optimization

In networking, AIoT can analyze network traffic patterns, identify anomalies, and optimize routing to enhance performance and reduce congestion (Herrera & Rivas-Lalaleo, 2020).

Automation is a pivotal force in revolutionizing networking productivity for the digital age. It involves deploying intelligent algorithms and systems to streamline and optimize network operations, reducing manual intervention and enhancing overall efficiency. By automating routine tasks such as device provisioning, configuration management, and troubleshooting, organizations can allocate resources to more strategic endeavors, accelerating deployment times and reducing the risk of human error (Otasek D. et al., 2019).

2.3 Unraveling the Potential of AIoT in Networking

AIoT possesses significant potential for networking optimization:

2.3.1 Self-Healing Networks

AIoT has the capability to create self-healing networks that autonomously detect and address issues. For instance, in the event of undersea cable damage, AIoT can reroute traffic to prevent service disruption (Lipinski, Oliveira, & Botha, 2018).

In today's digital age, automation is vital for managing IoT data streams. Intelligent algorithms allocate bandwidth, prioritize traffic, and enable self-healing for uninterrupted critical data processing (Gómez & Lambert, 2006).

2.3.2 Enhanced Security

AIoT bolsters network security by detecting anomalies and responding in real-time. In banking, it identifies suspicious transactions, triggering fraud alerts. Secure

Bootstrapping provisions IoT devices securely, preventing unauthorized access. AI-driven analytics enable real-time data processing, informed decision-making, and bolstered security against cyber threats. Dynamic scalability adapts to changing demands, while data-driven insights personalize user experiences and inform strategic decisions (Al-Shaer et al., 2018; Sun et al., 2006).

3.0 AIoT STRATEGY FOR NETWORKING PRODUCTIVITY

3.1 Designing the AIoT Strategy

Crafting an effective AIoT strategy requires alignment with organizational goals and adaptability to tech evolution. Start with a thorough analysis of current infrastructure, scalability, security, and performance benchmarks. Key components include assessing existing IoT capabilities, defining clear business objectives, and fostering a culture of continuous learning. Stay updated with AI, IoT, and networking advancements to ensure strategy relevance amidst evolving challenges and opportunities (Costa et al., 2022).

3.1.1 Clear Objectives

Clearly defined objectives drive AIoT implementation, such as reducing network downtime or enhancing security. AIoT transforms networking productivity through automation, optimization, and predictive maintenance, minimizing downtime and enabling swift, data-informed decision-making. Enhanced security, powered by AI-driven threat detection, strengthens defenses against cyber threats (Alahi & Mukhopadhyay, 2023).

3.1.2 Resource Allocation

Effective resource allocation, considering budget, talent, and technology, is crucial. Organizations determine AIoT investment for maximum value (Horita et al., 2017). Leveraging data drives networking productivity. Advanced analytics and AI insights extract valuable information for proactive network management, preemptive troubleshooting, and real-time resource allocation.

3.1.3 Integration Plan

The strategy should detail a plan for integrating AI with IoT, specifying which AI algorithms and IoT devices are suitable for the organization's specific needs.

Cyber-physical systems, at the intersection of physical processes and computational capabilities, form the backbone of AIoT. They facilitate the seamless interaction between the physical world and the digital realm, enabling the collection, processing, and analysis of real-time data from IoT devices. This integration empowers organizations to make data-driven decisions, optimize processes, and enhance operational efficiency (Suresh Babu, C. V. & Srisakthi, S., 2023).

3.2 IoT Device Deployment and Real-Time Data Collection

Efficient deployment of IoT devices and real-time data collection is a critical component of the AIoT strategy. IoT devices serve as the network's data collection nodes, continuously gathering data for analysis by AI systems.

According to Bibri & Allam (2022), real-time data collection is a cornerstone of the AIoT strategy, facilitating timely and informed decision-making. It commences with the deliberate selection of data types relevant to the defined objectives. For example, in the context of a smart building, this might encompass data related to temperature, humidity, occupancy, and energy consumption. Accurate sensor calibration is imperative to ensure the integrity of the collected data.

3.2.1 Device Selection

The careful selection of IoT devices is paramount. Devices must be compatible with the network, reliable, and capable of securely transmitting data. For instance, in a smart home environment, IoT devices like smart thermostats, cameras, and sensors can gather data on temperature, security, and environmental conditions (Atzori, Iera, & Morabito, 2010).

3.2.2 Data Quality

Maintaining data quality is of utmost importance. IoT devices should collect accurate and reliable data, as erroneous data can lead to incorrect AI analysis and decision-making (Al-Fuqaha et al., 2015).

As stated by Suresh Babu, C. V., Mahalashmi, J., et al. (2023), the system is designed to collect the necessary data using various IoT-based technologies. Collected data are trained and processed by machine learning algorithms to predict ways to enhance data quality.

3.3 Integration of AI for Analysis and Optimization

The integration of AI constitutes the cornerstone of AIoT. AI systems analyze the data collected by IoT devices, yielding valuable insights and enabling network optimization.

Data-Driven Insights and Reporting: AI-powered analytics derive valuable insights from the data gathered by IoT devices, supporting reporting, performance enhancement, and strategic decision-making. Organizations can extract valuable information about network performance through advanced analytics and AI-driven insights (Yaqub & Alsabban, 2023).

3.3.1 Machine Learning Algorithms

Incorporating machine learning algorithms into AI systems is essential for data analysis and pattern recognition. In agriculture, AI systems can analyze soil moisture data from IoT sensors to optimize irrigation practices (Mohanty, Hughes, & Zhang, 2016).

Machine learning finds applications across various domains and industries where extensive data handling is required. It is employed to process and analyze data from IoT devices, such as sensor data, to derive insights and make predictions. In agriculture, for instance, machine learning algorithms regularly monitor soil conditions, and assess nutrient levels, acidity, pH values, and other relevant factors to predict outcomes and provide recommendations for soil improvement (Suresh Babu, C. V et al., 2023).

3.3.2 Real-Time Decision-Making

AI should facilitate real-time decision-making. In manufacturing, AI algorithms can optimize production processes based on real-time data, reducing waste and enhancing efficiency (Zameer et al., 2017).

As mentioned by Tien, J. M. (2017), at the core of the Internet of Things (IoT) is an integrated system of computers that perform critical decision-oriented functions, encompassing data sensing, collection, fusion, data analysis, informed decision-making, and communication. This integrated system is pivotal for supporting informed decisions and future knowledge development.

3.3.3 Adaptive Learning

AI systems should possess adaptive learning capabilities, allowing them to continually enhance network performance over time. For example, AI can learn user preferences in a smart home context, adjusting heating and lighting based on individual preferences (Srivastava et al., 2015).

3.4 A Holistic Approach to Networking Enhancement

A comprehensive approach is essential to fully harness the benefits of AIoT in networking:

3.4.1 User Experience

Prioritizing user experience is vital in AIoT integration. It aims to optimize network performance and personalize services for seamless experiences. In enterprises, this involves aligning AIoT with existing operations, providing intuitive interfaces for employees and stakeholders to navigate easily.

4.0 KEY COMPONENTS OF THE AIoT STRATEGY

4.1 Proactive Network Monitoring

Proactive network monitoring is a pivotal element of the AIoT strategy, enabling organizations to preemptively identify and rectify issues before they lead to network downtime or disruptions. Through the utilization of AI algorithms for real-time data analysis, organizations can detect anomalies and potential issues, transforming network management from a reactive to a predictive approach. This proactive strategy ensures uninterrupted operations, enhances security, and optimizes resource allocation.

5.0 BENEFITS OF AIoT IMPLEMENTATION

5.1 Tangible Benefits of Implementing AIoT

Implementing AIoT offers a diverse range of concrete advantages to organizations across various industries. These benefits go beyond the conceptual and are evident in quantifiable improvements:

The fusion of Artificial Intelligence (AI) with the Internet of Things (IoT) empowers organizations to harness data-driven insights, enhancing efficiency, and enabling informed decision-making. AIoT's impact is palpable, resulting in improved productivity, cost savings, and enhanced customer experiences. It allows for service customization, predictive maintenance, and streamlined processes. These measurable outcomes underscore the pivotal role of AIoT in reshaping business paradigms, enabling organizations to thrive in an increasingly data-driven and interconnected world.

5.1.1 Enhanced Productivity

AIoT combines AI and IoT, automating routine tasks and processes, liberating human resources for more strategic and creative endeavors. This transition fosters a significant upsurge in productivity. By seamlessly integrating AI algorithms with IoT devices, routine operations, such as data analysis, maintenance, and monitoring, are executed autonomously and with precision. As a result, organizations can optimize resource allocation, reduce manual labor, and refocus human talent on tasks demanding ingenuity and critical thinking. This transition enhances efficiency and propels innovation in a competitive landscape driven by data and technology.

5.1.2 Reduced Downtime

The predictive maintenance capabilities of AIoT significantly reduce network downtime and prevent costly service disruptions. In manufacturing, for instance, machine breakdowns can be anticipated and maintenance scheduled proactively. AIoT technology merges AI with IoT, facilitating real-time analysis of data from connected devices and sensors to provide insights into machinery and equipment conditions. Consequently, maintenance activities can be scheduled in advance, averting unexpected downtimes, minimizing production interruptions, and curbing maintenance costs. AIoT enhances operational efficiency and cost-effectiveness across industries.

5.1.3 Optimized Resource Allocation

AIoT systems enable data-driven optimized resource allocation, prominently visible in agriculture. Real-time sensor data informs efficient allocation of resources like water and fertilizer, ultimately maximizing crop yield and resource utilization. AI's integration with IoT enables continuous monitoring of environmental conditions and crop health. AI algorithms analyze this data for informed decision-making on resource distribution. The result is tailored and dynamic allocation, ensuring crops receive the right resources precisely when needed. This leads to increased crop yield and resource conservation, promoting sustainable agricultural practices. AIoT drives efficiency and sustainability across various sectors.

5.2 Optimizing Networking Performance

AIoT excels in optimizing networking performance, providing organizations with a multitude of advantages. By utilizing real-time data analysis, predictive algorithms, and proactive management, AIoT equips organizations to achieve substantial improvements. The amalgamation of Artificial Intelligence (AI) with the Internet of Things (IoT) facilitates continuous network condition monitoring. AI algorithms process this data to predict potential issues and proactively address them before they escalate into disruptions. The outcome is enhanced network stability, reduced downtime, and improved overall performance. AIoT's capabilities are a cornerstone of efficient, reliable, and cost-effective networking, positioning it as a crucial tool for businesses aiming to stay competitive in the digital landscape.

5.2.1 Reduced Congestion

AIoT optimizes network traffic and mitigates congestion, resulting in faster data transfers and enhanced user experiences. AIoT leverages AI-driven algorithms to analyze network data in real-time, identifying traffic patterns and potential bottlenecks. This insights-driven approach allows AIoT systems to dynamically reroute data, allocate bandwidth efficiently, and prioritize critical traffic. As a result, network congestion is minimized, ensuring the smooth flow of data. This leads to expedited data transfers and an overall improved user experience. AIoT's ability to optimize network traffic and reduce congestion underscores its vital role in enhancing the performance and reliability of modern digital networks.

5.2.2 Improved Latency

The combination of edge computing and AI-driven network optimizations significantly reduces latency, a crucial advancement for latency-sensitive applications like autonomous vehicles and telemedicine. Edge computing brings processing closer to data sources, reducing data travel distances and mitigating latency. Simultaneously, AI-driven network optimizations continuously analyze and fine-tune data flow for efficient data transmission. This synergy ensures that data arrives swiftly and without delay, essential for real-time applications. For autonomous vehicles, low latency is imperative for split-second decision-making, while telemedicine relies on real-time data transfer for accurate remote diagnoses. The combined effect of edge computing and AI optimization significantly improves the viability and performance of these and other latency-critical applications.

5.3 Cost Reduction and Efficiency Gains

AIoT presents significant opportunities for cost reduction and operational efficiency:

5.3.1 Reduced Operational Costs

Predictive maintenance and automated network management in AIoT reduce manual interventions, lower operational costs, and optimize infrastructure management, especially in sectors like energy, enhancing efficiency and reliability (Gallagher et al., 2019).

5.3.2 Energy Efficiency

AIoT optimizes energy consumption in smart buildings by analyzing sensor data and making real-time adjustments to building systems, reducing utility costs and environmental impact. Its integration of AI with IoT drives economic and ecological benefits in urban areas.

5.4 Improving User Experience

AIoT has a direct impact on user experience across various domains:

5.4.1 Personalization

AI-driven insights facilitate the customization of services to individual preferences. For instance, in the realm of e-commerce, the delivery of product recommendations based on user behavior significantly enhances the shopping experience. This is powered by the fusion of Artificial Intelligence (AI) and the online marketplace, where AI algorithms continuously analyze user data, including past purchases and browsing habits. Consequently, personalized product suggestions are proactively offered to shoppers, elevating their online shopping experience. This approach not only augments user satisfaction but also boosts sales and customer engagement, showcasing AI's transformative influence in e-commerce.

5.4.2 Reliability

Enhanced network performance and reduced downtime lead to more reliable services. In the telecommunications sector, AIoT ensures superior call quality and connectivity. The convergence of AI and the Internet of Things (IoT) empowers networks to operate with greater reliability. AIoT systems continuously monitor network conditions, making real-time adjustments to optimize call quality and ensure seamless connectivity. This ensures that issues such as dropped calls, lag, and poor audio quality become rare occurrences, fostering a more satisfying user experience. The impact of AIoT in telecommunications demonstrates its pivotal role in providing dependable and superior service, underlining its potential to shape the future of communication technologies.

5.4.3 Security

AIoT enhances security in banking by monitoring user behavior and transaction patterns, swiftly detecting and responding to fraud. This proactive approach safeguards user accounts and assets, fostering trust and confidence in financial services (Suresh Babu & Srisakthi, 2023).

6.0 IMPLEMENTATION STEPS

6.1 Initiating the AIoT Strategy

Initiating the AIoT strategy is a crucial first step in realizing the potential benefits of AIoT for networking productivity. The process typically involves the following key actions:

Embarking on an AIoT strategy is a pivotal inaugural stride toward unlocking the manifold benefits of AIoT in enhancing networking productivity. This process entails a sequence of fundamental actions, crucial for achieving the desired outcomes.

6.1.1 Assessment and Planning

Organizations assess current infrastructure, align AIoT integration with goals, and identify areas for maximum value. This holistic approach streamlines implementation, enhancing efficiency and competitive advantage (Suresh Babu & Srisakthi, 2023).

6.2 Executing the Strategy in Real-World Scenarios

Executing the AIoT strategy in real-world scenarios involves the practical implementation of the plan:

6.2.1 IoT Device Deployment

This phase involves physically deploying IoT devices and sensors across the network, ensuring accurate placement and secure data transmission to AI systems. It forms the foundation for real-time insights and informed decision-making, optimizing network performance, and enhancing operational efficiency in AIoT implementations.

6.2.2 AI Integration

In this crucial phase, AI systems are integrated with data from IoT devices. Algorithms and machine learning models are established to process the influx of data, transforming it into actionable insights. This integration optimizes operations, enhances efficiency, and drives productivity, unlocking the full potential of AIoT strategies in organizations.

6.2.3 Testing and Validation

Thorough testing of the AIoT system ensures functionality, security, and alignment with goals. Real-world testing in operational environments uncovers potential issues. It validates performance, security, and reliability, safeguarding against vulnerabilities before deployment, ensuring optimal functionality, and alignment with organizational objectives.

6.3 Overcoming Implementation Challenges

Implementing AIoT can be complex, and organizations may encounter various challenges:

6.3.1 Data Privacy and Security

AIoT's data abundance raises privacy and security concerns. Organizations combat these risks with robust encryption, access controls, and governance policies. Regular audits ensure threat detection and response, fostering trust and regulatory compliance, thus maximizing AIoT's potential while safeguarding privacy and data integrity.

6.3.2 Integration Hurdles

Integrating AI and IoT technologies poses challenges, especially with legacy systems. Compatibility and data exchange issues may hinder seamless integration. Legacy systems lacking IoT and AI compatibility require retrofitting or middleware development to facilitate data exchange, crucial for realizing the synergistic potential of AI and IoT in a technologically advanced landscape.

6.3.3 Resource Constraints

Implementing AIoT requires significant investments in technology, talent, and infrastructure. Organizations must address resource constraints efficiently, balancing upfront costs with long-term returns. Efficient resource allocation and collaborative partnerships are crucial for optimizing investments and unlocking AIoT's full potential.

6.4 Measuring the Success of the AIoT Initiative

Measuring the success of an AIoT initiative is vital for evaluating its impact and making necessary improvements:

6.4.1 Key Performance Indicators (KPIs)

Define KPIs aligned with the objectives of the AIoT strategy. KPIs may include network uptime, cost savings, user satisfaction, and efficiency gains.

Setting Key Performance Indicators (KPIs) that are in harmony with the AIoT strategy's objectives is paramount. These KPIs may encompass network uptime, cost savings, user satisfaction, and efficiency gains, among other relevant metrics. The establishment of such measurable benchmarks serves as a yardstick for gauging the success and effectiveness of the AIoT implementation. By closely tracking these KPIs, organizations can ascertain whether the AIoT strategy is delivering the intended outcomes, and they can swiftly pinpoint areas that require adjustment or enhancement.

6.4.2 Data Analytics

Continuously analyze the data collected by AIoT systems to assess performance and identify areas for improvement.

A critical component of AIoT management is the continuous analysis of data collected by AIoT systems. This analysis is instrumental in evaluating performance and identifying areas that demand improvement. Regular data scrutiny allows organizations to determine how well their AIoT strategies align with their desired outcomes. This ongoing examination not only reveals insights that contribute to process refinement and enhanced efficiency but also serves as an early detection mechanism for issues or anomalies. Embracing this practice ensures that organizations remain adaptable and well-prepared to harness the full potential of AIoT, guaranteeing that their systems evolve to meet changing needs and remain at the forefront of innovation.

6.4.3 Feedback and Iteration

Gathering feedback from diverse stakeholders, including end-users and administrators, is vital for refining the AIoT strategy. This feedback loop provides insights for user-centric improvements, system manageability, and alignment with organizational objectives. Continuous feedback fosters adaptability, promoting ongoing enhancement and optimization of AIoT systems to deliver maximum value and efficiency.

7.0 CASE STUDIES

7.1 Real-world Examples of AIoT Success Stories

Real-world case studies provide concrete evidence of AIoT's benefits in enhancing networking productivity. Here are some compelling examples:

7.1.1 Smart Cities

Barcelona, Spain, stands as a shining example of a pioneering smart city. Within the urban landscape, IoT sensors are deployed to monitor critical aspects such as traffic patterns, waste management, and environmental data. The data collected is then subjected to analysis by AI algorithms. This AI-driven approach aims to optimize traffic flow, reduce congestion, enhance mobility, and minimize waste collection costs through efficient scheduling. Additionally, it supports environmental initiatives to improve air quality. Barcelona's smart city initiatives serve as a testament to the transformative potential of IoT and AI technologies in urban living enhancement (Kuğuoğlu, Janssen, et al., 2021).

7.1.2 Manufacturing

Bosch, the renowned global engineering and electronics company, has seamlessly integrated AIoT within its manufacturing facilities. On production lines, strategically positioned sensors collect data that AI systems subsequently use to predict maintenance requirements. This strategic implementation has resulted in a remarkable 10% reduction in downtime, concurrently boosting production efficiency by a substantial 15%. Bosch's utilization of AIoT underscores its capacity to deliver tangible benefits by optimizing manufacturing processes, minimizing disruptions, and enhancing operational efficiency.

7.1.3 Healthcare

The Cleveland Clinic, a major U.S. healthcare provider, has incorporated AIoT in patient care. Wearable devices continuously monitor patients' vital signs, with real-time analysis conducted by AI algorithms. This innovative approach has led to significantly improved patient outcomes. Early detection of deviations or anomalies triggers timely intervention, enhancing patient safety and healthcare delivery. The Cleveland Clinic's adoption of AIoT showcases its potential to revolutionize patient care and advance the healthcare industry.

7.2 Organizations That Have Successfully Leveraged AIoT

Numerous organizations have effectively harnessed AIoT to optimize their operations, resulting in impressive outcomes:

7.2.1 General Electric (GE)

GE's aviation division has adopted AIoT for predictive maintenance of aircraft engines. Equipped with sensors, these engines continuously gather real-time data. AI algorithms, purpose-built for this task, analyze the data to predict maintenance needs in advance, substantially reducing unplanned downtime. GE's innovative use of AIoT in aerospace not only enhances engine performance but also ensures increased reliability, safety, and cost-efficiency within the aviation industry.

7.2.2 Netflix

Netflix, a global streaming powerhouse, employs AIoT to transform its content recommendation system. IoT devices positioned within users' homes meticulously capture data regarding viewing habits and user interactions. AI algorithms, operating in real-time, interpret these patterns and preferences to offer highly personalized content recommendations. This strategic fusion of AI and IoT optimizes user experiences, resulting in elevated engagement, customer satisfaction, and retention, reshaping the content delivery and entertainment landscape.

7.2.3 Siemens

Siemens, a prominent global technology and engineering leader, has embraced AIoT for advanced energy management within smart buildings. Through the use of sensors that monitor energy consumption, AI systems dynamically optimize heating, cooling, and lighting based on real-time data and user behavior. This initiative has yielded a substantial 30% reduction in energy costs, exemplifying AIoT's potential to drive significant energy efficiency gains, reduce operational expenses, and contribute to sustainability and intelligent infrastructure management.

7.3 Lessons Learned from Case Studies

Analyzing AIoT case studies offers profound insights and valuable lessons for organizations contemplating similar implementations:

7.3.1 Alignment With Goals

Successful AIoT implementations hinge on well-defined strategic objectives. Clear, specific, and measurable goals guide technology deployment, aligning efforts with the organization's mission, enabling performance evaluation, and enhancing buy-in and resource optimization.

7.3.2 Data Quality

The bedrock of effective AIoT lies in high-quality, accurate data. Safeguarding data integrity and reliability is paramount. AIoT thrives on the precision and consistency of data, serving as the bedrock for informed decision-making, actionable insights, and the successful operation of AI algorithms. Ensuring data integrity encompasses protecting data against tampering or unauthorized access and maintaining data quality through rigorous validation and cleansing processes. Organizations must establish robust data collection and storage protocols to prevent corruption or loss. The dependability of AIoT systems hinges on this commitment to data quality, as they draw their transformative power from the trustworthiness and precision of the data they ingest, making it an indispensable prerequisite for success.

7.3.3 Scalability

The capacity for AIoT systems to scale is essential to accommodate growth. This flexibility involves the system's ability to seamlessly expand, welcoming an increasing number of IoT devices and the accompanying surge in data volumes. Scalability is imperative to ensure that AIoT solutions can evolve and adapt in harmony with the organization's evolving needs. This adaptability facilitates the integration of new devices, sensors, and data sources without compromising system performance, reliability, or efficiency. By prioritizing scalability, organizations position themselves to capitalize on the transformative potential of AIoT and ensure that their technological infrastructure remains agile and responsive in a perpetually evolving digital landscape.

7.3.4 User Involvement

Actively involving end-users in design and feedback processes is crucial for successful AIoT adoption. Their perspectives shape user-centric AIoT systems, ensuring solutions meet real-world needs. Iterative refinements based on feedback enhance usability and satisfaction, fostering ownership, buy-in, and widespread acceptance of AIoT technologies within the organization.

8.0 FUTURE TRENDS AND CONSIDERATIONS

8.1 Exploring Emerging Trends in AIoT for Networking

The landscape of AIoT for networking continues to evolve, with several emerging trends reshaping the field:

8.1.1 Edge AI

Edge computing, powered by AI, is gaining prominence as a crucial technological paradigm. It enables data processing and decision-making closer to the data source, reducing latency and enhancing real-time analytics. Edge AI plays a pivotal role in applications such as autonomous vehicles and industrial IoT (IIoT) (Zhang, K., Leng, S., He, Y., Maharjan, S., & Zhang, Y., 2018).

8.1.2 5G Integration

The rapid deployment of 5G networks is driving the adoption of AIoT. 5G's low latency and high bandwidth capabilities enable faster data transmission, making it ideal for AIoT applications, including smart cities and augmented reality. This convergence between 5G and AIoT is reshaping the technological landscape (Santhi & Muthuswamy, 2023).

8.1.3 AI-Driven Networking

The integration of AI into network infrastructure is on the rise, ushering in AI-driven networking. This paradigm empowers dynamic network adjustments, self-configuration, and enhanced security, revolutionizing the way networks operate and ensuring robust security measures (Yaqub, M. Z., & Alsabban, 2023).

8.2 Potential Developments and Innovations

The future of AIoT for networking is brimming with possibilities:

8.2.1 Autonomous Networks

Autonomous networks represent a fundamental shift in the networking domain. They are characterized by self-optimization, self-healing, and self-protection mechanisms, allowing them to adapt to changing conditions and make real-time adjustments without human intervention. This concept is particularly promising in large-scale applications, where the complexity and scale of network management are substantial (Engels & Zhuang, 2013).

8.2.2 Quantum Computing

The advent of quantum computing has the potential to revolutionize AIoT. Quantum computing can process massive datasets and perform complex AI computations at speeds that were previously unimaginable with classical computing systems. Leveraging principles like superposition and entanglement, quantum computers can tackle complex problems exponentially faster, making them ideal for the intricate and data-intensive demands of AIoT applications (Santhi & Muthuswamy, 2023b). As quantum computing technology matures, it opens up new frontiers in real-time analytics, predictive modeling, and optimization, ushering in a new era of AIoT innovation.

8.2.3 Blockchain Integration

The integration of blockchain technology can significantly enhance the security and transparency of AIoT systems. Blockchain offers an immutable, decentralized ledger where data records are cryptographically linked, ensuring data integrity and provenance. This is especially critical in applications like supply chain management, where product origin and journey can be transparently traced, and in healthcare, where patient data integrity is paramount. The fusion of blockchain with AIoT underlines the potential to reinforce trust and security in industries reliant on data integrity (Santhi & Muthuswamy, 2023b).

8.3 Ethical and Privacy Considerations in AIoT

As AIoT continues to advance, ethical and privacy considerations take on a growing significance:

8.3.1 Data Privacy

The substantial volume of data collected by AIoT systems gives rise to legitimate concerns about data privacy. Organizations bear the responsibility of implementing robust data protection measures to ensure the security and confidentiality of user information. These measures encompass stringent encryption protocols, access controls, and authentication mechanisms, collectively shielding data from unauthorized access or tampering. Simultaneously, comprehensive data governance policies and compliance frameworks must be established to guarantee the lawful and ethical handling of sensitive information. Regular security audits and vigilant monitoring further contribute to proactive threat detection and response. The effective implementation of these stringent measures not only fosters data security but also instills confidence and trust among stakeholders, enabling organizations to fully harness the potential of AIoT while upholding privacy and regulatory compliance.

8.3.2 Bias and Fairness

The utilization of AI algorithms within AIoT systems introduces the risk of inheriting biases from the training data. Ensuring fairness and reducing bias in AIoT decision-making processes is a primary concern. This endeavor entails comprehensive data screening to identify and rectify biased or discriminatory patterns in the training data. Furthermore, organizations must adopt fairness-aware AI models designed to mitigate bias, promote equitable outcomes, and uphold ethical standards. Inclusivity and diversity in both data and model development play a crucial role in mitigating bias. By proactively addressing this issue, AIoT technologies can ensure fairness and equitable treatment, mitigating the potential perpetuation of societal biases in automated decision-making, fostering a technology landscape that aligns with ethical, legal, and moral principles.

8.3.3 Security

The interconnected nature of AIoT systems introduces a complex web of potential vulnerabilities. Implementing robust cybersecurity measures is paramount to protect against cyber threats. These multifaceted systems are exposed to a spectrum of cyber threats that can compromise data integrity, privacy, and overall system functionality.

To safeguard against these perils, organizations must establish comprehensive security protocols, including stringent access controls, encryption, intrusion detection systems, and regular security audits. Additionally, proactive monitoring and swift response mechanisms are imperative to thwart emerging threats. As AIoT continues to evolve, cybersecurity remains a cornerstone of its success, preserving the trust of stakeholders and assuring the integrity of interconnected systems in an era marked by increasing digitization and data exchange.

9.0 CONCLUSION AND IMPLICATIONS

9.1 Summarizing Key Takeaways

As we conclude this exploration of AIoT for networking productivity, it's essential to summarize the key takeaways from this journey:

9.1.1 Convergence of AI and IoT

AIoT represents the convergence of Artificial Intelligence (AI) and the Internet of Things (IoT), combining AI's analytical capabilities with IoT's vast data streams to optimize networking and various industries. AIoT enhances the intelligence of IoT devices, enabling them to collect, comprehend, process, and act autonomously on data. This leads to smarter, self-optimizing systems that facilitate data-driven decision-making, operational efficiency, security enhancement, and personalized services. In essence, AIoT is reshaping technology, ushering in a new era of interconnected, intelligent, and data-driven applications across diverse domains.

9.1.2 Tangible Benefits

The implementation of AIoT offers tangible benefits across sectors such as healthcare, manufacturing, and smart cities. It enhances productivity, reduces downtime, and results in cost savings. In healthcare, it enables predictive maintenance and personalized patient care, while in manufacturing, it streamlines operations. In smart cities, it enhances services and improves the quality of life. AIoT is catalyzing a paradigm shift, enhancing efficiency, and delivering substantial real-world advantages.

9.1.3 Emerging Trends

AIoT continues to evolve, with emerging trends like edge AI, 5G integration, and autonomous networks reshaping the landscape and offering new opportunities for innovation. Edge AI enables real-time data processing, 5G integration accelerates data transmission, and autonomous networks introduce self-optimizing systems. These trends present new possibilities for AIoT to advance efficiency, connectivity, and intelligence across industries, opening up avenues for future innovations.

9.2 The Implications of Adopting AIoT for Networking Productivity

The implications of adopting AIoT are far-reaching and transformative:

9.2.1 Competitive Advantage

Organizations that embrace AIoT gain a competitive edge by optimizing operations, reducing costs, and delivering superior services. The infusion of AIoT technologies leads to operational optimization through data-driven insights, cost reduction through predictive maintenance, and enhanced user experiences through personalization. AIoT empowers organizations to pioneer the next wave of innovation, enabling them to thrive in an increasingly dynamic and digital landscape.

9.2.2 Data Governance

The massive data generated by AIoT requires robust data governance strategies encompassing data security, privacy, and ethics. Robust data security protocols are indispensable to safeguard data against unauthorized access and cyber threats. Ensuring data privacy and ethical considerations are also vital to maintaining trust and ethical practices.

9.2.3 Workforce Transition

The integration of AIoT necessitates upskilling the workforce to manage and extract value from these technologies. Organizations must invest in equipping their teams with the requisite knowledge and skills in data analysis, AI algorithms, IoT device management, cybersecurity, and data privacy awareness. Upskilling empowers the workforce to harness the full potential of AIoT.

9.3 Encouraging a Forward-Looking Approach

To fully harness the potential of AIoT for networking productivity, organizations are encouraged to adopt a forward-looking approach:

9.3.1 Innovation Culture

Nurturing a culture of innovation within an organization involves fostering an environment that champions experimentation, agility, and continuous learning. Such a culture encourages individuals to explore new ideas, take calculated risks, and think beyond conventional boundaries, ultimately driving creativity, progress, and competitiveness.

9.3.2 Collaboration

Collaborating with experts in AI, IoT, and other relevant fields is crucial for staying at the forefront of technological advancements. These partnerships provide access to cutting-edge knowledge, diverse perspectives, and facilitate the integration of the latest technological breakthroughs into the organization's operations.

9.3.3 Regulatory Compliance

Remaining informed about evolving regulations related to AIoT is essential to ensure compliance and responsible implementation. Ongoing monitoring, proactive adaptation, and active engagement with regulatory authorities and industry standards are key to upholding ethical and legal obligations, mitigating risks, and preserving organizational integrity and reputation.

REFERENCES

Abdelmaboud, A., Ahmed, A. I. A., Abaker, M., Eisa, T. E., Albasheer, H., Ghorashi, S., & Karim, F. K. (2022). Blockchain for IoT Applications: Taxonomy, Platforms, Recent Advances, Challenges and Future Research Directions. *Electronics (Basel)*, *11*(4), 630. doi:10.3390/electronics11040630

Abdollahpouri, H., Burke, R., & Mobasher, B. (2017). Context-aware event recommendation in event-based social networks. *User Modeling and User-Adapted Interaction*, *27*(4), 715–751.

Abid, M. A., Afaqui, N., Khan, M. A., Akhtar, M. W., Malik, A. W., Munir, A., Ahmad, J., & Shabir, B. (2022). Evolution towards Smart and Software-Defined Internet of Things. *AI*, *3*(1), 100–123. doi:10.3390/ai3010007

Adir, O., Poley, M., Chen, G., Froim, S., Krinsky, N., Shklover, J., Shainsky-Roitman, J., Lammers, T., & Schroeder, A. (2019). Integrating Artificial Intelligence and Nanotechnology for Precision Cancer Medicine. *Advanced Materials*, *32*(13), 1901989. Advance online publication. doi:10.1002/adma.201901989 PMID:31286573

Al-Fuqaha, A., Guizani, M., Mohammadi, M., Aledhari, M., & Ayyash, M. (2015). Internet of Things: A Survey on Enabling Technologies, Protocols, and Applications. *IEEE Communications Surveys and Tutorials*, *17*(4), 2347–2376. doi:10.1109/COMST.2015.2444095

Al-Shaer, E., Assi, C., & Zhang, N. (2018). Artificial intelligence for wireless communication systems: A comprehensive survey. *IEEE Communications Surveys and Tutorials*, *20*(4), 2759–2788.

Alahi, E. E., Sukkuea, A., Tina, F. W., Nag, A., Kurdthongmee, W., Suwannarat, K., & Mukhopadhyay, S. C. (2023). Integration of IoT-Enabled Technologies and Artificial Intelligence (AI) for Smart City Scenario: Recent Advancements and Future Trends. *Sensors (Basel)*, *23*(11), 5206. doi:10.3390/s23115206 PMID:37299934

Alsmadi, I., Tawalbeh, L. A., Alshraideh, M., & Jararweh, Y. (2014). A review of quality of service in mobile computing environments. *TheScientificWorldJournal*, *2014*, 16.

Atzori, L., Iera, A., & Morabito, G. (2010). The Internet of Things: A survey. *Computer Networks*, *54*(15), 2787–2805. doi:10.1016/j.comnet.2010.05.010

Bibri, S. E., Allam, Z., & Krogstie, J. (2022). The Metaverse as a Virtual Form of Data-Driven Smart Urbanism: Platformization and Its Underlying Processes, Institutional Dimensions, and Disruptive Impacts. *Computational Urban Science*, *2*(1), 24. Advance online publication. doi:10.1007/s43762-022-00051-0 PMID:35974838

Chang, Z., Liu, S., Xiong, X., Cai, Z., & Tu, G. (2021). A Survey of Recent Advances in Edge-Computing-Powered Artificial Intelligence of Things. *IEEE Internet of Things Journal*, *8*(18), 13849–13875. doi:10.1109/JIOT.2021.3088875

Chen, D., Xue, Y., & Yao, L. (2018). A survey of big data architectures and machine learning algorithms in large scale data processing. *Journal of King Saud University. Computer and Information Sciences*.

Chen, M., Saad, W., & Yin, C. (2018). Virtual Reality over Wireless Networks: Quality-of-Service Model and Learning-Based Resource Management. *IEEE Transactions on Communications*, *66*(11), 5621–5635. doi:10.1109/TCOMM.2018.2850303

Chen, S., Gu, X., Wang, J., & Zhu, H. (2021). AIOT Used for COVID-19 Pandemic Prevention and Control. *Contrast Media & Molecular Imaging*, *2021*, 1–23. doi:10.1155/2021/8922504 PMID:34729056

Cisco. (2019). *Cisco Visual Networking Index: Forecast and Trends, 2017–2022*. Cisco.

Cleveland Clinic. (2021). *How Cleveland Clinic's home hospital brings care to you*. Retrieved from https://my.clevelandclinic.org/patient/guides/hospital-at-ho me

Costa, I., Riccotta, R., Montini, P., Stefani, E., De Goes, R., Gaspar, M. A., Martins, F. S., Fernandes, A. A., Machado, C., Loçano, R., & Larieira, C. L. C. (2022). The Degree of Contribution of Digital Transformation Technology on Company Sustainability Areas. *Sustainability (Basel)*, *14*(1), 462. doi:10.3390/su14010462

Davenport, T. H., Harris, J., & Shapiro, J. (2010). *Competing on analytics: The new science of winning*. Harvard Business Press.

Deora, V., Shao, J., Gray, W. A., & Fiddian, N. J. (2003). A Quality of Service Management Framework Based on User Expectations. In Lecture Notes in Computer Science (pp. 104–114). doi:10.1007/978-3-540-24593-3_8

Dutta, A. (1996). Integrating AI and Optimization for Decision Support: A Survey. *Decision Support Systems*, *18*(3–4), 217–226. doi:10.1016/0167-9236(96)00026-7

Engels, A., Reyer, M., Xu, X., Mathar, R., Zhang, J., & Zhuang, H. (2013). Autonomous Self-Optimization of Coverage and Capacity in LTE Cellular Networks. *IEEE Transactions on Vehicular Technology*, *62*(5), 1989–2004. doi:10.1109/TVT.2013.2256441

Gallagher, B., McLean, T., Phillips, L. A., Warren, M., & Morrow, P. (2019). Predictive maintenance in the industry 4.0 era: A comprehensive survey. *IEEE Transactions on Reliability*, *68*(3), 1406–1420.

Gebraeel, N., Lawley, M. A., & Cahoon, P. (2016). Data-driven predictive maintenance using deep learning and embedded system technology. *IISE Transactions*, *48*(6), 494–505.

General Electric (GE). (2021). *Predictive maintenance for aviation*. Retrieved from https://www.ge.com/research/projects/predictive-maintenance-aviation

Ghahremani-Nahr, J., Nozari, H., & Sadeghi, M. E. (2021). Green Supply Chain Based on Artificial Intelligence of Things (AIoT). *International Journal of Innovation in Management Economics and Social Sciences*, *1*(2), 56–63. doi:10.52547/ijimes.1.2.56

Goethals, T., Volckaert, B., & De Turck, F. (2021). Enabling and Leveraging AI in the Intelligent Edge: A Review of Current Trends and Future Directions. *IEEE Open Journal of the Communications Society*, *2*, 2311–2341. doi:10.1109/OJCOMS.2021.3116437

Gómez, C., & Paradells, J. (2010). Wireless Home Automation Networks: A Survey of Architectures and Technologies. *IEEE Communications Magazine*, *48*(6), 92–101. doi:10.1109/MCOM.2010.5473869

Güngör, V. Ç., & Lambert, F. (2006). A Survey on Communication Networks for Electric System Automation. *Computer Networks*, *50*(7), 877–897. doi:10.1016/j.comnet.2006.01.005

Han, Z., Kumar, U., & Singh, M. (2013). An overview of Smart Grid: Concepts, benefits and challenges. *Electric Power Components and Systems*, *41*(14), 1453–1489.

Herrera, D., & Rivas-Lalaleo, D. (2020). Artificial intelligence of things (AIoT) applied to optimize wireless sensors networks (WSNs). *Sensors (Basel)*, *20*(19), 5417.

Hogg, S., Kurose, J., & Faloutsos, M. (2016). Network management protocols. In *Computer Networks and Internets* (6th ed., pp. 644–674). Pearson.

Horita, F. E. A., De Albuquerque, J. P., Marchezini, V., & Mendiondo, E. M. (2017). Bridging the Gap Between Decision-Making and Emerging Big Data Sources: An Application of a Model-Based Framework to Disaster Management in Brazil. *Decision Support Systems*, *97*, 12–22. doi:10.1016/j.dss.2017.03.001

IEEE Computer Society. (2017). *IoT Technical Committee*. IEEE. Retrieved from https://www.computer.org/technical-committees/internet-of-things/

Jadad, A. R. (2019). Promoting innovation in healthcare: The role of AI, IoT, and blockchain. *Journal of Medical Internet Research*, *21*(2), e16286.

Kamble, S. S., Oza, P., & Dongare, S. B. (2019). Internet of Things (IoT): A review of enabling technologies, challenges, and open research issues. *Computer Communications*.

Kim, J., Kim, I., Kim, J., & Kim, Y. (2020). An efficient deep learning-based network intrusion detection system for smart grids. *IEEE Access : Practical Innovations, Open Solutions*, *8*, 120087–120097.

Kim, Y., Tan, R., & Xu, L. (2017). Network security. In *Network Science and Cybersecurity* (pp. 259–279). Springer.

Kuğuoğlu, B., Van Der Voort, H., & Janssen, M. (2021). The Giant Leap for Smart Cities: Scaling Up Smart City Artificial Intelligence of Things (AIoT) Initiatives. *Sustainability (Basel)*, *13*(21), 12295. doi:10.3390/su132112295

Lea, R., Cater, A., & Vo, Q. (2018). Artificial intelligence and the internet of things for sustainable agriculture: A comprehensive review. *IEEE Access : Practical Innovations, Open Solutions*, *6*, 38335–38355.

Letaief, K. B., Shi, Y., Lu, J., & Lu, J. (2022). Edge Artificial Intelligence for 6G: Vision, Enabling Technologies, and Applications. *IEEE Journal on Selected Areas in Communications*, *40*(1), 5–36. doi:10.1109/JSAC.2021.3126076

Lipinski, P., Oliveira, L. B., & Botha, R. A. (2018). AIoT: Artificial intelligence as a service over IoT. *Future Generation Computer Systems*, *86*, 941–951.

Lopez, J., Sánchez, H., López, T., & Cerrada, M. (2018). IoT-based system for real-time monitoring and management of photovoltaic facilities. *IEEE Transactions on Industrial Informatics*, *14*(6), 2624–2633.

López-Granados, F., Jurado-Expósito, M., & Peña-Barragán, J. M. (2011). Quantifying efficacy and limits of unmanned aerial vehicle (UAV) technology for weed seedling detection as affected by sensor resolution. *Sensors (Basel)*, *11*(1), 119–131.

Lu, Z., Qian, P., Bi, D., Ye, Z., He, X., Zhao, Y., Su, L., Li, S., & Zheng-Long, Z. (2021). Application of AI and IoT in Clinical Medicine: Summary and Challenges. *Current Medical Science*, *41*(6), 1134–1150. doi:10.1007/s11596-021-2486-z PMID:34939144

McKenzie, J. E., Bossuyt, P. M., Boutron, I., Hoffmann, T., Mulrow, C. D., Shamseer, L., Tetzlaff, J., Akl, E. A., Brennan, S., Chou, R., Glanville, J., Grimshaw, J., Hróbjartsson, A., Lalu, M. M., Li, T., Loder, E., Mayo-Wilson, E., McDonald, S., . . . Moher, D. (2021). The PRISMA 2020 Statement: An Updated Guideline for Reporting Systematic Reviews. BMJ, 71. doi:10.1136/bmj.n71

McKeown, N., Anderson, T., Balakrishnan, H., Parulkar, G., Peterson, L., Rexford, J., ... Shenker, S. (2008). OpenFlow: Enabling innovation in campus networks. *Computer Communication Review*, *38*(2), 69–74. doi:10.1145/1355734.1355746

Meng, X., & Eneh, T. I. (2021). A survey on the impact of 5G network on edge computing and IoT. *IEEE Access : Practical Innovations, Open Solutions*, *9*, 75590–75606.

Menon, D., Anand, B., & Chowdhary, C. L. (2023). Digital Twin: Exploring the Intersection of Virtual and Physical Worlds. *IEEE Access : Practical Innovations, Open Solutions, 11*, 75152–75172. doi:10.1109/ACCESS.2023.3294985

Mohanty, S. P., Hughes, D. P., & Zhang, H. (2016). Using machine learning for crop yield prediction and climate change adaptation. *Computers and Electronics in Agriculture, 123*, 234–246.

Nassar, M. M., Soares, A. L., & Yadav, S. (2017). Internet of Things (IoT) and big data: A review of recent developments. *IEEE Internet of Things Journal, 4*(5), 1032–1043.

Netflix. (2021). *How Netflix works: The (hugely simplified) complex stuff that happens every time you hit Play.* Retrieved from https://netflixtechblog.com

Nguyen, G., Dlugolinský, Š., Tran, V., & García, Á. L. (2020). Deep Learning for Proactive Network Monitoring and Security Protection. *IEEE Access : Practical Innovations, Open Solutions, 8*, 19696–19716. doi:10.1109/ACCESS.2020.2968718

Nikitas, A., Michalakopoulou, K., Njoya, E. T., & Karampatzakis, D. (2020). Artificial Intelligence, Transport, and the Smart City: Definitions and Dimensions of a New Mobility Era. *Sustainability (Basel), 12*(7), 2789. doi:10.3390/su12072789

Otasek, D., Morris, J. H., Bouças, J., Pico, A., & Demchak, B. (2019). Cytoscape Automation: Empowering Workflow-Based Network Analysis. *Genome Biology, 20*(1), 185. Advance online publication. doi:10.1186/s13059-019-1758-4 PMID:31477170

Pan, L., Jiang, M., Wei, D., Chao, H. C., & Zhang, Y. (2018). Artificial intelligence of things (AIoT): A comprehensive survey. *Future Generation Computer Systems, 90*, 128–146.

Pereira, A. (2016). Plant Abiotic Stress Challenges from the Changing Environment. *Frontiers in Plant Science, 7*. Advance online publication. doi:10.3389/fpls.2016.01123 PMID:27512403

Perera, C., Liu, C. H., & Jayawardena, S. (2017). The emerging internet of things marketplace from an industrial perspective: A survey. *IEEE Transactions on Emerging Topics in Computing, 5*(2), 150–174.

Pise, A. A., Almuzaini, K. K., Ahanger, T. A., Farouk, A., Pant, K., Pareek, P. K., & Nuagah, S. J. (2022). Enabling Artificial Intelligence of Things (AIOT) Healthcare Architectures and Listing Security Issues. *Computational Intelligence and Neuroscience, 2022*, 1–14. doi:10.1155/2022/8421434 PMID:36911247

Ponemon Institute. (2016). *Cost of data center outages*. Ponemon Institute, LLC.

Qin, J., Liu, Y., & Yung, M. (2019). A survey of IoT key technology and applications in 5G. *IEEE Access : Practical Innovations, Open Solutions, 7*, 78714–78727.

Qin, Z., Zhang, H., Meng, S., & Choo, K. R. (2020). Imaging and Fusing Time Series for Wearable Sensor-Based Human Activity Recognition. *Information Fusion, 53*, 80–87. doi:10.1016/j.inffus.2019.06.014

Qiu, T., Chi, J., Zhou, X., Ning, Z., Atiquzzaman, M., & Wu, D. (2020). Edge Computing in Industrial Internet of Things: Architecture, Advances, and Challenges. *IEEE Communications Surveys and Tutorials, 22*(4), 2462–2488. doi:10.1109/COMST.2020.3009103

Rahmani, R., Adabi, S., & Javadi, H. H. S. (2018). Edge computing for the Internet of Things: A case study. *IEEE Internet of Things Journal, 5*(3), 1275–1284.

Rajasegarar, S., Leckie, C., Palaniswami, M., Bezahaf, M., & Bishop, A. (2012). Anomaly detection in wireless sensor networks. *IEEE Wireless Communications, 19*(5), 74–80.

Salah, K., Rehman, M. H. U., Nizamuddin, N., & Al-Fuqaha, A. (2019). Blockchain for AI: Review and Open Research Challenges. *IEEE Access : Practical Innovations, Open Solutions, 7*, 10127–10149. doi:10.1109/ACCESS.2018.2890507

Santhi, A. R., & Muthuswamy, P. (2023). Industry 5.0 or Industry 4.0S? Introduction to Industry 4.0 and a Peek into the Prospective Industry 5.0 Technologies. *International Journal on Interactive Design and Manufacturing, 17*(2), 947–979. doi:10.1007/s12008-023-01217-8

Shi, W., Cao, J., Zhang, Q., Li, Y., & Xu, L. (2016). Edge computing: Vision and challenges. *IEEE Internet of Things Journal, 3*(5), 637–646. doi:10.1109/JIOT.2016.2579198

Siemens. (2021). *Digitalization in building management*. Retrieved from https://new.siemens.com/global/en/products/buildings/topic-a reas/intelligent-infrastructure.html

Srivastava, N., Hinton, G., Krizhevsky, A., Sutskever, I., & Salakhutdinov, R. (2015). Dropout: A simple way to prevent neural networks from overfitting. *Journal of Machine Learning Research, 15*(1), 1929–1958.

Stoyanova, M., Nikoloudakis, Y., Panagiotakis, S., Pallis, E., & Markakis, E. K. (2020). A Survey on the Internet of Things (IoT) Forensics: Challenges, Approaches, and Open Issues. *IEEE Communications Surveys and Tutorials*, 22(2), 1191–1221. doi:10.1109/COMST.2019.2962586

Sun, Y., Shi, W., Xu, D., & Liu, Y. (2016). When mobile blockchain meets edge computing. *IEEE Network*, 30(2), 96–101.

Sun, Y., Yu, W., Han, Z., & Liu, K. (2006). Information Theoretic Framework of Trust Modeling and Evaluation for Ad Hoc Networks. *IEEE Journal on Selected Areas in Communications*, 24(2), 305–317. doi:10.1109/JSAC.2005.861389

Suresh Babu, C. V. N. S., A., P., M. V., & Janapriyan, R. (2023). IoT-Based Smart Accident Detection and Alert System. In P. Swarnalatha & S. Prabu (Eds.), Handbook of Research on Deep Learning Techniques for Cloud-Based Industrial IoT (pp. 322-337). IGI Global. doi:10.4018/978-1-6684-8098-4.ch019

Suresh Babu, C. V., Mahalashmi, J., Vidhya, A., Nila Devagi, S., & Bowshith, G. (2023). Save Soil Through Machine Learning. In M. Habib (Ed.), *Global Perspectives on Robotics and Autonomous Systems: Development and Applications* (pp. 345–362). IGI Global. doi:10.4018/978-1-6684-7791-5.ch016

Suresh Babu, C. V., Monika, R., Dhanusha, T., Vishnuvaradhanan, K., & Harish, A. (2023). Smart Street Lighting System for Smart Cities Using IoT (LoRa). In R. Kumar, A. Abdul Hamid, & N. Binti Ya'akub (Eds.), *Effective AI, Blockchain, and E-Governance Applications for Knowledge Discovery and Management* (pp. 78–96). IGI Global. doi:10.4018/978-1-6684-9151-5.ch006

Suresh Babu, C. V., & Srisakthi, S. (2023). Cyber Physical Systems and Network Security: The Present Scenarios and Its Applications. In R. Thanigaivelan, S. Kaliappan, & C. Jegadheesan (Eds.), *Cyber-Physical Systems and Supporting Technologies for Industrial Automation* (pp. 104–130). IGI Global. doi:10.4018/978-1-6684-9267-3.ch006

Sverko, M., Grbac, T. G., & Mikuc, M. (2022). SCADA Systems With Focus on Continuous Manufacturing and Steel Industry: A Survey on Architectures, Standards, Challenges and Industry 5.0. *IEEE Access : Practical Innovations, Open Solutions*, 10, 109395–109430. doi:10.1109/ACCESS.2022.3211288

Thibaud, M., Chi, H., Zhou, W., & Piramuthu, S. (2018). Internet of Things (IoT) in high-risk Environment, Health and Safety (EHS) industries: A comprehensive review. *Decision Support Systems*, 108, 79–95. doi:10.1016/j.dss.2018.02.005

Tikhamarine, Y., Souag-Gamane, D., Ahmed, A. N., Kisi, Ö., & El-Shafie, A. (2020). Improving artificial intelligence models accuracy for monthly streamflow forecasting using grey Wolf optimization (GWO) algorithm. *Journal of Hydrology (Amsterdam)*, *582*, 124435. doi:10.1016/j.jhydrol.2019.124435

Triantafyllou, A., Sarigiannidis, P., & Λάγκας, Θ. (2018). Network Protocols, schemes, and Mechanisms for Internet of Things (IoT): Features, open challenges, and trends. *Wireless Communications and Mobile Computing*, *2018*, 1–24. doi:10.1155/2018/5349894

Wang, C., He, T., Zhou, H., Zhang, Z., & Lee, C. (2023). Artificial intelligence enhanced sensors - enabling technologies to next-generation healthcare and biomedical platform. *Bioelectronic Medicine*, *9*(1), 17. Advance online publication. doi:10.1186/s42234-023-00118-1 PMID:37528436

Wang, X., & Gill, C. (2018). Integration of blockchain and IoT. In *Blockchain applications* (pp. 39–44). Springer.

Wei, W., Zhou, F., Hu, R. Q., & Wang, B. (2020). Energy-Efficient resource allocation for secure NOMA-Enabled mobile edge computing networks. *IEEE Transactions on Communications*, *68*(1), 493–505. doi:10.1109/TCOMM.2019.2949994

Xiang, Z., Zheng, Y., He, M., Shi, L., Wang, D., Deng, S., & Zheng, Z. (2021). Energy-effective artificial internet-of-things application deployment in edge-cloud systems. *Peer-to-Peer Networking and Applications*, *15*(2), 1029–1044. doi:10.1007/s12083-021-01273-5

Yaqoob, I., Hashem, I. A. T., Inayat, Z., Mokhtar, S., Gani, A., & Ullah Khan, S. (2019). Internet of things forensics: Recent advances, taxonomy, requirements, and open challenges. *IEEE Internet of Things Journal*, *6*(4), 6358–6370.

Yaqub, M. Z., & Alsabban, A. (2023). Industry-4.0-Enabled Digital Transformation: Prospects, instruments, challenges, and implications for business strategies. *Sustainability (Basel)*, *15*(11), 8553. doi:10.3390/su15118553

Yiğitcanlar, T., Desouza, K. C., Butler, L., & Roozkhosh, F. (2020). Contributions and Risks of Artificial Intelligence (AI) in Building Smarter Cities: Insights from a Systematic Review of the Literature. *Energies*, *13*(6), 1473. doi:10.3390/en13061473

Zameer, A., Anwar, S. M., Warsi, M. B., & Guergachi, A. (2017). Machine learning for Industry 4.0: A comprehensive survey. *Computers & Industrial Engineering*, *136*, 1–17.

Zhang, K., Leng, S., He, Y., Maharjan, S., & Zhang, Y. (2018). Mobile edge computing and networking for Green and Low-Latency Internet of Things. *IEEE Communications Magazine*, *56*(5), 39–45. doi:10.1109/MCOM.2018.1700882

Zhang, X., Cao, Z., & Wei, D. (2020). Overview of edge computing in the agricultural Internet of Things: Key technologies, applications, challenges. *IEEE Access : Practical Innovations, Open Solutions*, *8*, 141748–141761. doi:10.1109/ACCESS.2020.3013005

Zhang, Y., Chen, C., Zhang, Y., Han, Z., Xu, C., & Liu, K. (2014). Edge computing in the Internet of Things: A multidisciplinary survey. *IEEE Access : Practical Innovations, Open Solutions*, *6*, 6900–6944.

Zhang, Y., Yu, H. Z., Zhou, W., & Man, M. (2022). Application and research of IoT architecture for End-Net-Cloud Edge Computing. *Electronics (Basel)*, *12*(1), 1. doi:10.3390/electronics12010001

Zhang, Z., Zheng, Y., Sun, D., & Du, X. (2019). A survey of network anomaly detection techniques. *Journal of Network and Computer Applications*, *60*, 19–31.

Zhong, R. Y., Xu, X., Klotz, E., & Newman, S. T. (2017). Intelligent Manufacturing in the context of Industry 4.0: A review. *Engineering (Beijing)*, *3*(5), 616–630. doi:10.1016/J.ENG.2017.05.015

Chapter 6

Harnessing AIoT in Transforming the Education Landscape:
Opportunities, Challenges, and Future

Mukesh Chaware

iD https://orcid.org/0009-0008-6060-5438
Indian Institute of Management, Kozhikode, India

ABSTRACT

Artificial intelligence (AI) and internet of things (IoT) have been combined to create the artificial intelligence of things (AIoT), which could revolutionize education. Education has opportunities and difficulties from this growth. The benefits include tailored learning, real-time feedback, and immersive learning. Data privacy, security, and accessibility are problems beyond the digital divide. Strategic decision-making, communication, and stakeholder engagement are needed to maximise productivity and results using AIoT in education. The chapter provides a plan for ethical and equitable usage of the technology while avoiding hazards. This study investigates the role of AIoT in education to provide a much-needed understanding of future perspectives and ramifications. Responsible and ethical AIoT implementation improves learning and outcomes, according to the chapter.

1 AIoT IN EDUCATION: WHAT IS IT ALL ABOUT?

Let us start by peeping a bit into history and philosophical aspects of education instead of diving straight into the latest trends of artificial intelligence and its overlap with education.

DOI: 10.4018/979-8-3693-0993-3.ch006

According to well-known philosopher of yesteryears, Immanuel Kant, the human being is the sole living being which needs pretty dedicated levels of guidance and education to start becoming an effective member of its community. As such according to Kant, the notion of education holds significant importance in addressing a fundamental inquiry in philosophy: "What is THE human being?". Education serves as the mechanism via which the cognitive faculties that define our human nature are realized and nurtured. This is the process by which individuals undergo a transformation to attain their full humanity. Simultaneously, it is the process through which humanity endeavours to enhance and ultimately achieve perfection. This pursuit aims to enable us to live in accordance with moral principles within a universal realization of a society governed by ethical standards.

There is an interesting research work carried out wherein analysis was on the extent to which the integration of artificial intelligence (AI) in educational methodologies may compel us to reconsider Kant's above standpoint about human beings. There is a debate in academic circles, whether systems incorporating AI should be perceived solely as tools for enhancing educational processes or as autonomous entities capable of independent thought and agency. While today it is well known that advent of AI can allow a system to self-learn (machine learning) based on historical data patterns and can even predict further by 'deep learning' and as such if someone is adhering to the fundamental tenets of Kant's perspective, then their viewpoint necessitates significant modifications, thereby acknowledging that education is not solely confined to human interactions as Kant asserts. Thus, it can be suggested that certain nonhuman entities have the capacity to do both, educate and be educated and as such, the advent of AI necessitates a re-evaluation of the fundamental inquiry posed by Kant, namely, "What constitutes the essence of the human being?" The author of this research work ultimately concludes that despite the shift in perspective about essence of being a human, due to change in circumstances presented by emergence of AI, the viewpoint of Kant remains significant, albeit perhaps requiring additional support and modification in consideration of post- and transhumanist viewpoints (Kornilaev, 2022).

The overarching aim of this entire chapter, and not just this section, is to discuss the potential of AIoT in education sector with related analysis of opportunities, threats, and future possibilities. This chapter does not deal with the technical aspects of AIoT and neither does it adopt any research methodology like qualitative interviews or test experiments through controlled groups or survey based quantitative interpretations or alike. The author instead expects the reader to view it through the mindset of an 'all round review' and read it with an intent of 'sense making of things' before embarking on full-fledged transformation to using AIoT in education sector.

1.1 Introduction

The education industry finds itself on the cusp of a dramatic shift in the 21st century, driven by the rapid progress of digital technologies. Two revolutionary technologies, Artificial Intelligence (AI) and the Internet of Things (IoT), provide the foundation of the digital transformation in education. The combination of these technologies signifies the emergence of the Artificial Intelligence of Things (AIoT), a convergence that holds the potential to significantly transform educational methodologies and practices.

The fascination surrounding the integration of AIoT in the field of education extends beyond the realm of technologists and academic researchers, as it deeply resonates with a wider audience. Educators, who are enthusiastic about utilizing technological advancements to enhance their teaching approaches; policymakers, who are responsible for creating forward-thinking educational policies that incorporate technology; tech developers, who are leading the way in the architectural and functional development of AIoT; and lastly, the students, who are the main recipients of these advancements, all find themselves engaged in the discourse surrounding AIoT in education.

This study aims to provide a comprehensive examination of the role, possibilities, and challenges of AIoT in modern educational environments. This chapter provides an exploration of current implementations and potential future advancements in the field of AIoT inside educational institutions worldwide. It can act as a guide, helping to navigate the vast landscape of AIoT in the context of education sector which is getting increasingly globalised in its outlook whereas in contrast the investing stakeholders expect localised benefits and long serving outcomes.

With such prevailing context, it is necessary to understand the significance of our study. In the work (Holmes et al., 2023) titled "Artificial Intelligence in Education: Promises and Implications for Teaching and Learning," the authors articulate the notion that the integration of AI within educational contexts extends beyond mere system optimization. Instead, it serves to enhance the capacity of individuals to engage in creative thinking, critical analysis, and empathetic understanding. Therefore, the comprehension of AIoT assumes a crucial role, not only in terms of academic competence, but also in shaping a more educated, equitable, and empowered global community.

1.2 Convergence of AI and IoT

The integration of Artificial Intelligence (AI) being predominantly based on software and the Internet of Things (IoT) being predominantly hardware, has led to the emergence of the composite system called Artificial Intelligence of Things (AIoT), a

Figure 1. Block Diagram of AIoT

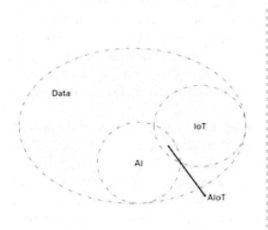

DATA: represents both types of data – real time data points generated by IoT devices plus the earlier prevailing historical data patterns used to train the AI model

IoT: it has the ability to enhance efficiency by streamlining repeated processes, hence reducing the amount of time required to do them.

AI: the AI algorithms are designed to gather and analyze large quantities of data to produce significant, practical and actionable insights. The data stream coming from IoT devices also gets analyzed to find trends.

AIoT: it involves the synthesis of raw data with AI-generated insights to effectively deploy data-driven solutions in real-time. And takes proactive measures to address the requirements of the end user.

technological convergence that is transforming various sectors and industries, which is represented through a simple block diagram, as shown in figure 01.

The education landscape is no exception, as AIoT is revolutionizing the way learning is delivered and perceived. Here we explore further on how, AIoT combines the predictive, cognitive capabilities of AI with the connectivity, ubiquitous computing potential of IoT, based on their backend data.

- **Data:** The term "data" refers to a collection of facts, statistics, or information that is unprocessed, including factual details, numerical values, and statistical measures, which can further manifest in diverse formats, including textual content, numerical representations, visual imagery, video recordings, or any other kind of organized or unorganized data. Data serves as the fundamental basis for any technology or process that relies on information. Data can be gathered, saved, manipulated, and examined in order to extract significant findings.
- **Artificial Intelligence (AI):** AI is a field within the realm of computer science that is primarily concerned with the development of systems or machines capable of executing activities that traditionally necessitate human intellect. These tasks encompass a wide range of abilities, including but not limited to comprehending natural language, discerning patterns, making informed judgments, and acquiring knowledge through experiential learning. AI systems leverage algorithms and data to generate predictions, automate tasks,

and enhance their performance through iterative learning processes. Data plays a pivotal role in the field of AI as machine learning and deep learning algorithms heavily depend on extensive datasets to facilitate the training of models and enable informed decision-making processes. Such extensive data sets, used for training AI models are also at times considered as Big Data (BD), however usage of BD is not restricted to training of AI models alone and it is well evolved sector within itself in the context of information technologies.

- **The Internet of Things (IoT):** This encompasses a network of networked physical devices, sensors, and objects that has the capability to gather and share data over the internet or other communication networks. IoT devices encompass a wide spectrum of technological artifacts, varying from rudimentary sensors to intricate entities such as intelligent thermostats, wearable fitness trackers, and self-driving automobiles. The IoT produces substantial quantities of data as these devices persistently gather data from the physical environment and transmit it to centralized systems or other devices for the purpose of processing and analysis.

Now let us further examine the overlapping portions between these constituent blocks of AIoT – Data, IoT and AI as below:

- **Data & IoT**: In a given ecosystem, the IoT devices produce a substantial volume of data through the utilization of sensors and interconnected devices. The dataset may encompass several types of information, such as temperature measurements, geographical coordinates, user engagements, and more data points. The collection and analysis of data obtained from IoT devices can provide valuable insights and enable well-informed decision-making. For instance, this data can be utilized to optimize energy consumption inside a smart building or to monitor the operational health of industrial machinery.

- **Data and AI**: An AI system is highly dependent on the availability and utilization of vast amounts of data. The Machine Learning (ML) algorithms, which are a subset of any artificial intelligence (AI) system, need substantial datasets in order to undergo training based on prevailing historical patterns. Once adequately trained, those AI models start possessing the capability to analyse and understand data in order to execute various tasks, such as natural language processing, picture recognition, and predictive analytics. AI plays a crucial role in the extraction of important insights and patterns from data sets that may possess a level of complexity beyond the capabilities of human manual analysis.

- **AI and IoT**: An AI system has the potential to augment the functionalities of IoT systems through its ability to swiftly process and analyse the vast amounts of data generated by IoT devices in real-time. As an illustration, AI algorithms possess the capability to identify anomalies within sensor data, forecast instances of equipment malfunction, and enhance the allocation of resources within smart urban environments. As such, AI facilitates the capability of IoT systems to make decisions autonomously and intelligently by leveraging the data they gather.

As one can notice, the interconnection between data, AI and the IoT, is highly evident in the contemporary technological milieu. Data plays a fundamental role as the primary resource for AI, which may then be utilized to analyse and extract significant insights from the data produced by IoT devices. The integration of data, AI, and the IoT has resulted in significant progress across many sectors, including healthcare, manufacturing, transportation, and smart cities. The integration of AI algorithms with the data collected from IoT devices leads to the emergence of AIoT. This AIoT convergence experience allows real-time decision-making and automation of numerous activities, enabling the creation of innovative digital services that improve end-user experiences and ecosystem performance. (Dwivedi et al., 2021; Javaid et al., 2022; Nayal et al., 2022; Salah et al., 2019; Sleem & Elhenawy, 2023).

The subsequent section examines the evolution of AIoT and the extent to which AIoT is capable of facilitating substantial transformations within the domain of education.

2 AIoT IN EDUCATION: HOW HAS IT EVOLVED?

The objective of this chapter is to conduct a comprehensive analysis of the influence of AIoT on education, delving into its diverse ramifications, potential advantages, and the obstacles it presents. The aim of this study is to demonstrate the transformative impact of AIoT on conventional pedagogical approaches, hence facilitating the emergence of individualized, engaging, and efficacious learning encounters for students. Moreover, this work will analyse the ramifications of this integration of technology for educators, students, and educational institutions, as well as other stakeholders within the ecosystem, in its entirety.

2.1 Understanding the Present

The AIoT provides a comprehensive experience involving intelligent and interconnected systems capable of autonomous decision-making, evaluation of decision outcomes, and continuous improvement over time. The integration of AI with the IoT to form AIoT can be comprehended by drawing parallels to the operations of the human brain in two fundamental manners:

- **Cloud AI**: One perspective suggests that the brain might be conceptualized as a centralized system responsible for processing various stimuli and making decisions. In this particular scenario, we are discussing a cloud-based system that serves as a central hub for receiving telemetry data and executing corresponding actions. Cloud AI refers to the utilization of artificial intelligence in a cloud computing environment, where the backend data is collected from diverse IoT devices, and the decision making is predominantly centralised. Example – Educational institutions can profit from Cloud-AI-based AIoT. Developing pupils' computational thinking skills is a major benefit as AIoT courses combine sensing technologies and problem-solving to improve students' computational thinking skills which is becoming increasingly relevant in digital age (Y.-S. Lin et al., 2021).

- **Edge AI**: A known crucial aspect of human neurological system is reflexes. Reflexes are involuntary responses that are generated by the nervous system without actually transmitting sensory information all the way up to the central processing unit, commonly known as the brain. These decisions are made in the peripheral, in close proximity to the source from where the data originated. The term used to refer to this concept is known as Edge AI, which stands for Artificial Intelligence at the Edge. Example – Smart water control systems can intermittently stop field valves when they detect leaks, decreasing water losses. It also alerts the central system, which can activate valves to switch water flow circuits, thus facilitating many such "predictive" maintenance options beyond "reactive" maintenance. In context of education ecosystems, there are known examples of successful implementations of Edge AI, like federated learning in edge computing is another customizable framework for intelligent systems. This architecture enables tailored AIoT learning through cloud-edge collaboration (Li & Wang, 2023; Wu et al., 2020).

The below table number 1, gives a good tabulated summarized representation of the key comparative aspects of Cloud AIoT versus Edge AIoT.

Edge AI is a burgeoning field holding significant promise and its intelligent solutions play a crucial role in facilitating real-time decision making, bolstering security measures, and ensuring a high level of reliability and bring certain advantages as enlisted below:

Table 1. Comparative Aspects of Cloud and Edge AIoT

No	Differentiation Parameter	Explanation for Cloud AIoT	Explanation for Edge AIoT	Example in context of Education Ecosystem
1	Energy consumption	High energy requirements	Low energy requirements	When using augmented or virtual reality for learning high energy consumption is expected as end users engage in live manner.
2	Connectivity	Needs to be continuous	Can be intermittent	Online proctoring would be needing continuous feed of live exam hall conditions hence continuous connectivity, whereas mobile based learning of specific course chapters can be downloaded offline, and connectivity can be intermittent.
3	Processing ability	Practically unlimited processing power	Edge AI generally requires more work to upgrade than cloud-based AI. Without frequent upgrades, Edge-enabled devices' processing power can suffer.	Mobile based course work apps are example of Edge AIoT in education ecosystem and they may become vulnerable to cyber-attacks or malfunction if they aren't getting regular security updates, as applicable.
4	Latency considerations	When leveraging the cloud for AI, data must be transported vast distances to centralized servers. These servers can be thousands of miles away, delaying service.	Edge AI processes on-device or close to it, reducing communication delays.	Due to connectivity constraints, a typical cloud AIoT might get over burdened with the usually large volume of data generated today – example when all students start to upload their course work submissions simultaneously near to cut-off date-time or want download their results around the same time, when they get declared.
5	Security protection	Cloud AI provides high security, but it is more vulnerable to being hacked when data is in transit.	Edge AI is suitable for locally processing sensitive data like student or entity identity records	Keeping students and ecosystem players identity data on servers owned, governed by the educational institute is ideal for sensitive data.

- Cloud AI offers a comprehensive analysis procedure that considers the entire system, while Edge AI offers rapid response time and autonomy.
- Edge AI's cost and latency appeal to businesses. As such, enterprises may prefer Edge AI over Cloud AI due of delay. Edge AI is better for remote AI workloads than Cloud AI. Edge AI excels offline.
- However, most AI workloads are cloud-based. They seek operational Cloud AI benefits. Companies desire AI, robotics, ML, IoT, VR, and AR. IoT investments in legacy assets and sensors will increase bandwidth needs for

organizations. Edge AI processes on Edge-enabled IoT devices for real-time insights.

- Edge AI minimizes Wi-Fi traffic and doesn't choke the network, making it ideal for business IoT devices. As enterprises manage AI performance, security, latency, bandwidth, and cost, edge AI is growing.
- Decision-making is real-time with Edge AI. In contrast, Cloud AI improves performance and AI learning through model learning.

However, similar to the human body, it is important to note that these two modes of response aren't mutually exclusive, but rather have the potential to complement each other. As such, specifically choosing only Edge or Cloud AI is pointless, as a studied implementation of both Edge and Cloud AI can let companies maximize both technologies.

AIoT has the potential to transform education by providing personalized learning experiences, real-time feedback, immersive and interactive learning experiences, and streamlined administrative processes.

- **Personalized learning pathways**: AIoT can collect data on students' learning styles and preferences and use this data to create personalized learning pathways that are tailored to each student's individual needs (Dignum, 2021).
- **Real-time feedback**: AIoT can provide students with real-time feedback on their work, AR technology lets AIoT give real-time feedback. A study indicated that AIoT learning with AR technology improved students' problem-solving, comprehension, application planning, and design capabilities. AIoT and AR allow students to get real-time feedback and make modifications (Y.-S. Lin et al., 2021).
- **Immersive and interactive learning experiences**: AIoT can be used to create immersive and interactive learning experiences that engage students and help them to learn more effectively (Tsai et al., 2022).
- **Streamlined administrative processes**: AIoT can be used to automate administrative tasks, such as grading papers and scheduling classes and support staff recruitment. This can free up teachers' time so that they can focus on teaching. Intelligent Tutoring Systems (ITS), are computer-based systems that give students tailored teaching and feedback, can automate paper grading with AIoT. These systems assess student responses and grade automatically using predetermined criteria. This saves teachers time and gives pupils rapid feedback, improving learning (Xiang et al., 2022).
- **Online proctoring**: AIoT helps in bringing in more reliable outcomes from test evaluations and zeroing down the possibilities of academic dishonesty and fraudulent practices (Fidas et al., 2023; Yaghi et al., 2022).

Overall, AIoT has the potential to transform classrooms into intelligent learning environments, tailor learning to individual students' needs and preferences, enhance teacher-student interactions, and assist in administrative decision-making. The aforementioned examples illustrate a limited selection of the potential effects that AIoT can have on the field of education. The ongoing development of AIoT is anticipated to exert a more substantial influence on the education sector.

2.2 Opportunities and Benefits

This portion of the chapter will explore the various prospects and benefits that the integration of AIoT offers within the educational domain. This study aims to examine the potential impact of integrating AI and IoT technologies on teaching and learning experiences. The focus will be on how this integration might contribute to enhanced learning outcomes, more efficient administrative operations, and an overall improvement in the effectiveness of educational systems. The implementation of learner-centric techniques facilitated by AIoT contributes to the advancement of inclusion, engagement, and accessibility within the realm of education. This study aims to thoroughly analyse the possible advantages that AIoT might offer to various stakeholders in the education sector, including teachers, students, administrators, and policymakers. Specifically, it focuses on how AIoT can facilitate learner-centric methodologies that foster inclusion, engagement, and accessibility in educational settings.

One of the methods by which AIoT might enhance accessibility for students with disabilities is through the utilization of intelligent devices. Smart technologies, enabled by the IoT, have the capability to facilitate uninterrupted contact with those who experience speech problems, visual impairments, hearing impairments, and similar conditions. This can facilitate the inclusion of students with impairments in class discussions and the fulfilment of their assignments. Furthermore, AI can be leveraged to enhance accessibility in the realm of testing, in addition to its applications in smart gadgets. AI has the potential to facilitate the development of adaptive examinations that are customized to cater to the unique requirements of individual students. This measure can contribute to the promotion of equal opportunities for academic success among students, irrespective of their disability. The utilization of assistive technology can serve as a method to augment accessibility within the domain of AIoT. According to the research conducted by Freitas and colleagues (de Freitas et al., 2022), the incorporation of AIoT holds promise in enabling the development of innovative solutions that address the specific needs of students with disabilities. To exemplify, the incorporation of AIoT has the potential to enhance the advancement of wearable gadgets that provide prompt feedback and assistance to students with mobility impairments. These technologies have the capacity to aid students in navigating their environment, carrying out tasks, and augmenting their communication skills..

AI has the potential to develop tools that enhance the accessibility of learning settings. One use of AI is the utilization of AI technology to create virtual reality (VR) simulations. These simulations have the potential to provide students with disabilities the opportunity to engage in diverse locations and scenarios. This intervention has the potential to enhance students' learning efficacy and foster the acquisition of essential skills necessary for success in practical contexts (Y.-S. Lin et al., 2021).

The AI for Accessibility initiative, developed by Microsoft, serves as an illustrative instance of the utilization of artificial intelligence (AI) in the creation of solutions aimed at addressing the many physical and cognitive obstacles encountered by those with disabilities in both professional and personal contexts. The overarching objective of this program is to foster social inclusion within society. The program is now engaged in the development of artificial intelligence (AI)-based solutions that aim to enhance communication, learning, and societal engagement for individuals with impairments (*MustardTek: An Educational Platform for Co-Designing Assistive Technologies*, n.d.).

Overall, AIoT has the potential to make a significant impact on the accessibility of education. By providing students with disabilities with the tools and support they need, AIoT can help to ensure that all students have an equal opportunity to succeed.

3 AIoT IN EDUCATION: WHY SHOULD IT MATTER?

In the book, "The Fourth Education Revolution Reconsidered: Will Artificial Intelligence Enrich Or Diminish Humanity?" (Seldon et al., 2020), the authors make a fervent appeal that the imminent revolution of AI and its associated technologies, bring out few issues of paramount concern. Their literary work makes a fervent plea to educators worldwide, spanning across various education-based communities, urging them to be cognizant of the imminent developments that lie ahead as adoption of such an approach ensures a higher probability that our future will be influenced by our collective efforts, serving the best interests of the entire population, failing which the decision-making power will be left in the hands of other entities such as multinational technology corporations and governmental bodies, who will prioritize their own interests. In such a scenario, the literary work suggests that we will bear sole responsibility for the consequences that may arise.

3.1 Historical Concerns

Continuing our reference to the earlier quoted book, "The Fourth Education Revolution Reconsidered: Will Artificial Intelligence Enrich Or Diminish Humanity?" (Seldon et al., 2020), the authors argue that numerous countries possess highly commendable primary and tertiary educational systems which have been very effectively catering to the demands of the previous era. The pedagogical approach of the third revolution was unsuccessful in effectively addressing issues of inequity and the lack of specialized training. The advent of artificial intelligence (AI) has brought about significant transformations in various aspects of our everyday lives, professional pursuits, and interactions with the environment. Therefore, it is indisputable that artificial intelligence (AI) will have a significant influence on our educational establishments. Nonetheless, there exists a limited timeframe within which we can exert our influence on this transformation. The revolution in question should not be regarded as a universal remedy, as its potential to gradually erode our creativity, personal beliefs, and capacity for empathy necessitates prompt intervention.

Artificial intelligence, particularly machine learning (ML), has exceptional proficiency in the identification and analysis of patterns within extensive datasets. According to forecasts from multiple leading consultancies like PWC and McKinsey, it is projected that AI will contribute an estimated $13 trillion to $16 trillion to the global economy by the year 2030 (Chui & Yee, 2023; Rao Anand & Verweij, 2017). Numerous intellectuals, including the renowned physicist Stephen Hawking, have posited that artificial intelligence, namely in its fourth iteration (AI/4.0), will constitute the most significant advancement in the annals of human civilization. The esteemed physicist had rather expressed serious concerns on the potential ramifications of developing an AI system that possesses the capability to equal or surpass human capabilities, yet acknowledging the significant utility demonstrated by early iterations of AI. According to Stephen Hawkin's statement in 2014, the renowned scientist expressed caution that AI could one day get sophisticated to an extent wherein it would undertake autonomous decisions on its own and even start reconfiguring or re-designing itself at an alarming rate' (Cellan-Jones, 2014).

So, like it happens with every new promising technology, the initial rapid progress starts to appear as elixir of life, hence it is important to use caution from the beginning as deeper scrutiny about AI in its current form, reveals some gaps and red flags. The contributing potential of AI in addressing the Covid-19 pandemic, which is often regarded as the most significant health catastrophe of the current century, did not materialize to the extent anticipated by its proponents. The availability of data required for AI applications is often limited, as evidenced by the profound impact of the Covid-19 pandemic, which has revealed the absence of a comprehensive and dependable database documenting individuals' mobility patterns.

The algorithms that underpin AI lack the ability to duplicate human intelligence and are contingent upon the capabilities of their human creators. The enhanced quality of data, the increased sophistication of algorithms, and the heightened computational capabilities of computers all contribute to a heightened vulnerability to potential misuse by authoritarian regimes and manipulation by profit minded market driven technological firms.

Hence one cannot be definitive about AI enhancing the human productivity and proving to be of resounding assistance as it couldn't stand up to the expected level during Covid-19 pandemic when it mattered the most.

3.2 Challenges Ahead

AIoT has the capacity to significantly transform the field of education. Nevertheless, the use of AIoT in educational settings necessitates the resolution of several issues and concerns. One of the foremost challenges is to the domain of data privacy and security. The utilization of AIoT systems in gathering and analysing extensive quantities of student data raises concerns regarding the potential for unlawful use of said data. Ensuring the incorporation of privacy and security considerations is crucial in the design of AIoT systems.

Another significant issue that needs to be addressed is the aspect of accessibility. It is imperative that AIoT systems are designed and implemented in a manner that ensures equal accessibility for all students, irrespective of their disability. This implies that the design of AIoT systems should prioritize inclusivity for all students, encompassing those with neurological issues in addition to physical disabilities, without excluding other groups.

Lastly, the matter of technical inequality must be addressed. The availability of technology varies across schools, so posing a potential risk of exacerbating the disparity between institutions that possess the necessary resources to adopt AIoT and those that lack such capabilities. Ensuring the implementation of AIoT in a manner that does not worsen pre-existing disparities is of paramount significance.

3.3 Promising Area

Online proctoring is considered to be a very promising domain for the integration of AIoT in the field of education. Online proctoring refers to the utilization of technological tools to supervise and oversee students during online examinations, with the primary objective of preventing academic dishonesty. The utilization of AIoT has the potential to enhance the precision and effectiveness of online proctoring. For example, AIoT can be used to identify students who are using unauthorized materials or who are behaving suspiciously.

Consider a hypothetical situation in which students from various geographical locations participate in tests without the physical presence of human invigilators, but simultaneously experiencing personalized instructional interactions. An AI proctored test refers to an examination in which a system employs a web camera attached to it to remotely monitor candidates. The browser activities system is capable of overseeing the entirety of exam invigilation in a manner similar to offline exam invigilators through the use of advanced technologies such as facial recognition, tracking candidate movements, and monitoring screen activities. The use of AI-enabled online educational assessment technologies has the potential to significantly improve the structure and integrity of online examinations, hence reducing the likelihood of academic dishonesty. The integration of AI-supported measures, which encompass both hardware and software elements, aims to discourage students from participating in acts of academic dishonesty during examination periods (Nigam et al., 2021).

AIoT based proctoring platforms leverage AI to oversee online examinations. They utilize video analysis techniques to identify any occurrences of cheating, thereby ensuring the preservation of exam integrity. However, in addition to the act of proctoring, the incorporation of AIoT introduces distinct supplementary advantages by supporting the enhancement of online education quality. The AIoT ecosystem has the capacity to offer personalized feedback that is precisely tailored to the performance of each individual student, effectively pinpointing specific areas that want attention. In addition, it enables the advancement of immersive educational experiences, such as simulations using virtual reality, that foster experiential learning. The capabilities of AIoT play a fundamental role in supporting the Educational Intelligent Economy (EIE), with a particular focus on utilizing data-driven personalization to improve and optimize learning outcomes. According to Jules and Salajan (Salajan & Jules, 2020), the ongoing progress of this technology possesses limitless possibilities to transform the domain of education.

4 AIoT IN EDUCATION: WHERE DOES THIS LEAD TO?

4.1 Strategic Decision-Making and Implementation

The implementation of AIoT in education is a complex process that requires careful planning and strategic decision-making. Educational institutions need to consider a number of factors, such as the technological infrastructure, the professional development needs of teachers, and the engagement of stakeholders.

Outlined below are certain ideas that contribute to the successful application of AIoT in the field of education (Cavanagh et al., 2020):

- To commence, it is imperative to establish a well-defined and unambiguous vision. What are the precise objectives of implementing AIoT in your institution?
- Evaluate your existing talents. Could you please provide information about your technological infrastructure? What are the requisite training requirements for your educators?
- Foster active involvement of relevant stakeholders. It is imperative to ensure that all relevant stakeholders agree and actively involved in the implementation of the AIoT initiative.
- Commence with modest actions. It is advisable to refrain from attempting to adopt AIoT in its entirety in a single endeavour. Commence with a preliminary undertaking, commonly referred to as a pilot project, and subsequently expand its scope as additional knowledge and insights are acquired.

4.2 Navigating Challenges

Several obstacles must be overcome in order to fully harness the potential of AIoT in the field of education. Some of the issues that arise include:

- The consideration of privacy and security is crucial in the design of AIoT systems.
- Ensuring Accessibility: It is imperative that AIoT systems are designed and implemented in a manner that guarantees equal access for all students, irrespective of their disability.
- Technological disparity: It is imperative to guarantee that the implementation of AIoT is carried out in a manner that does not further amplify pre-existing inequities.
- The widespread implementation of AIoT in educational environments presents various ethical quandaries. It is of utmost significance to possess an understanding of these ethical predicaments and to formulate ethical principles for the deployment of AIoT.

In order to tackle the aforementioned difficulties pertaining to data privacy and security, accessibility, and technological disparity in the field of AIoT, a range of tactics and solutions can be employed as follows:

- First and foremost, it is imperative to acknowledge the significance of resolving privacy problems inside machine learning systems. It is imperative to establish a connection between the machine learning and privacy communities in order to effectively confront privacy risks in the context of AIoT (Al-Rubaie &

Chang, 2019). Privacy-preserving machine learning methodologies, such as the utilization of differential privacy, can be implemented to safeguard confidential information while simultaneously enabling efficient data analysis and informed decision-making (Dwork & Roth, 2013).

- Secondly, the establishment of robust security measures for the IoT is crucial in ensuring the integrity and reliability of AIoT systems. The implementation of standardized protocols and practices can have a substantial impact on enhancing the security of IoT devices and networks (Keoh et al., 2014). The integration of robust authentication methods, such as blockchain-based solutions, has the potential to significantly bolster the levels of data security and privacy inside AIoT frameworks (Wazid et al., 2021).

- Thirdly, it is essential to prioritize the resolution of technological discrepancies in order to guarantee fair and equal access to AIoT technologies. The utilization of edge computing and the optimization of distributed message queuing systems have been identified as potential strategies for addressing obstacles associated with system circumstances, fluctuations in the number of devices, and latency issues (Sui & Liu, 2023; Xie et al., 2023). Furthermore, the implementation of artificial intelligence accelerators at the edge has the potential to enhance the Quality of Experience (QoE) and effectively tackle the obstacles encountered in the context of AIoT applications (Sun et al., 2023).

In order to address the issues associated with AIoT, it is imperative that research and development endeavours prioritize the enhancement of intelligent capabilities in devices, the establishment of standardized protocols, the implementation of privacy-conscious resource management strategies, and the adoption of secure resource sharing schemes. Furthermore, the implementation of agile methodologies in the context of AIoT can effectively tackle various issues and enhance the efficiency and effectiveness of developing and deploying AIoT systems (Slama, 2023).

4.3 Roadmap for the Future

The trajectory for the future of Artificial Intelligence of Things (AIoT) in the field of education is influenced by several factors and opportunities that have been highlighted in extent research literature.

The incorporation of AI and IoT technology in the field of education has the potential to augment the learning experience through the provision of personalized and adaptive learning environments. The utilization of AI algorithms for the analysis of student data and the provision of customized recommendations and feedback has been identified as a viable approach (Zhang & Tao, 2021). Furthermore, the integration of AIoT has the potential to facilitate the establishment of intelligent classrooms that are equipped with advanced equipment capable of supporting and enhancing teaching and learning activities (Y. Lin et al., 2021).

In addition, the integration of AIoT holds the capacity to significantly transform the field of education through the facilitation of intelligent campus environments. The process entails the incorporation of diverse IoT devices and sensors for the purpose of gathering data on campus activities and infrastructure. This data can then be subjected to analysis through AI algorithms, with the aim of optimizing resource allocation, enhancing security measures, and improving general administration of the campus (Rong et al., 2021).

Within the realm of assistive technology, the integration of AIoT possesses significant potential in providing essential assistance to students who have disabilities. Assistive technologies can be built to offer individualized help and accommodations for students with diverse needs, such as vision or hearing impairments, through the utilization of AI algorithms and IoT devices (Freitas et al., 2022).

Nevertheless, the adoption of AIoT in education is not without its obstacles and barriers that must be effectively handled. The aforementioned concerns encompass matters pertaining to the safeguarding of data privacy and security, the seamless integration of diverse IoT devices and platforms, and the imperative for possessing specialized expertise and understanding in the domains of AI and IoT (Ishengoma et al., 2022). Addressing these problems necessitates the collaborative efforts of educators, researchers, and industry stakeholders to establish comprehensive frameworks and standards for the integration of Artificial Intelligence of Things (AIoT) in the field of education (Lei et al., 2020).

The potential for AIoT in the field of education is promising. Through meticulous planning and strategic execution, the integration of AIoT possesses the capacity to fundamentally transform the field of education, hence enhancing the academic achievements of students across the board. In addition to the aforementioned issues and concerns, there exist some other factors that necessitate consideration during the implementation of AIoT in the field of education. Such factors are:

- The cost of AIoT systems.
- The availability of training and support for teachers.
- The need for clear policies and procedures for data privacy and security.
- The need to address ethical dilemmas raised by use of AIoT in educational settings.

Notwithstanding these limitations, the potential advantages of AIoT in the field of education are considerable. The integration of AIoT holds promising prospects for enhancing the educational achievements of pupils, irrespective of their diverse backgrounds or individual capabilities. Additionally, it has the potential to enhance the accessibility and inclusivity of education while mitigating any inherent biases. In summary, the strategic plan for the future of AIoT in the field of education encompasses the utilization of AI and IoT advancements to establish tailored and adaptable learning environments, construct intelligent educational campuses and supportive technologies, and tackle the obstacles associated with data privacy and security. Through the active adoption of these prospects and the successful surmounting of associated obstacles, the integration of AIoT holds the capacity to revolutionize the field of education and amplify the overall educational encounter for pupils.

5. CONCLUSION

The potential for AIoT in the field of education is promising. Through meticulous planning and strategic execution, the integration of AIoT possesses the capacity to fundamentally transform the field of education, hence enhancing the academic achievements of students across the board. In addition to the aforementioned issues and concerns, there exist some other factors that necessitate consideration during the implementation of AIoT in the field of education.

The convergence of technology and pedagogy has led to the emergence of an education system based on AIoT, which holds great potential in reshaping the landscape of learning. The integration of AI and the IoT holds the potential to not only optimize operational processes but also enhance and customize the learning experience, hence addressing current educational disparities. Nevertheless, the presence of immense potential necessitates a corresponding level of substantial responsibility. In light of the imminent arrival of this paradigm-shifting period, it is crucial to adopt a cautious approach towards the integration of AIoT. This entails acknowledging and embracing its potential, but also maintaining a vigilant stance towards the obstacles it presents. The prioritization of data privacy, equity, and the preservation of human touch in pedagogy will be of utmost importance.

The primary objective of AIoT in the field of education should be to enhance human capabilities, cultivating an educational setting that not only transmits information but also cultivates creativity, critical reasoning, and empathy. As we navigate this uncharted domain, it is important to bear in mind that technology, fundamentally, serves as a tool. Its genuine worth is in the manner in which we utilize it to enhance the educational experience for all those engaged in the process of learning.

REFERENCES

Al-Rubaie, M., & Chang, J. M. (2019). Privacy-Preserving Machine Learning: Threats and Solutions. *IEEE Security and Privacy*, *17*(2), 49–58. doi:10.1109/MSEC.2018.2888775

Cavanagh, T., Chen, B., Lahcen, R. A. M., & Paradiso, J. (2020). Constructing a Design Framework and Pedagogical Approach for Adaptive Learning in Higher Education: A Practitioner's Perspective. *International Review of Research in Open and Distance Learning*, *21*(1), 172–196. doi:10.19173/irrodl.v21i1.4557

Cellan-Jones, R. (2014). Stephen Hawking warns artificial intelligence could end mankind. *BBC News*. https://www.bbc.com/news/technology-30290540

Chui, M., & Yee, L. (2023). AI could increase corporate profits by $4.4 trillion a year, according to new research | McKinsey. In *McKinsey Report (originally appeared in Fast Company)*. https://www.mckinsey.com/mgi/overview/in-the-news/ai-could-increase-corporate-profits-by-4-trillion-a-year-according-to-new-research

de Freitas, M. P., Piai, V., Farias, R. H., Fernandes, A. M. da R., Rossetto, A. G. de M., & Leithardt, V. R. Q. (2022). Artificial Intelligence of Things Applied to Assistive Technology: A Systematic Literature Review. *Sensors (Basel)*, *22*(21), 8531. Advance online publication. doi:10.3390/s22218531 PMID:36366227

de Freitas, M. P., Piai, V. A., Farias, R. H., Fernandes, A. M. R., de Moraes Rossetto, A. G., & Leithardt, V. R. Q. (2022). Artificial Intelligence of Things Applied to Assistive Technology: A Systematic Literature Review. *Sensors (Basel)*, *22*(21), 8531. doi:10.3390/s22218531 PMID:36366227

Dignum, V. (2021). The role and challenges of education for responsible AI. *London Review of Education*, *19*(1). Advance online publication. doi:10.14324/LRE.19.1.01

Dwivedi, Y. K., Hughes, L., Ismagilova, E., Aarts, G., Coombs, C., Crick, T., Duan, Y., Dwivedi, R., Edwards, J., Eirug, A., Galanos, V., Ilavarasan, P. V., Janssen, M., Jones, P., Kar, A. K., Kizgin, H., Kronemann, B., Lal, B., Lucini, B., ... Williams, M. D. (2021). Artificial Intelligence (AI): Multidisciplinary perspectives on emerging challenges, opportunities, and agenda for research, practice and policy. *International Journal of Information Management*, *57*, 101994. doi:10.1016/j.ijinfomgt.2019.08.002

Dwork, C., & Roth, A. (2013). The Algorithmic Foundations of Differential Privacy. *Foundations and Trends® in Theoretical Computer Science*, *9*(3–4), 211–407. doi:10.1561/0400000042

Fidas, C. A., Belk, M., Constantinides, A., Portugal, D., Martins, P., Pietron, A. M., Pitsillides, A., & Avouris, N. (2023). Ensuring Academic Integrity and Trust in Online Learning Environments: A Longitudinal Study of an AI-Centered Proctoring System in Tertiary Educational Institutions. *Education Sciences*, *13*(6), 566. Advance online publication. doi:10.3390/educsci13060566

Holmes, W., Bialik, M., & Fadel, C. (2023). Artificial intelligence in education. In *Data ethics : building trust : how digital technologies can serve humanity* (pp. 621–653). Globethics Publications. doi:10.58863/20.500.12424/4276068

Ishengoma, F., Shao, D., Alexopoulos, C., Saxena, S., & Nikiforova, A. (2022). *Integration of Artificial Intelligence of Things (AIoT) in the Public Sector: Drivers, Barriers and Future Research Agenda*. Digital Policy Regulation and Governance. doi:10.1108/DPRG-06-2022-0067

Javaid, M., Haleem, A., Singh, R. P., & Suman, R. (2022). Artificial Intelligence Applications for Industry 4.0: A Literature-Based Study. *Journal of Industrial Integration and Management*, *07*(01), 83–111. doi:10.1142/S2424862221300040

Keoh, S. L., Kumar, S. S., & Tschofenig, H. (2014). Securing the Internet of Things: A Standardization Perspective. *IEEE Internet of Things Journal*, *1*(3), 265–275. doi:10.1109/JIOT.2014.2323395

Kornilaev, L. (2022). Kant's doctrine of education and the problem of artificial intelligence. *Journal of Philosophy of Education*, *55*(6), 1072–1080. doi:10.1111/1467-9752.12608

Lei, L., Tan, Y., Zheng, K., Liu, S., Zhang, K., & Shen, X. (2020). Deep Reinforcement Learning for Autonomous Internet of Things: Model, Applications and Challenges. *IEEE Communications Surveys and Tutorials*, *22*(3), 1722–1760. Advance online publication. doi:10.1109/COMST.2020.2988367

Li, F., & Wang, C. (2023). Artificial intelligence and edge computing for teaching quality evaluation based on 5G-enabled wireless communication technology. *Journal of Cloud Computing (Heidelberg, Germany)*, *12*(1), 1–17. doi:10.1186/s13677-023-00418-6

Lin, Y., Chen, S.-Y., Tsai, C.-W., & Lai, Y.-H. (2021). Exploring Computational Thinking Skills Training Through Augmented Reality and AIoT Learning. *Frontiers in Psychology*, *12*, 640115. Advance online publication. doi:10.3389/fpsyg.2021.640115 PMID:33708166

MustardTek: An Educational Platform for Co-Designing Assistive Technologies. (n.d.). Retrieved October 30, 2023, from https://sway.office.com/lwPjb0dnfmYLkCUd?ref=Link

Nayal, K., Kumar, S., Raut, R. D., Queiroz, M. M., Priyadarshinee, P., & Narkhede, B. E. (2022). Supply chain firm performance in circular economy and digital era to achieve sustainable development goals. *Business Strategy and the Environment*, *31*(3), 1058–1073. doi:10.1002/bse.2935

Nigam, A., Pasricha, R., Singh, T., & Churi, P. (2021). A Systematic Review on AI-based Proctoring Systems: Past, Present and Future. *Education and Information Technologies*, *26*(5), 6421–6445. doi:10.1007/s10639-021-10597-x PMID:34177348

Rao, & Verweij. (2017). PwC's Global Artificial Intelligence Study Report. *PwC Publications*. https://www.pwc.com/gx/en/issues/data-and-analytics/publications/artificial-intelligence-study.html

Rong, G., Xu, Y., Tong, X., & Fan, H. (2021). An Edge-Cloud Collaborative Computing Platform for Building AIoT Applications Efficiently. *Journal of Cloud Computing (Heidelberg, Germany)*, *10*(1), 1–14. doi:10.1186/s13677-021-00250-w

Salah, K., Rehman, M. H. U., Nizamuddin, N., & Al-Fuqaha, A. (2019). Blockchain for AI: Review and Open Research Challenges. *IEEE Access : Practical Innovations, Open Solutions*, *7*, 10127–10149. doi:10.1109/ACCESS.2018.2890507

Salajan, F. D., & Jules, T. D. (2020). Exploring Comparative and International Education as a Meta-Assemblage: The (Re)Configuration of an Interdisciplinary Field in the Age of Big Data. doi:10.1108/S1479-367920200000039014

Seldon, A., Abidoye, O., & Metcalf, T. (2020). *The Fourth Education Revolution Reconsidered: Will Artificial Intelligence Enrich Or Diminish Humanity?* (A. Seldon, O. Abidoye, & T. Metcalf, Eds.; 2nd ed.). University of Buckingham Press.

Slama, D. (2023). Agile AIoT. In *The Digital Playbook* (pp. 293–319). Springer International Publishing. doi:10.1007/978-3-030-88221-1_23

Sleem, A., & Elhenawy, I. (2023). Survey of Artificial Intelligence of Things for Smart Buildings: A closer outlook. *Journal of Intelligent Systems and Internet of Things*, *8*(2), 63–71. doi:10.54216/JISIoT.080206

Sui, Q., & Liu, X. (2023). Edge computing and AIoT based network intrusion detection mechanism. *Internet Technology Letters*, *6*(5), e324. Advance online publication. doi:10.1002/itl2.324

SunK.WangX.ZhaoQ.(2023).A Review of AIoT-based Edge Devices and Lightweight Deployment. *Preprint - Techrxiv.Org.*

Tsai, C.-C., Chung, C.-C., Cheng, Y.-M., & Lou, S.-J. (2022). Deep learning course development and evaluation of artificial intelligence in vocational senior high schools. *Frontiers in Psychology*, *13*, 965926. Advance online publication. doi:10.3389/fpsyg.2022.965926 PMID:36211841

Wazid, M., Das, A. K., & Park, Y. (2021). Blockchain-Envisioned Secure Authentication Approach in AIoT: Applications, Challenges, and Future Research. *Wireless Communications and Mobile Computing*, *2021*, 1–19. doi:10.1155/2021/3866006

Wu, Q., He, K., & Chen, X. (2020). Personalized Federated Learning for Intelligent IoT Applications: A Cloud-Edge Based Framework. *IEEE Open Journal of the Computer Society*, *1*, 35–44. doi:10.1109/OJCS.2020.2993259 PMID:32396074

Xiang, Z., Zheng, Y., He, M., Shi, L., Wang, D., Deng, S., & Zheng, Z. (2022). Energy-effective artificial internet-of-things application deployment in edge-cloud systems. *Peer-to-Peer Networking and Applications*, *15*(2), 1029–1044. doi:10.1007/s12083-021-01273-5

Xie, Z., Ji, C., Xu, L., Xia, M., & Cao, H. (2023). Towards an Optimized Distributed Message Queue System for AIoT Edge Computing: A Reinforcement Learning Approach. *Sensors (Basel)*, *23*(12), 5447. doi:10.3390/s23125447 PMID:37420614

Yaghi, M., Basmaji, T., Alamri, D., Hussein, N., Hammoudi, M., & Ghazal, M. (2022). Student Authentication and Proctoring System Using AI and the IoT. *2022 2nd International Conference on Computing and Machine Intelligence, ICMI 2022 - Proceedings.* 10.1109/ICMI55296.2022.9873661

Zhang, J., & Tao, D. (2021). Empowering Things With Intelligence: A Survey of the Progress, Challenges, and Opportunities in Artificial Intelligence of Things. *IEEE Internet of Things Journal*, *8*(10), 7789–7817. Advance online publication. doi:10.1109/JIOT.2020.3039359

ADDITIONAL READING

Abulibdeh, A., Zaidan, E., & Abulibdeh, R. (2024). Navigating the confluence of artificial intelligence and education for sustainable development in the era of industry 4.0: Challenges, opportunities, and ethical dimensions. *Journal of Cleaner Production*, *140527*, 140527. Advance online publication. doi:10.1016/j.jclepro.2023.140527

Rahiman, H. U., & Kodikal, R. (2024). Revolutionizing education: Artificial intelligence empowered learning in higher education. *Cogent Education*, *11*(1), 2293431. doi:10.1080/2331186X.2023.2293431

Wu, R., & Yu, Z. (2024). Do AI chatbots improve students learning outcomes? Evidence from a meta-analysis. *British Journal of Educational Technology*, *55*(1), 10–33. doi:10.1111/bjet.13334

Chapter 7
Machine Learning and Decision-Making

Félix Oscar Socorro Márquez
Complutense University of Madrid, Spain

ABSTRACT

At the end of the 1950s, the possibility of machine learning in computers began to be discussed. After more than half a century, thanks to technological advances, machine learning is more than a reality; it is a tool that supports different fields, such as education, business, science, and management. This chapter explores how machine learning takes place in decision-making, especially in the managerial field, and the advantages that it offers when studying scenarios, choosing strategies, exploring the possible consequences of our actions, and/or predicting the likely responses of the parties involved in a specific situation.

INTRODUCTION

The industrial revolution created new sources of employment, new markets, and new companies and, at the same time, inspired the imagination and creativity of many.

An example of this can be found in the work «The rebellion of the machines» by Romain Rolland, written in 1921, or in the 1927 film «Metropolis» directed by Fritz Lang, where anticipating by decades the concept of Artificial Intelligence (AI).

But those ideas did not remain trapped in science fiction, it was mathematically possible that a machine could process information and make decisions, therefore, in the following decades, formal science was dedicated to proving it.

For example, it is known that between 1936 and 1938 the first programmable computer was built, called Z1, created by the German Konrad Zuse (Russo, 2021).

DOI: 10.4018/979-8-3693-0993-3.ch007

Additionally, the Bombe computer was created in 1939 by Alan Turing in order to decrypt the messages of the German army and thus anticipate his strategy. It was an improved version of a device designed in 1938 by the Polish cryptologist Marian Rejewski. Additionally, Bombe was the forerunner of the digital electronic programmable computer (Ministerio Cultura Argentina, 2020).

In 1943, Warren McCulloch and Walter Pitts proposed, based on their studies of the nervous system, a formal neuron model implemented through electrical circuits, thus laying the foundations for what would later become AI, since the term had not yet been introduced at that time (Moya et al, 1998).

Also, in 1943, the USA Army approved the construction of the first general purpose computer based on electrical circuits, the ENIAC, designed by John W. Mauchly (1907-1980) and John Presper Eckert (1919-1995) (Pérez et al, 2017).

And, finally, in 1956 —13 years later— the term artificial intelligence was born, proposed by the scientist John McCarthy, who had the help of Marvin Minsky and Claude Shannon. This concept refers to the science and engineering dedicated to «making intelligent machines» (Redacción España, 2019). Since then, incredible progress has been made in the development and application of artificial intelligence.

It is known that intelligence and decision making are related, mainly because the first allows to be more successful using the second. Therefore, one way to measure the intelligence of a «machine» is how successful it can be in making decisions.

This chapter explores machine learning and its relationship with decision making, especially in the field of management, and why it is considered one of the tools that are changing the way of seeing, understanding and acting in the different administrative scenarios, where machine learning is used.

DECISION-MAKING

Making decisions is a constant and inherent activity in life. Everything we do —except for some biological functions— and how we do it has gone through a decision-making process. The aforementioned process can occur in fractions of seconds or be the result of long periods studying variables and probable scenarios, according to the possible decisions that are made.

Three elements are used to explain decision-making: process conception, choice of course of action, and the solution of problems or situations of organizational opportunity (Arévalo & Estrada 2017, p. 253).

However, in general, it can be said that the decision-making process is made up of a decision-maker —could be more than one—, the variables to be considered, the possible actions, and the study of their possible consequences.

Depending on the approach, the decision-maker can be rational, satisfying, organizational procedure approach, and political.

Simon (1980), defines the figure of the rational decision maker. In this case, the possible alternatives are identified and listed, the consequences derived from each one is analysed and these consequences are assessed and compared. As for the decision maker, he/she must describe his/her utility function, that is, his/her preference for different consequences (Canós et al, n.d., p.2).

Canós et al (n.d.) also state that the aforementioned behaviour is related to the «economic man» understood as the entity that seeks to maximize its performance and therefore tends to seek the best alternative.

Concerning the satisfier approach, this type of decision-maker is inclined to satisfy more than to optimize the solutions of the problem. This type of decision-maker —apparently— is usual in organizations where a satisfactory level of performance is accepted instead of a maximum one.

According to Cyert & March (1965) the organizational procedure approach focuses on the analysis of communication channels, the formalization of processes and the distinction between formal and informal structure (Canós et al, n.d., p.3).

And the final approach, the political one, describes people's decision making to satisfy their own interests. In this approach, preferences based on selfish personal goals seldom change as new information is acquired. In it, the definition of the problems, the search and collection of data, the exchange of information and the evaluation criteria are only methods used to predispose the result in favour of the decision maker (Chacín, 2010, p.2).

Nevertheless, other styles of decision-makers have also been identified, especially associated with the way of facing risks in decision making, such as the risk-prone decision-maker, the risk indifferent decision-maker, and the risk-averse decision-maker. See Table 1.

Table 1. Decision-Makers Characteristics

DECISION MAKER	CHARACTERISTICS
PRONE TO RISK	For this type of decision maker, the profits will vary proportionally more than the results vary.
INDIFFERENT TO RISK	For this type of decision, the profits will vary proportionally like what the results vary.
RISK ADVERSE	For this type of decision maker, profits will vary proportionally less than the results will vary.

Based on Bartolomeo (2015, p.1)

Such positions may respond to aspects associated with experience and information management, although subjective elements associated with hunches or simple intuition cannot be ruled out — both associated with human nature—, which may lead the decision-maker to be for or against the risk, or to be indifferent to it.

MACHINE LEARNING

Before talking about machine learning, it is important to explain what is meant by learning.

Learning refers to a broad spectrum of situations in which the learner increases his/her knowledge or abilities to accomplish a task. Learning applies inferences to certain information to build an appropriate representation of some relevant aspect of reality or of some process (Moreno et al, 1994, p.6).

There are seven types of learning process: habituation, association, conditioning, trial and error, latent learning, imitation, experience or imprint. These types of learning have been studied in animals and humans. As shown Table 2.

Table 2. Types of Learning Process

TYPE OF LEARNING	CHARACTERISTICS
Habituation	It consists of a response that decays before a set of repeated (or continuous) stimuli not associated with any type of reward or reinforcement. It can be characterized as associated with a specific stimulus and its relative permanence distinguishes it from temporary manifestations such as fatigue or sensory adaptation.
Association	This learning occurs when an event allows us to precede, with some confidence, the occurrence (or non-occurrence) of another event.
Conditioning	This learning occurs when the ability to respond to a given stimulus is acquired with the same reflex action with which one would respond to another conditioning stimulus (reinforcement or reward) when both stimuli are presented concurrently (or overlapped in a sequence) a certain number of times.
Trial-and-error	It requires the existence of reinforcement (or reward) to encourage the selection of the appropriate response from among a variety of behaviours in a given situation, until finally a relationship is established between the stimulus or situation and a response. correct answer to get a reward.
Latent learning	Latent learning is a type of associative learning that takes place in the absence of reward.
Imitation	It involves copying a new behaviour, action or expression that is impossible to learn if it is not copied from another individual. Imitation has frequently been considered as evidence of the existence of highly reflective behaviours.
Imprinting	It is an illustrative example of how a specific range of stimuli is capable of eliciting a response and can be limited and refined through experience.

Based on Moreno et al, 1994, pp.4-5

Now, the term «machine learning» refers to a wide spectrum of situations in which the learner increases their knowledge or skills to accomplish a task. Learning applies inferences to certain information to build an appropriate representation of some relevant aspect of reality or of some process. (Moreno et al, 1994, p.6).

Iberdrola (2022) pointed out that the term «machine learning» was first used in 1959. Nevertheless, the term has gained relevance in recent years due to the increase in computing power and the data boom known as Big Data. (p.1).

For its part, Big Data has been defined as a high-volume, high-velocity, and high-variety information asset, demanding cost-effective and innovative ways of processing information for enhanced insight and decision-making (Chen et al, 2016). Taking into account such demands, in terms of processing speed and data volume: it is easy to deduce that Big Data is closely linked to modern computers and, in a near future, to quantum computers. It is prudent to point out that the tests that are currently being carried out, related to quantum computing, promise data processing speeds significantly above what is offered today by the best of the computers that can be purchased on the market. According to the Clarín (2022), in June of 2022 it was announced that scientists had made progress to complete in just 36 microseconds a task that «classical» computers would take just over 9,000 years to complete.

Therefore, Machine Learning can also be understood as a discipline in the field of Artificial Intelligence (AI) that, through algorithms, gives computers the ability to identify patterns in massive data and make predictions (predictive analysis). This learning allows computers to perform specific tasks autonomously, that is, without the need to be programmed. (Iberdrola, 2022, p. 1).

In the previous definition, two concepts appear that must be explained: Artificial Intelligence (AI), and algorithms.

According to Saleh (2019) the Artificial Intelligence (AI) is a branch of computer science. It involves developing computer programs to complete tasks that would otherwise require human intelligence. AI algorithms can tackle learning, perception, problem-solving, language-understanding and/or logical reasoning. (p.3).

For Trigo (n.d.) an algorithm "is the sequence of steps that we must follow to solve a problem". (p.44).

MACHINE LEARNING AND DECISION-MAKING

It has been shown that, in the same way that a human being or a living being learns, it is possible to teach machines to learn and then, based on what they have learned —and the data collected— make decisions accordingly, or suggest actions that can lead to the achievement of a goal.

An example of the aforementioned aspect, is that recently, researchers from DeepMind, the Google company dedicated to the development of AI, have created a computational model to see if a deep learning system could acquire an understanding of certain basic physical principles from the processing of visual animations. The researchers state that it is easier than machines can do that if they are taught to learn like babies do, by studying their environment (Pascual, 2022).

Although machine learning is carried out in the same way that a child would face it, the time to achieve matches that allow predicting behaviours and making decisions is not the same as the time it would take for an infant to achieve it. This is an important advantage when exploring the ideal decisions in different scenarios through machine learning.

Decision making through machine learning is mainly based on predictive algorithms.

Predictive algorithms

A predictive algorithm is a mathematical representation of a factor that takes place in reality and that helps predict any future behaviour based on present knowledge (Onyx Soft, n.d.).

The combination of several predictive algorithms or the development of a complex algorithm results in a predictive model. The goal of a predictive model is to minimize the error between the real value and the predicted value, considering all the possible interference factors. (Joakin, 2021, p.46)

Machine learning is driven by «Deep Learning», a method that allows training an Artificial Intelligence to obtain a prediction given a set of inputs. (Joakin, 2021), this can be done in two main ways: supervised or unsupervised.

According to Rueda (2022) a supervised machine learning algorithms base their learning on a set of previously labelled training data. By labelling we mean that for each occurrence of the training data set we know the value of its target attribute. This will allow the algorithm to «learn» a function capable of predicting the target attribute for a new data set. (p.1)

In relation to unsupervised methods, Rueda (2022) explains that these are algorithms that base their training process on a set of data without previously defined labels or classes. That is, no target or class value is known, either categorical or numerical. Unsupervised learning is dedicated to grouping tasks, also called clustering or segmentation, where its goal is to find similar groups in the data set. (idem).

Data Processing and Decision Making

Using machine learning it is easier, more accurate and faster to analyse a set of data. This occurs because it is not necessary to program the computer to perform a task which it has already «learned» through deep learning. Therefore, finding patterns, coincidences, trends and/or contradictions will be much faster than if human talent were used for it.

By exploring the different scenarios that arise due to the data that is supplied to the system, using machine learning, computers can also create links between them, creating groups and subgroups with the same or similar trends, which facilitates the identification of patterns and establishing strategies to address each of them.

As the system lacks typical expressions of human beings, such as hunch or intuition, data processing occurs objectively, which offers an advantage when making decisions that favour the majority or, failing that, affect only a minority portion.

Levels of Decision Making Through Machine Learning

Like any other tool, the results obtained from it will depend on the complexity that it is allowed to achieve in a given process.

Regarding decision-making, depending on what the final decision-maker considers necessary, three levels can be found, according to Datision (2021): Fully automated decisions. Increased options available. Here, the solution offers those responsible one or several alternatives, and these are the ones who have the last word, and Decision support. It is the people who decide, but with the support of artificial intelligence solutions and the analysis they have done. (p.2)

Each of these levels offers a valid alternative when choosing the strategy that corresponds to a particular scenario or situation that must be addressed immediately or that, instead, requires greater analysis and in-depth study.

Another significant contribution of the aforementioned levels is that it allows management to focus on problems or events that require greater attention and care, allowing machine learning (if it is one of the objectives of the decision-maker) to be in charge of responding to those requirements., deviations or alterations whose data coincide with past experiences, previously studied and resolved and, therefore, following the same pattern, can be resolved without human intervention.

Machine Learning and Management Decisions

In the past, data management was not a problem, markets were relatively small, and the behaviour of customers, suppliers, and employees was easy to predict.

But in the contemporary world, due to globalization, the rise of information technologies, social networks, technological development and access to increasingly complex products and services, it is almost impossible to explore all possible scenarios without neglecting some detail, especially if it is left in the hands only of the human ability to capture, process and analyse the data that the markets offer.

The development of artificial intelligence (AI) and with it, the emergence of machine learning, has not only been useful for the pure sciences, it has also been useful for administration and management.

Given the complexity of the contemporary world, the need to give immediate answers to customers and suppliers, to predict market behaviour, and the possible actions of the competition, machine learning has positioned itself as a vital tool for companies that are committed to going a step forward and thereby guarantee their success.

Machine learning, as already mentioned, allows companies —according to the level assigned to it— to make decisions by itself and thereby maintain a level of responses to common problems, or offer possible alternatives to the scenarios that a trend can generate or choose the best response to a change that could have been predicted thanks to the processing of the data that has been captured.

Machine learning, as a management tool, also makes it possible to not worry about repetitive tasks, minimizing or eradicating human errors, when calculating taxes, controlling budgets, monitoring income or projecting sales. This allows managers to focus on aspects of greater interest and relevance that can significantly affect the business.

Among other thing, the use of machine learning in decision making helps to reduce bias, decreases subjectivity, avoids the linear analysis of complex data, allows human error to be detected, and broadens the scope of data study from different perspectives.

In terms of risk, regardless of the position of the decision maker (prone, indifferent, adverse), when machine learning is used, it is more likely to minimize losses and optimize profits, this is because predictive models can point to the «safest» course of action or the one with the least risk if chosen.

Thanks to machine learning, establishing the three most common scenarios when making a decision —such as the optimistic, the pessimistic and the trend— is faster, more accurate and more objective. This makes it easier to observe with greater scope the pros and cons that the company faces and act accordingly, establishing strategies that will respond to the registered and analysed behaviour of the market, customers, suppliers, the government, the institutions and/or the competition, according to the case.

MACHINE LEARNING, DECISION-MAKING, AND AIoT

An analysis between the relationship between Machine Learning and Decision Making could not be considered complete if another aspect associated with technological advances and the digital era is not addressed, such as Artificial Intelligence of Things, or AIoT.

AIoT is understood as the integration of new artificial intelligence (AI) with Internet of Things (IoT) communications to carry out precisely those operations with greater effectiveness and thereby improve communications between the user and the machine, achieving with This will increase the quality of information processing and management (Revanthy et al, 2020).

According to Corchado (2020), the term AIoT also refers to the new technological wave that combines, as already noted, Artificial Intelligence (AI) and the Internet of Things (IoT). Corchado (2020) warns that IoT devices produce significant amounts of data, so AI is required to handle these large volumes when processing, classifying and analysing them, calling this process «connected intelligence».

As would be expected, upon coming into contact with this «connected intelligence», aspects associated with the company, such as productivity and performance, would be significantly affected, thereby promoting the organizational transmission, moving from an open networked organization (ONO) —as Tapscott (1997) visualized it—, to an *artificially intelligent and constantly learning organization* (AICLO). And this makes sense.

In the transition from ONO to AICLO, the AIoT will have a leading role; it will be this tool that will drive the changes that the market and customers demand from companies, that the possible scenarios foresee and that the trends indicate. The traditional aspects of the 21st century company will be associated with technological changes, more frequently than what was experienced in the 20th century, therefore, organizations will experience constant changes in the way of seeing and doing things, produce and sell them.

If an organization makes appropriate use of artificial intelligence and incorporates it into its planning, marketing, sales, distribution, logistics and production processes —to name just a few—, to create strategies, campaigns, promotions, routes, optimal conditions for conservation and distribution of its products, in addition to the production itself of the goods or services it offers, respectively, it could be expected an increase in its efficiency and effectiveness rates, positively impacting the production process.

At present, it is not possible to imagine an organization, regardless of the market it serves, that is not tempted to include artificial intelligence in some of its operations, and that has been not only a consequence of the technological advances that have been in IoT and mobile devices, it has also been due to the COVID-19 pandemic and how confinement promoted the development of interconnection technologies and the consumption of that technology.

As reported by the website Cepymenews (2023), artificial intelligence experienced astonishing growth in adoption rates from 2017 to 2018, levelling off significantly between 2019 and 2021, before experiencing a 2.5-fold increase thereafter — approximately— in 2022, if compared with the figures for 2017. This information allows us to infer that to the extent that the usefulness of this tool materializes in organizations, impacting their means of production, trade and management, it will become even more popular and necessary, adding to this the contributions that it will undoubtedly offer in decision-making and the strengthening of machine learning tools, as already mentioned.

The Internet of Things (IoT) significantly expanded connectivity, making communication between humans and machines faster, fluid, intuitive and efficient, now, with the addition of artificial intelligence (AI) to the equation, it is not difficult to assume that results will contribute to making processes more effective, saving costs, reducing errors, maximizing productivity and guaranteeing the proper use of resources.

It is simple, among other things, AIoT allows organisations to improve the decision-making process because it analyses, processes, offers information in real-time, and makes use of different sources in a more efficient way, for this AIoT is supported by machine learning and processes of deep learning, which allow the creation of more optimal and reliable processing chains in each run, resulting in accurate and reliable outputs.

PRODUCTIVITY, ORGANIZATIONAL TRANSITION, AND MACHINE LEARNING

At this point there are two fundamental elements regarding the impact that Machine Learning, Decision Making and AIoT have on productivity and organizational transition.

To develop this point, it is necessary to explore the last two concepts aforementioned.

For Björkman (1992) productivity "is not a uniform universal concept; there exist many different definitions that are suitable under different conditions. This leads to (…) difficulties and problems when productivity is used as a practical tool or measurement". (p. 204). However, there are generally accepted considerations that allow us to understand the concept. For example, according to Yadav & Marwah (2015), productivity is understood as the "measure of the ability to produce a good or service" (p.192). They also explain that "productivity is the measure of how specified resources are managed to accomplish timely objectives as stated in terms

Figure 1. Organizational Transition
Source: *Alkaya & Hepaktan (2003), p. 51*

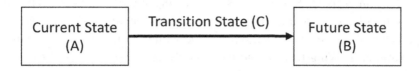

of quantity and quality" (idem). These two authors finally state that "productivity may also be defined as an index that measures output (goods and services) relative to the input (labor, materials, energy, etc., used to produce the output)" (idem). These three points of view could represent the general idea about the definition of productivity.

It can be said that productivity is the result of what we hope to produce and how it is produced, taking into account the quality and quantity in which it is produced.

On the other hand, organizational transition is understood as the changes that the organization undergoes when it moves from its current position to a different one, ideally better than the previous one (Aggarwal, 2020).

According to Alkaya & Hepaktan (2003), the approach proposed by Beckhard & Harris explains the transition through the implementation of a change, viewing it as a new organizational design, where the organization moves from its current position towards a desired future state, as can be seen in Figure 1.

Alkaya & Hepaktan (2003), explain that "the effective management of change involves developing an understanding of the current state, developing an image of the desired future state, and moving the organization through a transition period" (p.51).

Once this is understood, it can be stated that the relationship between productivity, organizational transition, and machine learning is bidirectional and mutually reinforcing on multiple levels. On the one hand, machine learning can boost productivity by optimizing processes, identifying areas for improvement, and automating repetitive tasks. On the other hand, the constant search to improve productivity can drive organizations to adopt new technologies, such as machine learning, as part of their transition process toward a more efficient and competitive model.

In terms of productivity, machine learning has the potential to fundamentally transform the way work is done and decisions are made within an organization. By leveraging sophisticated algorithms and advanced data analytics, companies can improve accuracy and efficiency in a variety of areas, from supply chain management to customer service. For example, in manufacturing, machine learning can optimize production scheduling, reduce downtime, and improve product quality, leading to greater productivity and profitability.

The adoption of machine learning is also driving significant changes in the way organizations manage organizational transitions. Companies that recognize the strategic value of this technology are proactively integrating machine learning into their operations, seeking to leverage its capabilities to quickly adapt to changing market demands and gain a sustainable competitive advantage. This may involve restructuring teams, investing in machine learning training and skill development, and reviewing processes to make the most of the capabilities of this emerging technology.

However, the successful adoption of machine learning and its integration into organizational transition is not without challenges. Implementing new technologies can face resistance from employees, require significant investments in infrastructure and training, and raise ethical and privacy concerns related to data use. Additionally, long-term success requires a deep understanding of how to effectively integrate machine learning into existing business processes and how to mitigate the risks associated with its implementation.

CONCLUSIONS

The 21st century began with profound changes thanks to the development of information technologies, robotics, data mining and the consolidation of artificial intelligence as a tool that is available to almost all fields of knowledge, society, technologies and the companies, among others.

Artificial intelligence is found in cell phones, video cameras, smart TVs, and other everyday gadgets; so, it is logical to understand that every day their presence becomes more common and necessary.

Just as artificial intelligence has taken over almost all everyday tools, it has also crossed the borders of hard sciences and military agencies to become an ally of administration and, therefore, of management.

As has already been pointed out, machine learning is a part of artificial intelligence, which offers the opportunity to teach machines to learn on their own and, thereby, respond to situations that require fast, precise, and objective attention.

The processing of high volumes of data, in fractions of seconds, as well as its categorization and interweaving, offers managers the opportunity to know, visualize and project the scenarios —through machine learning— that are required to make simple, moderate, or complex decisions, as needed, with a level of precision that the human brain can rarely offer.

Although it cannot yet be said that a high degree of artificial intelligence has been achieved that guarantees 100% accuracy in predictions of behaviour or possible responses to a present or future situation, without a doubt, the advances that have been achieved —until now— allow visualize the potential that AI has in the field of management and business, which will be increasingly complete and complex once quantum computing is present in companies and organizations like the technology that is present in them today.

Additionally, it is timely to highlight that the emergence of AIoT will end up changing the way in which companies develop, adapt, transform and evolve. But it is necessary to be careful, the adoption and consolidation of the frequent use of AIoT does not mean that there will be – in the immediate future – a total dependence on digital tools and very little human intervention, on the contrary, greater acuity will be required when analysing the results, greater critical thinking when exploring possible scenarios and leadership when choosing the strategies, tactics, changes and/or transformations that AIoT offers to organizations.

REFERENCES

Aggarwal, R. (2020, March). *Organisational Change*. Retrieved from https://www.arsdcollege.ac.in/wp-content/uploads/2020/03/OB-Change.pdf

Arévalo-Ascanio, J., & Estrada López, H. (2017). Decisionmaking. A review of the topic. In *Gerencia de las organizaciones: Un enfoque empresarial* (pp. 249–278). Ediciones Universidad Simón Bolivar.

Asil, A. L. K. A. Y. A. A., & Erdem Hepaktan, A. C. (2003). Organizational Change. *Yönetm ve Ekonom, 10*(1), 31-58. doi:https://dergipark.org.tr/tr/download/article-file/145822

Bartolomeo, J. (2015, April 29). *Tipo de decisor según su postura ante el riesgo*. Retrieved from https://administrarconsultora.wordpress.com: https://administrarconsultora.wordpress.com/2015/04/29/tipo-de-decisor-segun-su-postura-ante-el-riesgo/

Björkman, M. (1992). What is productivity? *IFAC Proceedings Volumes, 25*(8). 10.1016/S1474-6670(17)54065-3

Canós, L., Pons, C., Valero, M., & Maheut, J. P. (n.d.). *Toma de decisiones en la empresa: proceso y clasificación*. Retrieved from https://riunet.upv.es/bitstream/handle/10251/16502/TomaDecisiones.pdf

Cepymesnews. (2023, May 10). *Estadísticas de inteligencia artificial*. Retrieved from https://cepymenews.es/estadisticas-inteligencia-artificial-para-2023

Chacín, L. (2010). Information technology as support to decision making management process in organizations in electrical sector in Venezuela. *Espacios, 31*(2), 11-13. Retrieved from https://www.revistaespacios.com/a10v31n02/10310233.html

Chen, D. Q., Preston, D. S., & Swink, M. (2016). How the use of Big Data analytics Affects value creation in supply chain management. *Journal of Management Information Systems*, *32*(4), 4–39. doi:10.1080/07421222.2015.1138364

Clarín. (2022, June 3). *Una computadora cuántica resuelve en 36 microsegundos un problema de 9.000 años*. Retrieved from https://www.clarin.com/tecnologia/computadora-cuantica-resuelve-36-microsegundos-problema-9-000-anos_0_BzzaX42xoE.html

Corchado, J. M. (2020, November 23). *The role of the AIoT and deepint.net*. Retrieved from IEEE International Conference on Electronics Circuits and Systems: https://gredos.usal.es/bitstream/handle/10366/144254/ICECS.pdf

Datision. (2021, October 14). *La inteligencia artificial para la toma de decisiones empresariales*. Retrieved from https://datision.com/blog/inteligencia-artificial-decisiones-empresariales/

Hinestroza, D. (2018). *El machine learning a través de los tiempos, y los aportes a la humanidad*. Retrieved from https://repository.unilibre.edu.co/bitstream/handle/10901/17289/EL%20MACHINE%20LEARNING.pdf

Iberdrola. (2022). *Descubre los principales beneficios del 'Machine Learning'*. Retrieved from https://www.iberdrola.com/innovacion/machine-learning-aprendizaje-automatico

Joakin, I. (2021, September 5). *Aplicación de tecnologías de aprendizaje automático para predecir negocios y tomar decisiones empresariales*. Retrieved from http://sedici.unlp.edu.ar/bitstream/handle/10915/127018/Documento_completo.pdf?sequence=1

Ministerio Cultura Argentina. (2020, June 22). *Alan Turing, el padre de la inteligencia artificial*. Retrieved from https://www.cultura.gob.ar/alan-turing-el-padre-de-la-inteligencia-artificial-9162/

Moreno, A., Armengol, E., Béjar, J., Belanche, L., Cortés, U., Gavaldà, R., . . . Sànchez, M. (1994). *Aprendizaje automático*. Barcelona-España: Edicions de la Universitat Politècnica de Catalunya (UPC).

Moya Anegón, F. d., Herrero Solana, V., & Guerrero Bote, V. (1998). La aplicación de Redes Neuronales Artificiales (RNA): a la recuperación de la información. *Bibliodoc: anuari de biblioteconomia, documentació i informació*, 147-164.

Onyx Soft. (n.d.). *¿Qué son los algoritmos predictivos de planificación de procesos?* Retrieved from https://www.onyxerp.com/blog/algoritmos-predictivos

Pascual, M. G. (2022, July 11). *La inteligencia artificial de Google es capaz de aprender como un bebé.* Retrieved from https://elpais.com/tecnologia/2022-07-11/la-inteligencia-artificial-de-google-es-capaz-de-aprender-como-un-bebe.html

Pérez, T. E., Raya, R., & Santos, E. (2017, September 29). *Las chicas del ENIAC (1946-1955).* Retrieved from https://mujeresconciencia.com/2017/09/29/las-chicas-del-eniac-1946-1955

Redacción España. (2019, November 11). *Origen del concepto de Inteligencia Artificial.* Retrieved from https://agenciab12.com/noticia/origen-concepto-inteligencia-artificial

Revathy, R., Gopal, M., Selvi, M., & Periasamy, J. (2020, June 4). Analysis of artificial intelligence of things. *International Journal of Electrical Engineering and Technology*, *11*(4), 275–280. doi:10.34218/IJEET.11.4.2020.031

Rueda, J. F. (2022). *Aprendizaje supervisado y no supervisado.* Retrieved from https://healthdataminer.com/data-mining/aprendizaje-supervisado-y-no-supervisado/

Russo, M. (2021, February 2). *¿Cuándo se inventó el primer ordenador?* Retrieved from https://www.info-computer.com/blog/cuando-se-invento-el-primer-ordenador/

Saleh, Z. (2019). *Artificial Intelligence Definition, Ethics and Standards.* Retrieved from https://www.researchgate.net/publication/332548325_Artificial_Intelligence_Definition_Ethics_and_Standards

Tapscott, D. (1997). The Digital Economy: Promise and Peril. In *The Age of Networked Intelligence*. McGraw-Hill Education.

Trigo Aranda, V. (n.d.). Algoritmos. *Acta*, 43-50. Retrieved from https://www.acta.es/medios/articulos/matematicas/035041.pdf

Yadav, P., & Marwah, C. (2015). The Concept of Productivity. *International Journal of Engineering and Technical Research*, *3*(5), 192–196. https://www.erpublication.org/published_paper/IJETR032199.pdf

Chapter 8
The Effect of Artificial Intelligence Over the Educational System (School and University)

Saman Omidi

ⓘ https://orcid.org/0009-0002-1516-2865
University of Isfahan, Iran

Ali Omidi

ⓘ https://orcid.org/0000-0003-1882-0456
University of Isfahan, Iran

ABSTRACT

This chapter delves into the impact of artificial intelligence (AI) on the educational systems, spanning both schools and universities. The primary components of an educational system include students, teachers, monitoring and evaluation (M&E), and administrative organization. While there are undoubtedly other components, these mentioned items are the most crucial. Artificial intelligence significantly influences all of them. In this chapter, the authors thoroughly examine and analyze the effect of AI on each of the four educational components, employing a systematic review and library methodology. Consequently, after the introduction, the chapter is organized into four parts, corresponding to each component. The main finding indicates that AI technology has revolutionized education by offering personalized and adaptive learning, flexible teaching, as well as smart administration and swift assessment systems. This chapter also highlights the drawbacks associated with it.

DOI: 10.4018/979-8-3693-0993-3.ch008

INTRODUCTION

Artificial intelligence is a form of capability exhibited by machines, in contrast to the natural intelligence displayed by animals, including humans. Any system capable of comprehending its environment and taking actions to optimize its likelihood of achieving goals is considered an example of artificial intelligence. Among the most significant applications of artificial intelligence are its contributions to self-driving vehicles, including drones and autonomous cars, medical diagnoses, artistic creation, mathematical theorem proving, strategic gameplay, image and sound recognition (such as face recognition), energy storage, internet browsers, search engines, contract preparation, and judicial decision prediction. Introducing techniques that can effectively manage complexity should be positioned as the fundamental core of past, present, and future scientific and research endeavors across all branches of computer science, particularly in the realm of artificial intelligence. An essential role of artificial intelligence tools is assisting humans in optimally addressing exceedingly intricate issues.

The process of educating learners at different educational levels is considered a very complex issue in human life. Presently, humanity seeks to address this complexity by harnessing the power of artificial intelligence. University professors worldwide are increasingly utilizing software like essay grader to enhance their educational processes. These programs have the capability to respond to learners' queries, send reminders about homework, and assess essays and written assignments. Students interact with these virtual teaching assistants through email, often unaware that they are communicating with a computer program rather than a human being.

While humanoid robots may not assume the role of teachers in the immediate future, numerous ongoing projects leverage computer intelligence to assist both students and teachers in elevating the quality of education. This chapter explores the impact of artificial intelligence on transforming the nature of learning. This chapter underscores the influence of artificial intelligence in the realm of education across four key areas: the education audience (students), trainers and teachers, educational monitoring and evaluation, and educational administration. Consequently, the chapter is structured into four sections. The first section delves into the impact of artificial intelligence on learners, encompassing both schools and universities, and it explores the hypothesis of AI-driven personalization of education. The second part focuses on the ramifications of artificial intelligence on the changing role of teachers. The third part deals with the impact of artificial intelligence on Monitoring and Evaluation (M&E) in the educational context. In the fourth section, the influence of artificial intelligence on the smartness of educational administration is addressed. Finally, the chapter concludes by presenting the results and sources.

1. PERSONALIZATION OF EDUCATION

One significant effect of artificial intelligence is its role in facilitating personalized education. To grasp the essence of personalized learning, envision a classroom without a standardized teaching approach. In this setting, the teacher doesn't deliver identical lessons to all students but instead guides each learner along a unique educational path. The content, timing, location, and instructional method are customized to address each student's strengths, skills, needs, and interests. While diverse methods may be employed to acquire specific skills, the overarching expectation is that all students achieve a predefined level of proficiency.

Personalized learning is a pedagogical model grounded in the premise that each student should receive an educational program tailored to their learning style, current knowledge level, skills, and interests. This approach diverges from the conventional "one-size-fits-all" method predominant in many schools today. The ultimate aim of personalized education is to grant every student access to educational content aligned with their individual interests and talents. With the support of artificial intelligence, students' potential and distinctive talents can be more effectively identified, nurturing their personal growth and success. AI facilitates the maintenance of current and comprehensive records for learners, enabling a profound understanding of each student's strengths, needs, motivations, progress, and goals. These records are updated more frequently than traditional report cards, providing accurate information that empowers teachers to make decisions that positively influence student learning. Educators can harness AI technologies to gain valuable insights into students' abilities, interests, and learning styles. This capability empowers them to create personalized guidance and educational experiences tailored to the unique needs of each individual (Maghsudi, et al., 2021).

Additionally, access to personalized educational resources is often limited in traditional education methods. Historically, such opportunities were typically reserved for nobles, the children of sultans, and rulers. The integration of AI in education brings numerous benefits, including expanded access to a wide range of resources. Students can now avail themselves of digital libraries, online courses, e-books, research papers, and multimedia content from anywhere and at any time through AI-based platforms and applications. AI algorithms play a crucial role by adapting and recommending different resources based on individual learning preferences. This ensures that students have access to the best resources aligned with their interests and goals. By doing so, geographical barriers are eliminated, providing all students with an equal opportunity to learn and easily pursue their desired fields of study (Chen et al., 2020).

Artificial intelligence can analyze students' strengths, weaknesses, and learning styles, adjusting the educational path based on individual experiences. By harnessing

AI algorithms, educational institutions can scrutinize extensive student data to identify patterns and trends with the potential to forecast future academic performance. Consequently, schools can leverage these valuable insights to optimize the allocation of resources in a more targeted and streamlined manner.

In fact, an education system regulated by artificial intelligence takes into account students' characteristics when designing programs, making it less challenging for students. In a study conducted by Hirankerd and Kittisunthonphisarn (2020), they developed a management system that incorporates artificial intelligence and utilizes augmented, virtual, and mixed-reality technologies. This system was designed to monitor the progress of students' learning and assign adaptive tasks accordingly. QuillBot, as an AI-powered proofreader, serves as a second pair of eyes by providing real-time feedback to students, assisting them in enhancing the quality of their written assignments, and making a strong impression. By offering instantaneous suggestions and corrections, QuillBot expedites the editing and proofreading process, acting as a valuable time-saving tool. This feature not only helps refine students' work but also saves time that would otherwise be spent on manual revisions (York, 2023).

Additionally, chatbots and AI-based virtual assistants can serve as proficient tutors, addressing students' questions, providing explanations, and simplifying complex concepts. Essentially, AI-based programs function as personalized tutors. Virtual tutors can operate around the clock, offering immediate support to students and eliminating space and time limitations. Intelligent educational systems, such as Carnegie Learning, operate as tutors, personalizing students' learning experiences by considering their strengths and weaknesses (Guilherme, 2019). These systems can also monitor students' progress and generate insightful reports for teachers, enabling the creation of relevant content for each student (Tapalova & Zhiyenbayeva, 2022).

AI can also enhance education by making it more engaging and interactive. Artificial intelligence algorithms can design virtual reality and augmented reality programs that enable students to explore virtual environments, conduct experiments, and experience real-world scenarios in a simulated setting. For instance, students can embark on virtual tours, immersing themselves in environments typically only found in textbooks. This feature is particularly beneficial for subjects like history, geography, and experimental sciences. By providing practical experiences rather than mere memorization of concepts, students gain a better understanding of different subjects. They can even engage in conversations with virtual assistants to discuss various educational topics (Sharma et al., 2021). Along with this adaptive teaching strategy, Munawar et al. (2018) developed an intelligent virtual laboratory that caters to students' needs by assigning laboratory tasks at an appropriate difficulty level. Similarly, ChatGPT is a form of artificial intelligence known as GenAI, functioning as a chatbot capable of generating human-like conversation. Its purpose is to enhance students' learning abilities by locating and presenting information in an engaging narrative style. This has the potential to enable students to acquire knowledge more efficiently through various approaches. However, it's important to note that ChatGPT, like web pages, is not entirely reliable and may contain errors or produce inaccurate results (Chiu, 2023).

Despite the advancements in science and technology, teaching disabled individuals remains a challenge. However, artificial intelligence technologies can create suitable learning environments for these individuals in educational institutions and organizations. Many disabled students require special resources to learn. For example, visually impaired students often rely on Braille books as their primary source of information. Screen readers equipped with artificial intelligence can read various sources aloud for these students. The same technology can be used to convert spoken language into text for students with hearing impairments, enabling their participation in discussions, lectures, and classes. In other words, artificial intelligence can personalize each person's learning path and adapt the pace of education to their specific abilities and needs. Furthermore, as previously mentioned, artificial intelligence can provide 24-hour support for these students. By utilizing these technologies, no individual is deprived of education due to physical or mental disabilities (Cromby et al., 1996).

Therefore, personalization of education helps each student choose an educational path that matches their progress, motivations, and goals. For example, a school might create a schedule for the student with weekly updates on their academic progress and interests. Each student's program is unique to them, but it probably includes several teaching methods. This combination may include project-based instruction with a small group of peers, independent work on specific skills or complex tasks, and one-on-one instruction by an instructor. A personalized learning path allows students to work on different skills at different speeds, but that doesn't mean the school will let them fall behind schedule. Teachers closely monitor each student's work and provide additional activities if needed. This type of system also plays a significant role in how learners interact with information in their personal and professional lives, and it can change the way they search for information and use it in school. Therefore, with newer and more integrated technologies, students in the future will have different experiences in searching for information and finding the right answers compared to today's students.

The impact of artificial intelligence on the education industry owes a large part to the use of applications made using this technology. In fields that require interactive methodologies, such as learning foreign languages or treating patients, AI technology has afforded facilities for learners to communicate and interact with the audience. Below are some examples of these applications that students, teachers, and institutions can utilize (Vazquez-Cano et al., 2021). For instance, Duolingo is a language learning application based on artificial intelligence that teaches more than

30 living languages of the world. Duolingo boasts more than 300 million active users. This application enables its users and learners to engage in various exercises and test themselves at their convenience, regardless of time or location (Astarilla, 2018). Brainly is a platform with 150 million users who utilize it to find answers to their questions and exercises. Leveraging artificial intelligence, this platform compiles answers contributed by students, teachers, and professors, easily accessible through a quick search. It serves as an excellent platform where students can collaborate and find the solutions they need (Billy, 2020). Odysee is a fantastic app that aids children in learning geography in a fun way. In Odysee, users can explore different parts of the world using artificial intelligence algorithms. Additionally, the platform allows users to take custom tests and assess their progress (Bonnett, 2023).

Dr. Michael Co Tiong-hong from HKUMed's LKS Faculty of Medicine and Dr. John Yuen Tsz-hon from HKU's Department of Computer Science have collaborated to create Hong Kong's inaugural diagnostic application called 'AI virtual patients,' aimed at training medical students. Through the utilization of generative AI technology and actual surgical cases, the research team has developed 'humanized' AI virtual patients with unique personalities and medical traits. These virtual patients enable medical students to effectively engage with patients in a virtual environment. This groundbreaking initiative significantly enhances students' professional skills and their ability to accurately gather patients' medical histories (news-medical.net, 2023).

2. AI AND TEACHERSHIP

Artificial intelligence (AI) stands as one of the technological advancements that have completely transformed various facets of life and industries, including education, communication, transportation, health, manufacturing plants, and even the world of art. The potential for AI to replace numerous jobs has led to concerns across various sectors, with teachers in the field of education being among those apprehensive about job displacement. Professor Anthony Seldon, Vice-Chancellor of the University of Buckingham, was quoted in the Times in 2018, suggesting that AI could provide "an Eton education for all" and that machines could potentially replace the inspirational role of teachers, becoming "extraordinarily inspirational" (Schools Week, 2018). However, this paper challenges the hypothesis that AI can truly replace teachers in schools or universities.

While AI tools are advancing rapidly in terms of speed, accuracy, and efficiency, their primary function is to assist humans. With the aid of AI, teachers, and trainers can become more efficient, thus dramatically transforming the education industry. AI technologies have significantly enhanced classroom management for teachers by uploading, assigning, and distribution of learning materials and assignments across various subjects such as physical education and language education. Additionally, AI-powered tools are capable of vocalizing text-based problems, further aiding in teaching and learning processes. These applications have effectively boosted efficiency in educational settings (Chiu et al, 2023). In the future, AI can play a valuable role by providing real-time information and up-to-date data to teachers, enabling them to personalize their teaching methods. Furthermore, AI can make education more accessible to a wider population, allowing individuals who do not have access to traditional schools and educational facilities to participate in virtual classrooms through artificial intelligence. AI can support teachers in the educational process in various ways:

Creating FAQ: Teachers can leverage AI applications to design various questions and generate accurate, comprehensive answers. This resource can assist students in understanding key concepts and enhance their learning. AI programs have the potential to boost efficiency and productivity in educational settings by addressing common and straightforward questions that students may have. Through the use of chat-based AI, students can interactively seek satisfactory answers to lower-order inquiries, such as definitions or procedures, and receive immediate responses. This approach, in turn, frees up teachers' time, enabling them to concentrate on addressing higher-order questions and providing guidance (EDU Support, 2023).

Adaptive teaching: AI can replace complex and vague sentences with precise words, enhancing the clarity and accuracy of instructional materials to facilitate student comprehension. Teachers can input their handout text into AI platforms such as ChatGPT, which possesses natural language processing capabilities, to simplify the tone. In this regard, Luo (2018) and Standen et al. (2020) utilized artificial intelligence (AI) systems that incorporated multimodal sensor data to recognize the emotional states of students. This recognition helps educators determine the most effective ways to present content, employ teaching methods, and utilize communication strategies (Chiu et al., 2023).

Decreasing failure percentage: Educators can harness AI to reduce educational failure. For instance, Ivy Tech Community College in Indiana conducted a pilot study utilizing data from 10,000 course sections. By identifying 16,000 students at risk of failing within the initial two weeks of the semester, the college took steps to address non-academic challenges impeding their progress. By the end of the semester, 3,000 students were rescued from failing, with an impressive 98% of the contacted students achieving a grade of C grade or higher. Through their initiative called Project Student Success, Ivy Tech Community College has been able to support and aid a growing number of students, reaching a total of 34,712 and continuing to provide assistance (Verma, 2023).

Productivity and efficiency: AI can provide an interactive teaching experience for students. Platforms like ChatGPT enable teachers to offload tasks such as answering questions and explaining topics, while AI technologies create interactive scenarios to enhance understanding, fostering a personalized and engaging learning environment (Guilherme, 2019). Teachers often have various administrative duties, including communication with students and their families, assignment evaluations, and feedback provision. At times, teachers spend a significant amount of time on these administrative tasks instead of focusing on direct interaction with students. However, incorporating appropriate AI tools can automate or simplify these tasks, granting teachers more time for direct engagement with their students (Dené Poth, 2023). Moreover, teachers can receive swift feedback on their teaching weaknesses and make necessary corrections (Wongvorachan et al., 2022).

Nonetheless, it has been observed that some teachers lack a comprehensive understanding of how these technologies function. Without a clear understanding of the underlying mechanisms behind task assignments and teaching strategy recommendations, teachers have reported feeling that their control and main role over the teaching process is diminished as if they are diminishing the role of a "user". This perceived low role, potentially may discourage them from embracing AI as a tool for supporting their classroom teaching.

However, it is important to note that AI cannot fully replace teachers and professors. Instead, AI can serve as a teaching assistant, supporting and enhancing the work of educators. While the idea of replacing teachers with AI may initially seem appealing due to factors such as cost-effectiveness and availability, it should be recognized that education is not solely about imparting knowledge. The quality of interactions and the ability to shape minds are essential aspects of education that AI cannot replicate. Teachers possess the capacity to explain concepts repeatedly, cater to diverse audiences, and inspire critical thinking and innovation — tasks that AI cannot accomplish. While AI can significantly contribute to making education more effective, it cannot replicate the human connection and empathy that teachers provide in the classroom. The primary role of a teacher is to guide and inspire students, extending beyond the mere transmission of information (Felix, 2020).

As the co-author of this chapter, I began my university studies during the COVID-19 pandemic, experiencing three fully online academic semesters. It was only in the fourth semester when classes were held in person, that I truly realized the value of face-to-face human interaction and the role of teachers and professors in the learning process. It felt as though I had transitioned from a machine world to a human world, imbuing my life with greater meaning. While AI and machines can facilitate improved learning, they can never replace the depth of human face-to-face communication. Considering this, the most plausible future role for AI in education is to assist teachers in becoming more effective, rather than replacing them.

3. MONITORING AND EVALUATION (M&E)

Traditional evaluation methods primarily rely on summative assessments, such as tests and exams, which are typically administered at the end of instructional periods. These assessments, often standardized across institutions, focus on measuring students' factual recall and basic comprehension, often overlooking the development of crucial skills such as critical thinking, problem-solving, and creativity. In traditional evaluation methods, assessments are typically uniform, meaning that the same tasks or items are given to every student, irrespective of their prior knowledge, abilities, experiences, and cultural backgrounds. This approach may introduce bias into the assessment process, resulting in unequal opportunities for students to demonstrate their learning abilities (Swiecki et al., 2022).

In contrast, the advent of artificial intelligence has the potential to bring about significant changes in student evaluation. It not only possesses the ability to score tests efficiently but also assists in managing administrative tasks, thus alleviating a considerable burden on teachers (Diyer et al., 2020). For instance, platforms like ProProfs Quiz Maker offer online test software that enables personalized exams with automatic scoring and results. This allows teachers to enhance their productivity and allocate more time to teaching sessions. In recent years, electronic assessment platforms (EAPs) that facilitate online or offline exam administration have gained increasing popularity. EAPs offer several key advantages, such as the ability to present questions that are challenging or impractical to present on paper, including questions in a predetermined order, and providing prompt and tailored feedback to learners (Gokturk-Saglam, & Sevgi-Sole, 2023).

Covert assessment techniques aim to gather data that goes beyond assessing whether students answered questions accurately. The concept of 'stealth assessment' was introduced by Shute and Ventura to describe an approach that utilized automatically collected data from learners while they engaged in a digital game. Through the analysis of data generated in a digital physics game commonly used in schools, they were able to identify indicators of conscientiousness, creativity, and physics proficiency (Shute, & Ventura 2013).

Bayesian Knowledge Tracing (BKT) is widely recognized as the most prominent technique for estimating latent knowledge. This method employs four parameters to estimate a learner's ability to apply a knowledge component. The parameters include (a) the probability that the learner has already mastered the knowledge component, (b) the probability of learning the knowledge component after a learning opportunity, (c) the probability of providing a correct response even when the learner has not yet mastered the knowledge (guess), and (d) the probability of incorrectly applying a knowledge component despite having acquired it (slippage). BKT has been applied not only in subject-specific learning but also in assessing learners' self-regulation abilities and providing personalized visualizations (Martínez-Comesaña et al., 2023).

One critical component of automated assessment design is the creation of secure assessment tools based on artificial intelligence techniques. These tools generate questions and exercises for test-takers using a predetermined question bank designed by teachers. What sets these exams apart is their enhanced security compared to regular exams. They offer intelligent settings that can be applied when assigning tests to specific users. These settings allow for access restrictions and the creation of different groups by randomly selecting questions for each, thus preventing cheating. While these programs save considerable time for teachers, they also reduce their workload and mitigate the risk of human error. (Swiecki et al., 2022).

Coursera, established in 2012 by Stanford University computer science professors, Andrew Ng and Daphne Koller, is a prominent provider of massive open online courses (MOOCs) based in the United States. Collaborating with universities and organizations worldwide, Coursera offers a diverse range of online courses, certifications, and degrees across numerous subjects. As of 2023, Coursera boasts a catalog of over 4,000 courses, which are provided by more than 275 universities and companies. Assessments within Coursera encompass a variety of formats, including quizzes, assignments, projects, peer reviews, and exams. Some assessments may be graded automatically, while others may be graded by peers or instructors. The assessments may also be timed or untimed and may allow multiple attempts or only one attempt (Wei, & Taecharungroj, 2022, and Wikipedia, 2023). Likewise, Gradescope is a grading platform powered by artificial intelligence (AI) that offers instructors a streamlined approach to assessing different types of assignments, such as written work, multiple-choice questions, and programming assignments. With its AI capabilities, Gradescope has the ability to group similar answers together, enabling instructors to provide feedback more efficiently. Specifically for programming assignments, Gradescope utilizes an autograder system that automatically evaluates students' code using predefined test cases and criteria. This automated process helps expedite the grading process and provides consistent and objective assessments for programming assignments (tensorway.com, 2023).

The use of facial recognition technology to identify students through face scanning is one of the new applications of artificial intelligence. This technology not only eliminates the need for a student card but also serves various research, security, and administrative purposes. For example, these scanners can replace student cards when students want to borrow a book from the library. Furthermore, many schools worldwide utilize face-scanning technology to prevent cheating and verify student identities (Tsai & Li, 2021).

Certainly, the non-neutrality of AI-based assessments is a significant and valid point of discussion. Like any form of assessment, it can be argued that AI-based assessments inherently reflect the cultural, disciplinary, and individual norms, value systems, and hierarchies of knowledge of their creators. Moreover, these assessments may unintentionally reinforce certain norms, values, and hierarchies of knowledge among students. This can influence students' behavior as they might align their actions with what is algorithmically evaluated and rewarded. The potential biases embedded in AI systems, whether intentional or unintentional, raise important questions about fairness, equity, and the unintended consequences of relying on AI for assessment in educational settings. Addressing and mitigating these biases is crucial to ensure that AI-based assessments provide a fair and unbiased evaluation of students' capabilities (Si-Jing & Wang, 2018).

4. SMART ADMINISTRATION

Educational administrative management is defined as the general process of planning, organizing, directing, coordinating, monitoring, and evaluating the education process, with a primary focus on the hardware dimensions of training. AI systems have the potential to automate many tasks that were previously performed by humans, leading to fundamental changes in the education staffing system. This could result in certain jobs becoming obsolete or necessitating the creation of new ones. Numerous administrative tasks can be automated through AI, resulting in improved information flow between students and administrative staff. This includes accelerating the registration process, handling large volumes of data more efficiently, and reducing bureaucracy for the benefit of learners.

Envisioning universities and schools where data is entered only once, never lost, and interconnected to provide valuable services, AI has the potential to transform educational institutions into modern and digital entities. Such institutions may not necessarily require physical buildings and excessive bureaucracy, as AI integrates physical space, technology, and the virtual world to create the most engaging experience for its beneficiaries. Indeed, AI has revolutionized academic and administrative activities in various ways, spanning the admission process, counseling, library services, assessment, feedback, tutoring, course allocation, hostel management, staff recruitment, fee collection, and more.

Learning and teaching involve numerous administrative tasks, such as registering students in different courses or sessions, handling applications and admission processes, and filtering suitable and potential students. AI technologies have empowered educational administrators and management teams by providing evidence-based support for their decision-making processes. This has the potential to streamline operations, enhance efficiency, and ultimately contribute to a more effective and responsive education system. Through access to vast amounts of data, AI agents can accurately predict the likelihood of students discontinuing their courses, identify the key factors influencing student academic performance, and provide valuable assistance to students in selecting the most suitable courses (Duan et al., 2019).

One of the most appealing aspects of AI in education is its potential to make universal classrooms accessible to all students, irrespective of language barriers or disabilities. Tools like Presentation Translator, a free PowerPoint plugin, can provide teachers with simultaneous captioning, making teaching more inclusive for learners with hearing impairments or those who speak different languages. AI technology also opens opportunities for students who are unable to attend school due to illness, enabling them to continue learning at their own pace. For instance, Microsoft has developed an AI application for the blind that can define and describe objects, read text, provide product introductions, describe surroundings, identify currency, describe individuals, and even recognize emotions displayed on a person's face. Consequently, individuals with disabilities no longer need to travel long distances to access education. AI breaks down traditional barriers between schools and grade levels, fostering a more cohesive and inclusive learning experience. (Knox et al., 2019).

Virtual world conferences utilizing artificial intelligence offer opportunities for growth and collaboration among learners. These conferences enable instructors and learners from around the globe to interact in a virtual environment, exchanging ideas and learning from one another, thereby enhancing their skills and staying updated. In essence, a global educational and research network is being established, facilitating inclusive learning by overcoming traditional challenges pertaining to physical limitations and accessibility. It is important to note that this does not imply the complete disappearance of universities, colleges, and schools, but rather, depending on specific circumstances, these physical spaces will coexist alongside artificial intelligence (Rajasingham, 2009).

Artificial intelligence can help administrators predict enrollment trends, identify at-risk students, evaluate the effectiveness of different instructional strategies, and more. This data-driven approach can lead to more informed, objective, and effective decision-making. Therefore, one of the key advantages of a smart educational system based on artificial intelligence is its potential to reduce costs while increasing efficiency and accessibility. AI systems can manage admissions, registration, and course scheduling, thus reducing the workload of administrative staff. This reduction in variable costs, along with the potential for tailored curriculum delivery, increases access. Therefore, with the help of AL tools, the need for large research centers will decrease in the future, and the structure will move towards the core network. Collaborations will mainly occur through the Internet and online, and many research activities will be product-oriented. The universities of the future will have smaller campuses, fewer classes, and cramped dormitories. The library may not exist in them or will be very small. The student community of the university will be global. Of course, some universities will be purely local (in the form of a college) to respond to the needs of the surrounding community. On the other hand, workshop and laboratory activities will be outsourced (Rouhani, 2020).

CONCLUSION

AI-based technologies, such as virtual reality, augmented reality, and mixed reality, have a significant impact on enhancing the learning experience for students in both online and face-to-face classes. These technologies facilitate easier and faster access to relevant resources, allowing students to personalize their learning journey. Additionally, artificial intelligence (AI) can identify areas where students are struggling and provide appropriate methods to improve their performance. The education industry has been transformed by AI, with institutions and teachers benefiting from automation in tasks like administrative affairs, test evaluation, and grading. AI also contributes to enhanced security in online testing. Furthermore, AI technology has revolutionized the lives of students, offering personalized learning experiences and anytime, anywhere access to educational materials.

However, it is crucial to acknowledge that AI-based technology involve negative implications as well. Despite claims that educational programs based on artificial intelligence personalize education, they are constructed within specific cultural and educational frameworks, making it impractical to apply them universally with the same effectiveness. Another challenge with these programs is their complexity, making it difficult for teachers and students to quickly grasp their functionalities, potentially leading to disappointment for both groups. Additionally, while educational software based on artificial intelligence may be suitable for disciplines like engineering, it may not be as effective for humanities. Although interdisciplinary programs may be developed in the future, but we are currently distant from that point. Furthermore, the progression of artificial intelligence in education may induce philosophical despair among both teachers and students, as they fear being replaced by machines, leading to impact negatively the existential meaning of human life. The use of personal data as a criterion for developing new educational programs raises ethical and privacy concerns. Can we entrust artificial intelligence machines with the personal information of individuals to make educational decisions? Is there not a risk of potential exposure to hacking if such sensitive information is utilized in this manner?

One significant drawback is the potential roboticization of human life, including the learner's experience. Excessive reliance on AI may lead to decreased human interaction and a diminished sense of meaning in life. Human communication heavily relies on nonverbal cues such as tone and body language, which current AI systems lack such potentials. Moreover, as AI evolves, it absorbs the characteristics of the datasets it was trained on, making it prone to perpetuating biases present in those datasets. One common AI tool's error arises when it deals with facts or events that never really happened, while human life always experiences new developments. Nevertheless, it is unlikely that artificial intelligence will entirely replace teachers. Instead, AI can serve as a valuable tool for teachers in the future. Research findings demonstrate that positive interactions between students and teachers significantly impact student success—something that AI-related software cannot replace at present.

REFERENCES

Astarilla, L. (2018). University students' perception towards the use of duolingo application in learning english. *Prosiding CELSciTech*, *3*, 1–9.

BillyM. (2020). The influence of dynamic organizations and the application of digital innovations to educational institutions in the world during the COVID-19 pandemic. *Available at* SSRN 3588233. doi:10.2139/ssrn.3588233

Bonnett, A. (2023). *What is geography?* Rowman & Littlefield.

Chen, L., Chen, P., & Lin, Z. (2020). Artificial intelligence in education: A review. *IEEE Access: Practical Innovations, Open Solutions*, *8*, 75264–75278. doi:10.1109/ACCESS.2020.2988510

Chiu, T. K. (2023). The impact of Generative AI (GenAI) on practices, policies and research direction in education: A case of ChatGPT and Midjourney. *Interactive Learning Environments*, 1–17. doi:10.1080/10494820.2023.2253861

Chiu, T. K., Xia, Q., Zhou, X., Chai, C. S., & Cheng, M. (2023). Systematic literature review on opportunities, challenges, and future research recommendations of artificial intelligence in education. *Computers and Education: Artificial Intelligence*, *4*, 100118. doi:10.1016/j.caeai.2022.100118

Cromby, J. J., Standen, P. J., & Brown, D. J. (1996). The potentials of virtual environments in the education and training of people with learning disabilities. *Journal of Intellectual Disability Research*, *40*(6), 489–501. doi:10.1111/j.1365-2788.1996.tb00659.x PMID:9004109

Diyer, O., Achtaich, N., & Najib, K. (2020). Artificial Intelligence in Learning Skills Assessment. *Proceedings of the 3rd International Conference on Networking, Information Systems & Security*. 10.1145/3386723.3387901

Duan, Y., Edwards, J. S., & Dwivedi, Y. K. (2019). Artificial intelligence for decision making in the era of Big Data–evolution, challenges and research agenda. *International Journal of Information Management, 48*, 63–71. doi:10.1016/j.ijinfomgt.2019.01.021

Felix, C. V. (2020). The role of the teacher and AI in education. In *International perspectives on the role of technology in humanizing higher education* (pp. 33–48). Emerald Publishing Limited. doi:10.1108/S2055-364120200000033003

Gokturk-Saglam, A. L., & Sevgi-Sole, E. (Eds.). (2023). *Emerging Practices for Online Language Assessment, Exams, Evaluation, and Feedback*. IGI Global. doi:10.4018/978-1-6684-6227-0

Guilherme, A. (2019). AI and education: The importance of teacher and student relations. *AI & Society, 34*(1), 47–54. doi:10.1007/s00146-017-0693-8

Hirankerd, K., & Kittisunthonphisarn, N. (2020). E-learning management system based on reality technology with AI. *International Journal of Information and Education Technology (IJIET), 10*(4), 259–264. doi:10.18178/ijiet.2020.10.4.1373

Knox, J., Wang, Y., & Gallagher, M. (2019). Introduction: AI, inclusion, and 'everyone learning everything'. *Artificial intelligence and inclusive education: Speculative futures and emerging practices,* 1-13.

Luo, D. (2018). *Guide teaching system based on artificial intelligence.* https://online-journals.org/index.php/i-jet/article/view/9058

Maghsudi, S., Lan, A., Xu, J., & van Der Schaar, M. (2021). Personalized education in the artificial intelligence era: What to expect next. *IEEE Signal Processing Magazine, 38*(3), 37–50. doi:10.1109/MSP.2021.3055032

Martínez-Comesaña, M., Rigueira-Díaz, X., Larrañaga-Janeiro, A., Martínez, J. M., Ocarranza-Prado, I., & Kreibel, D. (2023). Impact of artificial intelligence on assessment methods in primary and secondary education: systematic literature review. Revista De Psicodidáctica. doi:10.1016/j.psicoe.2023.06.002

Munawar, S., Toor, S. K., Aslam, M., & Hamid, M. (2018). Move to smart learning environment: Exploratory research of challenges in computer laboratory and design intelligent virtual laboratory for eLearning technology. *Eurasia Journal of Mathematics, Science and Technology Education, 14*(5), 1645–1662. doi:10.29333/ejmste/85036

News Medical Life Science. (2023). *AI virtual patients' diagnostic application revolutionizes medical education in Hong Kong.* https://www.news-medical.net/news/20231117/AI-virtual-patients-revolutionize-medical-education-in-Hong-Kong.aspx

Rajasingham, L. (2009). *The impact of artificial intelligence (AI) systems on future university paradigms.* Victoria University of Wellington Wellington.

Rouhani, S. (2020). *AI and the Future of Universities.* Mehr News Agency. https://www.mehrnews.com/news/5035864

Sharma, U., Tomar, P., Bhardwaj, H., & Sakalle, A. (2021). Artificial intelligence and its implications in education. In Impact of AI Technologies on Teaching, Learning, and Research in Higher Education (pp. 222-235). IGI Global. doi:10.4018/978-1-7998-4763-2.ch014

Shute, V., & Ventura, M. (2013). *Stealth assessment: Measuring and Supporting learning in Video Games.* The MIT Press. doi:10.7551/mitpress/9589.001.0001

Si-Jing, L., & Wang, L. (2018). Artificial Intelligence Education Ethical Problems and Solutions. *2018 13th International Conference on Computer Science & Education (ICCSE).* 10.1109/ICCSE.2018.8468773

Standen, P. J., Brown, D. J., Taheri, M., Galvez Trigo, M. J., Boulton, H., Burton, A., & Hortal, E. (2020). An evaluation of an adaptive learning system based on multimodal affect recognition for learners with intellectual disabilities. *British Journal of Educational Technology, 51*(5), 1748–1765. doi:10.1111/bjet.13010

Support, E. D. U. (2023). *Educational uses for teachers.* https://edusupport.rug.nl/2430042249

Swiecki, Z., Khosravi, H., Chen, G., Martinez-Maldanado, R., Lodge, J. M., Milligan, S., Selwyn, N., & Gašević, D. (2022). Assessment in the age of artificial intelligence. Computers & Education. *Artificial Intelligence, 3,* 100075. doi:10.1016/j.caeai.2022.100075

Tapalova, O., & Zhiyenbayeva, N. (2022). Artificial Intelligence in Education: AIEd for Personalised Learning Pathways. *Electronic Journal of e-Learning, 20*(5), 639–653. doi:10.34190/ejel.20.5.2597

Tensorway. (2023). *AI in Higher Education: Impact of AI on Student Assessment.* https://www.tensorway.com/post/ai-for-student-assessment-in-higher-education

Tsai, M., & Li, M. (2021). Attendance Monitoring System based on Artificial Intelligence Facial Recognition Technology. *2021 IEEE International Conference on Consumer Electronics-Taiwan (ICCE-TW)*. 10.1109/ICCE-TW52618.2021.9603093

Vázquez-Cano, E., Mengual-Andrés, S., & López-Meneses, E. (2021). Chatbot to improve learning punctuation in Spanish and to enhance open and flexible learning environments. *International Journal of Educational Technology in Higher Education*, *18*(1), 1–20. doi:10.1186/s41239-021-00269-8

Verma, N. (2023). *How Effective is AI in Education? 10 Case Studies and Examples*. https://axonpark.com/how-effective-is-ai-in-education-10-case-studies-and-examples/

Week, S. (2018). *Seldon: AI can replace Ofsted inspectors within a decade*. https://schoolsweek.co.uk/seldon-ai-can-replace-ofsted-inspectors-within-a-decade/

Wei, X., & Taecharungroj, V. (2022). How to improve learning experience in MOOCs an analysis of online reviews of business courses on Coursera. *International Journal of Management Education*, *20*(3), 100675. doi:10.1016/j.ijme.2022.100675

Wongvorachan, T., Lai, K. W., Bulut, O., Tsai, Y. S., & Chen, G. (2022). Artificial intelligence: Transforming the future of feedback in education. *Journal of Applied Testing Technology*, *23*, 95–116.

York, A. (2023). *10 Educational AI Tools for Students in 2023*. https://clickup.com/blog/ai-tools-for-students/

Chapter 9
Theoretical Impact of AI on Working Hours and Wage Rate

Saumya Ketan Jhaveri
ⓘ https://orcid.org/0009-0004-2463-1111
School of Liberal Studies, Pandit Deendayal Energy University, India

Anshika Chauhan
School of Liberal Studies, Pandit Deendayal Energy University, India

Nausheen Nizami
School of Liberal Studies, Pandit Deendayal Energy University, India

ABSTRACT

This chapter assesses the correlation between AI expansion and its influence on diverse economic factors like labor supply, wage rates, and working hours. In the subsequent segment, the authors explore the connection between the utility function and the derived hours equation. Unlike the direct utility indicators of leisure and consumption, indirect utility functions are represented through income/wage rates or prices. These models and diagrams are developed by the authors to illustrate the correlation between income/consumption and its association with the balance between working hours and leisure hours. This chapter encompasses diverse models predicated on specific assumptions, contingent on market demand and supply scenarios.

INTRODUCTION

Throughout centuries and decades, humanity has borne witness to numerous revolutions that have brought both remarkable progress and advancements to our

DOI: 10.4018/979-8-3693-0993-3.ch009

world. Each industrial revolution has presented solutions to existing challenges while simultaneously giving rise to new concerns. With three industrial revolutions behind us and a fourth underway, the forthcoming industrial revolution is poised to leverage information and communication technologies across various sectors. This includes domains like 'Artificial Intelligence,' 'Automation,' 'Digitalization,' and 'Machine Learning.'

Artificial intelligence involves computers and machines replicating processed human intelligence. However, this growth in production and output has been accompanied by issues like unemployment, job displacement, market shifts, and price fluctuations. This analysis aims to investigate how the expansion of AI relates to different economic variables such as labor supply, wage rates, and working hours.

Historically, skepticism has surrounded the impact of technological change on labor demand. This skepticism dates back to Plato's Phaedrus (Frank et al., 2019, 2), which discussed how writing could supplant human memory. Similarly, during the industrial revolution, there was apprehension about machines replacing jobs. In the current context of the fourth industrial revolution, there's a hypothesis that AI and robots might lead to job displacement. Research suggests that while AI might initially bring about creative destruction, over time it could generate more employment opportunities.

Theoretical Models

This chapter assesses the correlation between AI expansion and its influence on diverse economic factors like labor supply, wage rates and working hours. In the subsequent segment, we explore the connection between the utility function and the derived hours equation. Unlike the direct utility indicators of leisure and consumption, indirect utility functions are represented through income/wage rates or prices.

These models and diagrams are developed by author to illustrate the correlation between income/consumption and its association with the balance between working hours and leisure hours (Koutsoyiannis, 1979). This will also reveal its influence on the labor supply (L) and wage rates. The variables dependent on the outcome are wage rates (W), Artificial Intelligence (A*), where AI functions as a curve-shifting factor. Other factors encompass working hours (WHr), leisure (LE), unemployment, output (Y), capital (K), and productivity. In figures 2 and 3, the X-axis represents leisure hours, while the Y-axis signifies consumption or income. The premise is that consumption equates to income. AE and BE symbolize budget constraints. U1 and U2 depict indifference curves that reflect similar levels of satisfaction, rendering individuals indifferent to specific combinations of working hours, leisure hours, and monetary income. P and R denote points of intersection with the budget constraints.

This chapter encompasses diverse models predicated on specific assumptions, contingent on market demand and supply scenarios.

IMPACT OF CHANGES IN INCOME/WAGE RATE ON WORKING HOURS AND LEISURE

Suppose an individual's utility function is denoted as U(C, L). In this case, utility is quantified in terms of dollars/rupee (a standardized unit) that a consumer allocates to consumption, while "L" represents the chosen leisure time. Alongside, one accounts for alternative sources of income labeled as "V," given a designated time frame "T" in hours. The vertical axis represents consumption, while the horizontal axis represents leisure hours ranging from 0 to 110 (right to left), and work hours ranging from 0 to 48 (left to right) are plotted. This approach is in line with the framework proposed by Krueger & Meyer (n.d.). As per the equation, an individual's wage rate hinges on both the hours worked and any supplementary income sources. The individual's budget constraint can be formulated as follows:

C= wh+ V

C= w(T-L)+ V

TANGENCY CONDITION: This pertains to the scenario where the marginal rate of substitution (MRS) is equivalent to the wage rate.

MRS=The pace at which a person is willing to trade off leisure time to gain additional goods or services.

Wage rate=The rate at which the market allows individuals to exchange an hour of leisure time for consumption. The assessment of leisure hours is influenced by the individual's inclinations and the current market circumstances.

$$w = -\frac{MU_L}{MU_C}$$

MODEL 1

The first model examines how the supply of labour will behave when a firm is catering to the full demand from the market i.e. whatever the existing demand is the

Figure 1. Budget Constraints

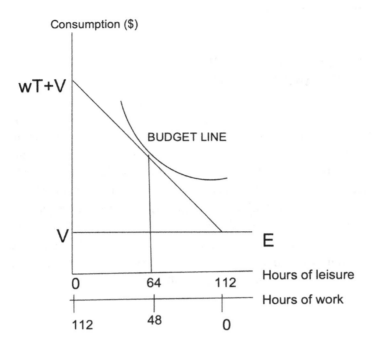

firm is able to fulfil it by their capability to deliver supply. The first graph considers the situation of a firm with the extensive use of AI and limited work distribution amongst the labourers.

As per the assumptions the firm is catering to full demand and because of that the existing worker will not get any incereament if he completes his work early because the firm does not have any additional work to give that employee. Therefore the employee will not be able to work more for extra income or because of AI his work hours will reduce and in the remaining hours that he got because of AI he will not be able to work. This theory will lead to constant income C. AI will increase productivity (Accenture) and it will lead to a decrease in the working hours of the labourer in graph 1 from T_1E to T_2E because the work in this model is limited and that labour has done his work with the use of AI in T_2E time which he used to do in T_1E time. The hours of work will decrease and hours of leisure will increase from OT_1 to OT_2. The Indifference curve U_2 will shift towards the right from P to R on different budget lines from AE to BE because of the impact of AI the working hours

Figure 2. Distributions Among Laborers

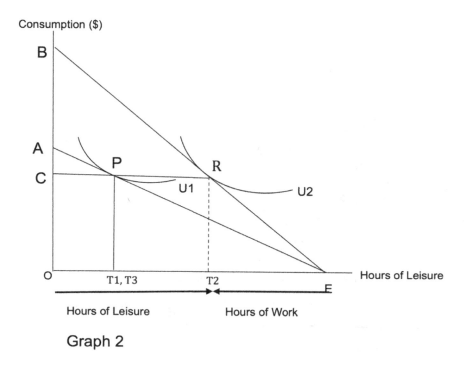

Graph 2

are decreased. In the short run, employment will be constant because it takes time to analyse the situation and take action against current employees. In the long run, this curve will have a leftward shift from R to P. It will come back to the previous equilibrium at P but with an unemployment effect. The employers will then know that the work is done by lesser employees so because of limited work they will fire people. Hours of work will increase again from T_2E to T_1E as some of the employees of this firm will get unemployed therefore the work will be added to the remaining employees. The wage rate will be constant at C because they will be working for same amount of hours which they used to do before AI. This has happened in previous technological change such as industrial revolution and for some time as there are no new job created, the unemployment rate increases. Jobs and businesses that fall under this model will experience unemployment and job displacement. 21% of jobs will be displaced in India due to AI. AI will also create 60% of new jobs giving a net impact of 40%. The global net impact on jobs is 25% (World economic forum, 2023) so it will create jobs. This data shows that there will be job displacement and unemployment which proves our theoretical model 1.

Figure 3. Demands for Goods and Labor

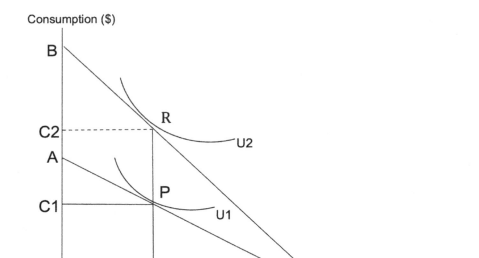

Graph 3

MODEL 2

This model shows the impact of demand for goods for a firm on the supply of labour i.e. the demand is abundant that the firm with existing technology and labour is not able to fullfil the supply. Here, productivity will increase as more demand in the market leads to more work done by labourers and an increase in their wage rate, simultaneously.

In the second model - graph 3, from the assumption that the firm is getting excess demand. Therefore, the labours will get extra time because their work which used to be completed in T hours will now complete somewhere between TE. But in this graph there won't be a rightward shift because the labour will get extra incentive to work and utilise the time he got left with the use of AI to increase their production. Therefore, the labour will get higher salary and the indifference curve U_1 will shift upwards to U_2 from point P to R and it will go from budget line AE to BE. The

Figure 4. Cost Decrease and Revenue Increase From AI Adoption by Function, 2021

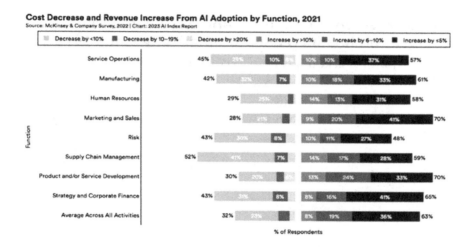

Cost Decrease and Revenue Increase From AI Adoption by Function, 2021

wage rate will also increase from C_1 to C_2 as the Marginal productivity of labour is increasing. Hours of work will be the same TE. This shift occurs with an impact on unemployment in the short run because the employer will not hire other people because he will not need another labour as capital(AI) is working with existing labour and work will get done what they used to receive but could not complete. The curve will remain at R in the long run.

- **Impact on revenue and cost functions**

There are evidences that point to rise in labour productivity due to AI and this may lead to falling cost and rising revenues (graph-4). However, it can be seen that how productivity will increase. Their wages may not increase as much as their productivity as a survey indicates (Economic Policy Institute, n.d.) graph 5.4 that there is a widening gap in the graph of wages and productivity. Also because of this widening gap companies are getting more profit which can be observe from Graph 4.

- **Gap between productivity and hourly compensation growth**

The compensation in wages and benefits that are incurred with the help of AI can be found (Graph 5). Dark blue lines mark productivity growth and light blue line marks the compensation growth rate. The time period from 1979-2021 shows total productive growth of 62.5% and compensation growth of 15.9%.

Figure 5. The Gap Between Productivity and Worker's Compensation

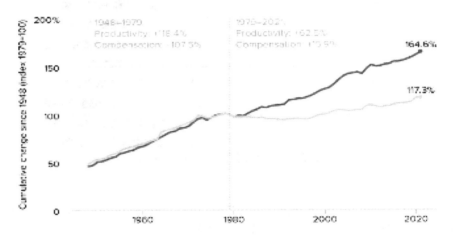

The gap between productivity and a typical worker's compensation has increased dramatically since 1979

Productivity growth and hourly compensation growth, 1948–2021

Notes: Data are for compensation (wages and benefits) of production/nonsupervisory workers in the private sector and net productivity of the total economy. "Net productivity" is the growth of output of goods and services less depreciation per hour worked.

Source: EPI analysis of unpublished Total Economy Productivity data from Bureau of Labor Statistics (BLS) Labor Productivity and Costs program, wage data from the BLS Current Employment Statistics, BLS Employment Cost Trends, BLS Consumer Price Index, and Bureau of Economic Analysis National Income and Product Accounts.

Economic Policy Institute

EMPIRICAL MATHEMATICAL METHODS

IMPACT OF AI ON PRODUCTIVITY/ MEASUREMENT OF TECHNICAL CHANGES (AI)

The research has analyzed directed change in labour supply models with exogenous labour supply and endogenous technology. Instead of giving one equation, sets of equations have been used to simplify the implications and effect of one on another. We formulated a change of impact of AI resulting in changes in production and output through cobb douglas production function, deriving the need for (new) skilled labours has been examined. Followed by the determination of wage rate from changes in labour supply due to AI in supply of labour, wage rate, work and leisure and the movements in graphs of various countries individually.

Figure 6. Output and Capital Labor

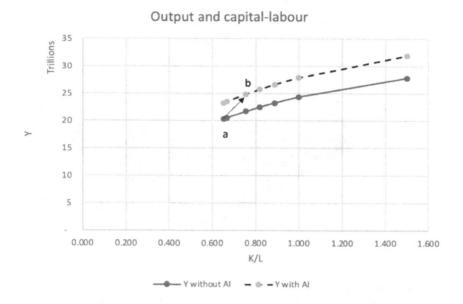

Our model considers the traditional method of measuring technical changes obtained from Cobb-Douglas production function, (Cottrell, 2019).

Y= AKαLβ (A1)

Where, Y stands for total output, 'A' refers to changes due to AI with respect to capital (K) and Labour (L), *α and β* are output elasticities, α represents how much capital contributes to the production of output. It measures responsiveness of output to change in levels of labour or capital respectively.

This graph (6) has been derived for output growth with and without AI in the case of US economy. Cobb douglas production function has been used to calculate the production using the elasticity of labour and capital with respect to AI.

The data was taken for total capital cost and labour compensation from (U.S. Bureau of Labor Statistics, 2023), the elasticity of capital intensity and labour productivity from (Marquetti, 2007) and growth of capital, and labour productivity from (OECD, 2022).

Capital/labour ratio will decrease, therefore because of higher capital intensity it the curve will shift upwards resulting in higher production(graph 6). The slope is increasing which depicts as capital increases the production will increase. Task performed by humans is 70% today which will decrease to 62% and task performed by machines is 30% which will increase to 38% by 2027(World economic forum, 2023). The dotted line depicts the Output after AI. The current production is at point

Figure 7. Capital Labor Ratio
- Cobb–Douglas production function. (2023, August 19). In Wikipedia.

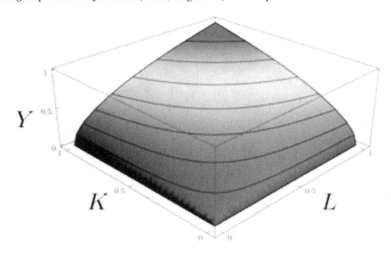

Figure 8. Technology Shock
Source- Cobb–Douglas production function. (2023, August 19). In Wikipedia

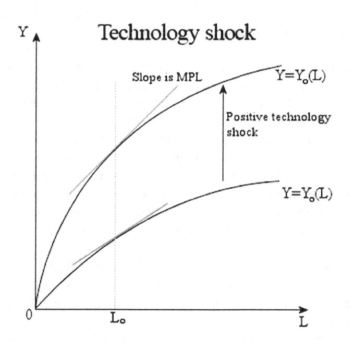

a and is projected to rise to point b as capital labour ratio will increase by 2027. From this graph we can see that labour share will decrease so K/L will increase.

Determination of Real Wage Rate From the Supply of Labour

In the previous section, we estimated the impact of changes in AI on the variable factors (K, L) was projected. It becomes necessary to derive the change in labour appointed which can further help us determine the wage rate. Wage rate helps us to connect technological change with the supply of labour and the decision to choose the limit of their working hours. For this, one needs to derive MPL (marginal productivity of labour) from the Cobb-Doughlas production function and then study its impact on the wage rate. The marginal productivity of labour (MPL) shows the relative change in output resulting from hiring an additional unit of Labour. Change in AI is likely to change MPL respectively. For example, increasing AI will contribute to higher productivity, which will increase the marginal productivity of Labour (the demand curve will shift to the right) which increases the number of employed workers and the increase in the wage rate. In the figure given below, the x-axis marks labour included and the y-axis marks the amount of output derived by the addition of a unit of labour, with and without AI.

DERIVATION OF MPL FROM COBB-DOUGLAS

$$Y = AK^{\alpha} L^{1-\alpha}$$

$$\frac{\partial Y}{\partial L} = (1-\alpha) AK^{\alpha} \left(L^{-\alpha} \right)$$

$$= (1-\alpha) A \left(\frac{K}{L} \right)^{\alpha}$$

$$\text{MPL} = (1-\alpha) \left(\frac{Y}{L} \right)$$

:- Real wage rate $= \dfrac{W}{p} = MRPL$ (Marginal revenue product of Labour)

Figure 9. Income-Leisure Indifference Curve
Source-Author's calculation

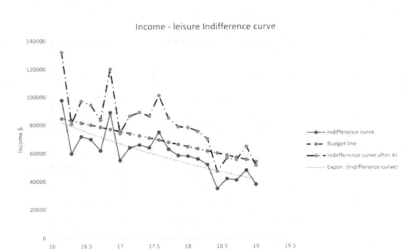

(MRPL= MPL*MR) where MPL= Marginal productivity of labour, MR= Marginal Revenue.

Modifications to the marginal productivity of labor can occur due to various factors. One such factor is technological advancements, which have the potential to enhance the marginal productivity of labor, resulting in a shift of the demand curve to the right. An illustrative example is the introduction of computer technology, which has effectively increased the productivity (marginal product) of numerous job categories. As a consequence, the marginal revenue product of labor for these occupations has risen, causing firms to adjust their demand for labor in an upward direction. This, in turn, leads to a twofold outcome: an expansion in the number of employed workers and an upward adjustment in the wage rate.

Empirical evidences for the leisure hours and income can be analysed from the data (Kolmar & American community survey). The indifference curve has been computed of leisure and income for now and after AI. Graph 9 is an empirical evidence of model 2. We have assumed 30$ as consumtion to prepare budget line.

CONCLUSION

The emergence of AI-driven technology, as seen in Model 1, suggests that there will be a rise in overall productivity and a decrease in costs. In the short term, there might be a reduction in employment due to displacement, but in the long term, employment opportunities are likely to expand with the emergence of new AI-related jobs. Model 2 indicates that the productivity gains from AI will ultimately lead to tasks becoming easier, less exhausting, and more profitable. Increased work output will result in higher wages. However, over time, this effect will become standard. The advancement of AI will lead to enhanced labor and capital productivity. As AI technology continues to advance, both labor and capital productivity will rise. Even from empirical model graph 6, we can see that when K/L ratio rises, output rises, but robots still can't accomplish everything on their own, necessitating human use, creation, management, innovation, and maintenance. The goal of humankind has always been to become more productive through various advancements. They have created technologies that have made their lives easier from the Stone Age through the Fourth Industrial Revolution. The environment can benefit and be harmed by this innovation. To establish policy, the effects on the social and environmental spheres are carefully examined. There is evidence that AI can both improve and harm the environment's health. There is evidence that AI can both improve and harm the environment's health. Since it is not the primary focus of this study, we won't go into great detail about it here. To minimize harm and maximize profit, policymakers must adapt employment policies to new technology like AI. The countries' policy makers have boosted their policies for employment (ILO, 2023).

REFERENCES

A., N. (2017). Automation and inequality The changing world of work in the global South Green economy. *International Institute for Environment and Development.* http://pubs.iied.org/11506IIED

Accenture. (n.d.). Where AI is Aiding Productivity. *Frontier Economics.* https://www.statista.com/chart/23779/ai-productivity-increase/

Cobb–Douglas production function. (n.d.). Wikipedia. Retrieved June 1, 2023, from https://en.wikipedia.org/wiki/Cobb%E2%80%93Douglas_production_function

Cottrell, A. (2019). *Economics 207, 2019.* The Cobb–Douglas Production Function. Retrieved June 1, 2023, from http://users.wfu.edu/cottrell/ecn207/cobb-douglas.pdf

Damioli, G., Roy, V. V., & Vertesy, D. (2021, January 21). The impact of artificial intelligence on labor productivity. *Eurasian Business Review.* doi:10.1007/s40821-020-00172-8

Economic Policy Institute. (n.d.). *The gap between productivity and a typical worker's compensation has increased dramatically since 1979*. Author.

Frank, M. R., Autor, D., & Bessen, J. E. (2019, April 2). Toward understanding the impact of artificial intelligence on labor. *National Academy of Sciences, 116*(14), 10. https://www.jstor.org/stable/10.2307/26698534

ILO. (2023). *Overview of national employment policies adoption*. ILO. https://shorturl.at/flAH0

Keane, I. (2023, March 26). *New York Post (AI impact)*. Retrieved June 1, 2023, from https://nypost.com/2023/03/26/up-to-80-percent-of-workers-could-see-jobs-impacted-by-ai/

Keane, I. (2023, March 26). *Up to 80 percent of workers could see jobs impacted by AI*. New York Post. Retrieved June 1, 2023, from https://nypost.com/2023/03/26/up-to-80-percent-of-workers-could-see-jobs-impacted-by-ai/

Kolmar, C., & American Community Survey. (n.d.). *An Interactive Exploration Of Earnings By Hours Worked*. Zippia. Retrieved June 1, 2023, from https://www.zippia.com/research/earnings-vs-hours-worked/#lessons

Koutsoyiannis, A. (1979). *Modern Microeconomics*. Macmillan. doi:10.1007/9781349160778

Krueger, A. B., & Meyer, B. D. (n.d.). *Labour supply*. Retrieved June 1, 2023, from https://scholar.harvard.edu/files/gborjas/files/lechapter2.pdf

Marquetti, A. (2007). A cross-country non parametric estimation of the returns to factors of production and the elasticity of scale. *Nova Economia Belo Horizonte, 4*(C14), 95-126.

Maslej, N., Fattorini, L., & Stanford University. (2023, April). The AI Index 2023 Annual Report. *Institute for Human-Centered AI*. https://aiindex.stanford.edu/report/

McKinsey. (2018, September). *McKinsey Global Institute*. Retrieved May 31, 2023, from https://www.mckinsey.com/mgi

NASSCOM. (2022). *Implications of AI on the Indian Economy*. AI Adoption Index.

OECD. (2022, December). *Growth in GDP per capita, productivity and ULC*. OECD Statistics. Retrieved June 1, 2023, from https://stats.oecd.org/Index.aspx?DataSetCode=PDB_GR#

PwC. (2017). *Human in the loop*. Pwc. www.pwc.com/AI

Sachs, G. (2023, April 5). *Generative AI Could Raise Global GDP by 7%*. Goldman Sachs. Retrieved June 1, 2023, from https://www.goldmansachs.com/intelligence/pages/generative-ai-could-raise-global-gdp-by-7-percent.html

U.S. Bureau of Labor Statistics. (2023, March 23). *News releases from the Office of Productivity and Technology*. Bureau of Labor Statistics. Retrieved June 1, 2023, from https://www.bls.gov/productivity/news-releases.htm

World Economic Forum. (2023, January 5). *Future of jobs Report 2023*. Weforum. Retrieved May 31, 2023, from https://www.weforum.org/reports/the-future-of-%20jobs-report-2023/

Chapter 10
Theories and Models in AIoT:
Exploring Economic, Behavioral, Technological, Psychological, and Organizational Perspectives

Surjit Singha

https://orcid.org/0000-0002-5730-8677
Kristu Jayanti College (Autonomous), India

Ranjit Singha

https://orcid.org/0000-0002-3541-8752
Christ University, India

ABSTRACT

AIoT, or artificial intelligence of things, is a transformative combination of artificial intelligence and the internet of things (IoT) that has far-reaching ramifications across multiple domains. This chapter examines the theories and models underlying its development and implementation. Businesses can assess the costs, benefits, and competitive advantages of AIoT by using economic models and market dynamics. Understanding human behaviour and trust is crucial for user acceptance, while ethical considerations underpin the development of accountable AIoT applications. Data management, security, and interoperability are technical facets that architectural frameworks address. The alignment of AIoT with human needs is enhanced by cognitive models and user experience, thereby fostering well-being. Change management and organizational learning are essential for effective implementation, which fosters innovation. AIoT promotes innovation and efficiency in manufacturing, healthcare, and smart cities.

DOI: 10.4018/979-8-3693-0993-3.ch010

INTRODUCTION

AIoT is a paradigm shift that signifies the fusion of two influential technologies, Artificial Intelligence (AI) and the Internet of Things (IoT). AIoT makes IoT devices sharper, more responsive, and capable of advanced data processing and decision-making through AI algorithms and machine learning techniques. IoT devices can analyze data, adapt to changing circumstances, and make intelligent, autonomous decisions without human intervention (Ye et al., 2023; Kuzlu et al., 2021). AIoT systems are designed to integrate seamlessly with IoT devices, sensors, and networks, allowing them to comprehend context, learn from data, and progress over time. This convergence of AI and IoT technologies is at the core of AIoT, and it can potentially revolutionize multiple domains, including smart cities, healthcare, manufacturing, and agriculture (Adli et al., 2023; Hung, 2021). The relevance of AIoT lies in its capacity to drive substantial progress and revolutionize diverse domains. In manufacturing, agriculture, and logistics industries, AIoT systems can optimize processes, predict and prevent faults, and automate duties, resulting in greater productivity. By enabling IoT devices to analyze and process data, AIoT allows organizations and individuals to make informed decisions in real-time, resulting in enhanced resource management and outcomes.

AIoT can substantially reduce operational costs and improve resource efficiency through predictive maintenance, energy management, and optimization. AIoT enables innovative and customized services (Kuzlu et al., 2021). For instance, it can facilitate remote patient monitoring and personalized treatment plans in healthcare. IoT addresses environmental challenges by optimizing energy consumption, reducing waste, and promoting sustainability in smart cities and industries. The integration of Artificial Intelligence of Things (AIoT) can enhance individuals' quality of life by providing cutting-edge and user-friendly solutions for everyday chores, such as smart homes and wearable gadgets. Integrating AIoT into various industries can stimulate economic growth by fostering innovation and generating new business opportunities. AIoT also raises significant concerns regarding data privacy, security, ethics, and the potential for job displacement, making it a subject of substantial societal and policy debates.

AIoT, or Artificial Intelligence of Things, is a profoundly impactful technology movement that can revolutionize industries, improve our lifestyles and productivity, and address intricate challenges. Its significance in the technological landscape is highlighted by its multifaceted effect on society and the economy. Artificial Intelligence of Things (AIoT) is a multidimensional field that encompasses a variety of domains and draws from numerous disciplines. AIoT's multidisciplinary nature reflects the diversity of its components and effects. AIoT is founded on the Internet of Things, which includes sensors, actuators, and communication networks. AI techniques such as machine learning, deep learning, and natural language processing are crucial for enhancing the intellect and decision-making capabilities of AIoT. AIoT generates enormous data, necessitating expertise in managing and analyzing large data sets. Decision-making in AIoT applications requires data-driven insights (Ye et al., 2023; Adli et al., 2023). The development of AI and IoT-integrated applications requires software engineering expertise. IoT systems must address security and privacy concerns, rendering cybersecurity expertise essential. AIoT's foundational aspect is designing and engineering IoT devices and sensors. Understanding the economic implications of AIoT applications and devising viable business models are imperative.

It is assessing the AIoT adoption's cost-effectiveness and financial benefits. Examining how individuals and businesses embrace and implement AIoT technologies. It considers the AIoT's ethical and societal implications, such as data ownership and privacy, designing user-friendly and intuitive AIoT interfaces (Adli et al., 2023; Wynsberghe, 2021) and understanding how AI and its applications affect human emotions and social interactions. It successfully navigated the organizational changes necessary for AIoT implementation and developed data management policies and procedures for AIoT ecosystems. It assessed the environmental impact and possible benefits of AIoT for sustainable solutions. The Internet of Things has implications for remote patient monitoring, medical diagnostics, and health administration. It is understanding the legal and regulatory environment surrounding data privacy and security and considering issues associated with AIoT patents and intellectual property rights. AIoT contributes to creating intelligent and more efficient urban environments, which requires urban planning and architecture expertise. AIoT's multidisciplinary nature emphasizes its all-encompassing impact on society, technology, and business. Collaboration across these diverse disciplines is essential for maximizing the AIoT's potential and addressing the challenges and opportunities it presents. This multidisciplinary approach is necessary for comprehending and harnessing the potential of AIoT for organizational transformation and productivity.

ECONOMIC THEORIES IN AIoT

When companies consider adopting AIoT (Artificial Intelligence of Things) technologies, they rely on various economic models and frameworks to make informed decisions. These models and frameworks provide structured methods for evaluating the economic viability and repercussions of implementing AIoT solutions. Rogers' model of the diffusion of innovations categorizes individuals who accept innovations into distinct groups: innovators, early adopters, early majority, late majority, and laggards. Understanding where an AIoT solution lies on this adoption curve facilitates targeting the appropriate audience and optimizing adoption strategies (Frei-Landau et al., 2022). CBA is a fundamental economic framework for evaluating the costs and benefits of AIoT adoption. It helps organizations quantify the financial implications by comparing the costs (investment, maintenance, operational expenses) against the anticipated benefits, which include increased efficiency, cost savings, and revenue generation (Vagdatli & Petroutsotou, 2022; Ortega, 2020). Return on Investment (ROI) is a commonly employed financial metric for calculating the ROI of AIoT projects. It compares the net profit or advantage garnered to the initial investment. A positive ROI is a crucial indicator of the economic viability of the undertaking.

Total Cost of Ownership (TCO) is a model that incorporates all direct and indirect costs associated with implementing and operating AIoT systems. It offers a comprehensive view of the financial implications throughout the system's lifecycle (Gadatsch, 2023). Innovation Diffusion Theory investigates the dissemination of innovations and the diffusion of AIoT solutions across markets and industries. It considers perceived innovation attributes, communication channels, and the influence of the social system on adoption (Baskerville et al., 2014). Economic Impact Assessments estimate the economic impact of AIoT projects on local economies, industries, and employment markets using financial models and frameworks. They aid policymakers and other parties in comprehending the broader economic implications. The Resource-Based View (RBV) framework highlights the significance of an organization's internal resources, capabilities, and competencies in determining its competitive advantage. It can be used to evaluate how adopting AIoT improves a company's resource base and competitiveness (Seriki, 2020; Nagano, 2020; Taher, 2011). Technology Acceptance Models (e.g. TAM and UTAUT) investigate user acceptance and adoption of technology, considering factors such as perceived ease of use and utility. Understanding these factors is essential for effectively implementing AIoT solutions (Archi & Benbba, 2023).

In conjunction with market research, economic models and frameworks are frequently employed to evaluate the demand for AIoT solutions, competitive forces, pricing strategies, and market positioning. The adoption of AIoT may also be affected by government policies, standards, and regulations, which can affect the economic factors associated with data privacy, security, and compliance. These financial models and frameworks provide a structured and systematic evaluation of the adoption of AIoT's costs, benefits, and economic implications. They are essential for organizations and decision-makers attempting to integrate AIoT technologies into their operations and strategies while making informed decisions. Cost-benefit analysis (CBA) and return on investment (ROI) are essential economic instruments used to evaluate the financial viability and implications of implementing AIoT (Artificial Intelligence of Things) solutions. They assist organizations in making informed judgments regarding adopting AIoT by quantifying the costs and anticipated benefits (Vagdatli & Petroutsotou, 2021).

A cost-benefit analysis is a methodical technique for assessing the economic viability of AIoT projects. It entails comparing the costs of implementing and operating AIoT systems to the anticipated returns. This cost includes acquiring AIoT hardware, software, sensors, and required infrastructure. Maintenance, data storage, energy consumption, and employee training are ongoing costs. AIoT can expedite processes, decrease downtime, and boost productivity. AIoT can generate new revenue streams, such as through the monetization of data or the provision of value-added services. Attribute monetary values to each expense and advantage. Assess the current and future worth of cash flows by applying time-value-of-money concepts. Present Value (NPV) is calculated by subtracting a project's total expenditures from its benefits. A positive NPV indicates the project's financial viability. Changes in assumptions, such as implementation costs or market conditions, can substantially affect a project's economic viability (Gaspars-Wieloch, 2017).

Return on Investment (ROI) is a financial metric that measures the profit or return generated relative to the initial investment in AIoT. In this context, Net Profit represents the total benefits (revenue, cost savings, and increased efficiency) minus the total costs (initial investment and operating expenses). AIoT benefits may accrue over time, and ROI can fluctuate as the system matures. Therefore, the timeframe for calculating ROI must be determined, considering the expected project lifecycle. Identifying the risks and uncertainties associated with IoT adoption and comprehending how they impact ROI is essential. ROI is a valuable instrument for comparing the financial performance of various AIoT projects or investments, thereby assisting with initiative prioritization. Organizations frequently employ industry benchmarks to determine if the ROI of their AIoT initiatives is competitive or needs to meet expectations. Post-implementation monitoring and evaluation of actual benefits and costs are required to ensure that the projected ROI matches reality. Cost-benefit analysis (CBA) and Return on Investment (ROI) are essential instruments for evaluating the financial aspects of AIoT adoption. These analyses aid organizations in making data-based decisions, effectively allocating resources, and ensuring that AIoT initiatives deliver the anticipated economic benefits.

In Artificial Intelligence of Things (AIoT), market dynamics and competitive advantage are essential factors. Understanding the dynamics of AIoT ecosystems and how to obtain a competitive advantage within them is necessary for organizations seeking success in this emerging field. The AIoT market can be segmented by industry, application, or geographic region. Understanding the specific market segments and their characteristics is necessary for market entry and positioning. Evaluate the degree of competition in AIoT markets. Is it a highly competitive environment with numerous participants, or is disruptive innovation possible? The competitive environment can impact pricing, innovation, and market strategy. Consider whether the AIoT ecosystem consists of a small number of dominant actors (oligopoly) or a large number of small players (perfect competition). Market concentration can influence pricing and market leverage. Positive network effects occur in the IoT when the ecosystem's value increases as more participants join. It can result in network effects in which larger ecosystems become more attractive, making it difficult for new entrants to enter the market. Understanding network effects can be essential for positioning strategies. AIoT solutions frequently enjoy economies of scale, in which the cost per unit of service or product decreases as the scope of operations increases. This concept is essential for effective cost management and market penetration.

The Porter's Five Forces framework is excellent for studying the competitive dynamics within AIoT ecosystems. The analysis considers the relative influence of suppliers and buyers, the potential for new competitors to enter the market, the presence of alternative products or services, and the level of rivalry within the sector. Organizations can devise strategies to counteract these forces effectively. Evaluate the entry barriers in the AIoT ecosystem, such as high initial investment, regulatory impediments, and proprietary technology. Understanding these obstacles can affect market entry strategies. Innovation is a significant factor in AIoT competitive advantage. Organizations that develop cutting-edge AIoT technologies, novel use cases, and services with added value are better positioned to obtain a competitive advantage. Differentiated AIoT products or services, such as customized solutions or distinctive features, can help a business stand out in a competitive marketplace. Companies can generate new revenue streams using the data generated by AIoT ecosystems. Data monetization is frequently a source of competitive advantage.

AIoT ecosystems may include device manufacturers, software developers, providers of data analytics, and service providers, among others. To identify partnership opportunities and competitive advantages, it is essential to comprehend the function and position of each entity within the ecosystem. The IoT is governed by various regulations concerning data privacy and security. Understanding and adhering to these regulations is crucial for maintaining market dynamics and competitive advantage. After implementing AIoT, continually assess performance and adapt to changing market conditions. It may involve modifying strategies, expanding services, or confronting new obstacles. Staying attuned to market dynamics and positioning strategically within AIoT ecosystems are crucial for organizations seeking a competitive edge in this swiftly evolving field. Continuous evaluation, innovation, and adaptability are essential success factors in the AIoT landscape.

BEHAVIORAL SCIENCE PERSPECTIVES IN AIoT

Understanding how AIoT (Artificial Intelligence of Things) systems influence human behaviour and decision-making is essential for designing user-friendly and effective AIoT solutions. People tend to search for information that validates their beliefs or expectations. It can manifest in AIoT systems as users favouring data or insights that accord with their preconceived notions. Individuals may rely significantly on the first information they encounter when making decisions. It can influence user decisions in the IoT when presented with initial data or AI-generated suggestions. Users may place excessive faith in AI recommendations, assuming the system is always accurate. It could contribute to an apathetic acceptance of decisions generated by artificial intelligence. Due to a lack of trust or comprehension of AI systems, users may underutilize or abuse AI capabilities. They may favour manual procedures because they have more faith in human judgment. In AIoT, consumers frequently have the option to override decisions made by AI. It is essential to comprehend when and why users disregard AI recommendations. It may take time for users to calibrate their Trust in AI. They understand the factors contributing to trust calibration, such as system performance and user experience. Complex IoT systems may impose a cognitive burden on end users. UX design is informed by behavioural science to reduce the cognitive burden and enhance user comprehension. Effective feedback mechanisms in AIoT systems can aid users in comprehending how their actions affect the system's behaviour and instilling confidence. Behavioural science enables designers to consider user emotions and satisfaction, ensuring that AIoT systems elicit positive emotional responses. A crucial factor is the perception of control and autonomy. Users may favour automation for routine or time-consuming tasks while retaining control over certain aspects of AIoT systems. When the decision-making process is transparent, users typically feel more at ease with AI recommendations—enhancing user confidence by revealing how the AI arrived at its recommendations.

Understanding these behavioural aspects of AIoT systems is crucial for designing interfaces and experiences corresponding to human preferences and decision-making patterns. AIoT solutions can be more effective and user-centric if they account for cognitive biases, foster Trust, and optimize user experiences. In addition, studying the evolution of human behaviour and decision-making as users acquire experience with AIoT can inform the ongoing refinement of these systems. AIoT (Artificial Intelligence of Things) technologies depend on user acceptability and adoption to be successful. Organizations and developers must comprehend the factors that influence the adoption of AIoT solutions by individuals and organizations.

The Technology Acceptance Model (TAM) was developed by Davis in 1989. According to this model, the intention of users to use a technology is determined by two primary factors: perceived simplicity of use and perceived usefulness. This model helps assess how users perceive AIoT systems and their willingness to implement them (Ursavaş, 2022; Maranguni & Grani, 2014) in the context of AIoT. The Unified Theory of Acceptance and Use of Technology (UTAUT) is an expanded version of the Technology Acceptance Model (TAM) incorporating several aspects that impact technology adoption. The components of this include anticipated performance, anticipated effort, social influence, and enabling conditions. This model can be utilized to enhance understanding of the adoption of AIoT technologies (Dwivedi et al., 2017). Users will be more likely to adopt AIoT technologies if they are intuitive and straightforward to interact with.

Designing intuitive and user-friendly interfaces can improve the perception of usability. Users must perceive the utility of AIoT systems. It is essential to demonstrate how AIoT can benefit their lives, increase productivity, or solve specific problems. In AIoT systems, user confidence in the dependability and capabilities of AI components is crucial. Building trust requires openness in decision-making, consistency in performance, and predictability (Yang & Wibowo, 2021). The operation of IoT systems must be reliable and consistent. Users are more likely to embrace AIoT technologies if they consistently perform as anticipated. Identifying behavioural barriers to adoption, such as dread of job loss or resistance to change, is essential. Training, change management, or communication strategies may be required as behavioural interventions to address these obstacles. Continuous user feedback is critical. AIoT systems should be designed to collect user feedback and adapt to user preferences and requirements. AIoT solutions should evolve via iterative development, integrating user feedback and enhancements to increase acceptance. Implementing AIoT technologies on a smaller scale enables users to become accustomed to the systems. Acceptance can be increased by positive pilot program experiences (Ahmetoglu et al., 2021).

Ensuring users are confident in their interactions with AIoT systems, adequate training and education programs can reduce user apprehension and increase acceptance. Peer, colleague, and social network influence can impact user acceptance. Positive recommendations and shared experiences can encourage adoption. Diverse cultures may have diverse perspectives on technology. Understanding cultural nuances is essential for customizing AIoT solutions for particular regions or demographics. Compliance with data privacy regulations and ethical standards is critical for establishing user confidence. When users perceive their data is handled responsibly, they are more likely to accept AIoT systems. Understanding and addressing these factors is essential for promoting the acceptance and successful adoption of AIoT technology. Organizations can increase the likelihood that individuals and organizations will embrace AIoT solutions by devising user-centric systems, providing training and support, and fostering Trust (Telliolu, 2021; Lin et al., 2021).

Ethical considerations and user trust are crucial in designing and implementing AIoT (Artificial Intelligence of Things) systems. Trust and ethical conduct are essential to ensure users feel safe and comfortable interacting with these technologies (Reinhardt, 2022; Choung et al., 2022; Hagendorff, 2020; Ryan, 2020). In the realm of AIoT (Artificial Intelligence of Things) systems, where extensive user data collection and processing is the norm, robust data protection measures, encryption, and the solicitation of user assent are crucial for maintaining user confidence. It is essential to clarify data ownership and utilization rights so that users know who owns the data, how it will be used, and who can access it. Giving users control over their data, including the ability to opt-out or delete it, is a crucial trust-building measure.

Fostering user comprehension of the functionality of AI algorithms and the reasoning behind specific decisions is crucial, highlighting the significance of transparent algorithms and decision-making processes. To foster confidence, AI systems must be designed to provide transparent justifications for their decisions, enabling users to comprehend the underlying logic. AIoT systems must undergo exhaustive testing for biases that may lead to discriminatory outcomes, necessitating robust bias detection and mitigation strategies to ensure equity. The principle of impartiality should be woven into the fabric of AIoT systems, encompassing training data, algorithms, and decision-making processes.

Defining roles and responsibilities for AIoT system stakeholders is essential for empowering users and ensuring they know where to seek redress for problems or concerns. Liability should be determined in the event of AI-generated errors or damage caused by AIoT systems, giving users recourse to adverse outcomes. Consent must be assiduously obtained for data collection and AI interactions, with users fully aware of how their data will be utilized and processed. Encouraging user participation in decision-making and customization of AIoT interactions, allowing them to configure preferences and affect system behaviour, can significantly increase Trust. Compliance with the ethical frameworks and codes of conduct that govern the design and operation of AIoT systems is essential. Establishing internal ethics committees or consulting with external experts to evaluate ethical implications is crucial. Educating users about AIoT technologies, their inner workings, and their capabilities and limitations is a trust-building measure because educated users are more likely to interact confidently with AIoT systems. Consideration of third-party auditing to evaluate the ethical practices and standards conformance of AIoT systems is a proactive measure. Implementing mechanisms for users to provide feedback on AIoT interactions to utilize this feedback for continuous development is of great value. In the end, integrating ethical considerations into the development and operation of AIoT is a commitment that must evolve in tandem with technology development. AIoT system developers and organizations can provide a positive and secure user experience by emphasizing ethical considerations and nurturing Trust. Ethical

AIoT interactions, in which data is handled responsibly, decisions are transparent, and users have control, are more likely to acquire and maintain user trust, thereby increasing the acceptance and adoption of these technologies.

TECHNOLOGICAL MODELS FOR AIoT

AIoT (Artificial Intelligence of Things) system architecture and protocols facilitate efficient data communication, processing, and interoperability. IoT architectures define an IoT system's structure's components, layers, and interactions. OSI is a standard reference architecture that divides IoT systems into physical, data link, network, and application layers. It is essential to comprehend IoT protocol frameworks, such as the TCP/IP protocol suite, and their function in facilitating data exchange between IoT devices and the cloud.

Edge computing facilitates data processing close to the source, minimizing latency and enabling instantaneous decision-making. Popular architectures for peripheral computing include edge servers, fog computing, and local processing nodes. Edge devices, which are frequently resource-constrained, require lightweight communication protocols such as MQTT (Message Queuing Telemetry Transport) and CoAP (Constrained Application Protocol) (Talebkhah et al., 2020). MQTT (Message Queuing Telemetry Transport) is a common, lightweight, and efficient protocol IoT and AIoT devices use to publish and subscribe to data. It is well-suited for environments with limited bandwidth and unreliable networks (Cao et al., 2020).

HTTP (Hypertext Transfer Protocol) and its secure version, HTTPS, transmit data between devices and servers in web-centric applications. CoAP (Constrained Application Protocol) is a lightweight UDP-based protocol designed for IoT devices with limited resources, making it suitable for AIoT applications (Jung et al., 2020; NkonoB, 2020). Fog computing expands the functionalities of the cloud to the outer edges of the network. Cisco's IOx and OpenFog are frameworks that define how data can be processed and acted upon at the network periphery. (Dizdarevi et al., 2019) Fog computing relies on communication protocols such as MQTT, HTTP, and standard IoT communication protocols to facilitate seamless data flow between edge and cloud resources. AIoT systems require data integration and analytics frameworks such as Hadoop, Spark, and Apache Flink to process and analyze large volumes of data generated by IoT devices. AIoT systems should include machine learning models and frameworks like TensorFlow and PyTorch for data analysis and predictive modelling. Adherence to interoperability standards such as IoTivity, oneM2M, and OCF (Open Connectivity Foundation) enables diverse AIoT devices and platforms to communicate and interoperate without difficulty (Lemus et al., 2022; Chen & Wu, 2021; Adi et al., 2020; Vermesan et al., 2020).

In constrained IoT environments, AIoT systems require robust security protocols, such as SSL/TLS for encrypted communication, OAuth for secure access control, and DTLS (Datagram Transport Layer Security) for secure data transmission. In AIoT systems, Blockchain Technology can be utilized for data integrity, traceability, and secure transactions, particularly in supply chain management and healthcare applications. Identity and Access Management (IAM) solutions ensure that only authorized users and devices can access the IoT system, improving security (Paris et al., 2023; Shen, 2023). Secure Boot and Updates to Firmware To maintain the integrity of AIoT devices and prevent vulnerabilities, specific launch processes and over-the-air firmware updates are essential. These architectural frameworks and protocols serve as the structural backbone for IoT systems, enabling them to function efficiently, securely, and in a manner that promotes interoperability and data exchange between devices and platforms. Choosing the appropriate architecture and protocols for each AIoT application depends on its requirements and constraints. AIoT (Artificial Intelligence of Things) systems are built around effective data management and analytics. These processes entail collecting, storing, processing, and analyzing the immense amounts of data generated by IoT devices to derive meaningful insights and facilitate informed decision-making (Razzaq, 2020; Gardaevi et al., 2016).

Effective data management and analytics are essential to maximizing the utility of AIoT (Artificial Intelligence of Things) applications, especially in light of the abundance of data produced by IoT devices. It begins with identifying and integrating data from various IoT devices, sensors, and sources, both at the periphery and in the cloud, while accommodating structured, semi-structured, and unstructured data formats. Implementing data ingestion mechanisms ensures the efficient, dependable, and secure accumulation of data, including batch processing, real-time streaming, and data transformation.

Data cleansing techniques correct errors, eliminate duplicates and improve data quality. Validation procedures are required to confirm the precision and completeness of the data. To accommodate the volume and variety of AIoT data, robust data storage solutions, such as databases, data lakes, cloud storage, or hybrid systems, are employed. Time-series databases' efficient storage and retrieval capabilities are advantageous to time-stamped and structured IoT data. Particularly for large datasets, data can be partitioned or sharded to optimize storage and retrieval performance. Preparing data for analysis entails converting it into a format suitable for AI algorithms, which includes data normalization, aggregation, and feature engineering. Dimensionality reduction techniques improve processing efficiency without compromising vital data. Big data technologies and analytics platforms like Hadoop, Spark, and Apache Flink efficiently process and analyze large volumes of AIoT data. Data processing and real-time analytics are essential for AIoT applications that require immediate insights and decisions. Machine learning models are developed and deployed to process and interpret AIoT data for predictive and prescriptive analytics, with tools such as TensorFlow and PyTorch instrumental in model development and deployment.

Establishing data governance policies and procedures is essential for ensuring data's ethical and compliant management. To protect AIoT data from unauthorized access and intrusions, stringent data security measures, such as encryption, access control, and audit logs, are implemented. Compliance with privacy regulations is a pillar of data governance, and user data must be handled carefully. Using data visualization tools and techniques enhances data interpretation by presenting insights and findings in a plain, actionable manner. User-friendly interfaces are created so that users can visually interact with and investigate AIoT data. A monitoring system is implemented to measure the performance and health of AIoT data pipelines, with anomaly and problem alerts configured. Data management and analytics processes are continuously optimized to increase efficiency and decrease latency. These data management and analytics considerations are essential for deriving actionable insights from the abundance of AIoT data, thereby supporting decision-making across various applications, including predictive maintenance, healthcare monitoring, and smart cities.

PSYCHOLOGICAL FACTORS IN AIoT

Understanding cognitive models and human-computer interaction (HCI) in the context of AIoT (Artificial Intelligence of Things) is essential for developing user-friendly and effective systems. Cognitive models enable us to comprehend how users perceive and make decisions when interacting with AIoT devices and systems (Schürmann & Beckerle, 2020; Howes, 1995). Users frequently have mental models or preconceived conceptions regarding the functionality of AIoT systems. Align AIoT system designs with these expectations to reduce cognitive dissonance. Important is system behaviour consistency. When AIoT systems behave predictably, users can construct more accurate mental models and make better decisions. Consider the cognitive burden imposed on users when AIoT systems require decisions or provide feedback. Systems should present information following the cognitive abilities of users (Hay et al., 2020; Marikyan et al., 2020).

The design process should incorporate several crucial factors to create a seamless user experience in AIoT (Artificial Intelligence of Things) systems. It is essential to minimize response time, as delays can disrupt users' thought processes and contribute to frustration. Additionally, feedback should be plain and understandable, avoiding ambiguity and mental strain. A user-centered design philosophy should underpin AIoT system development. This strategy incorporates iterative design, user feedback, and a commitment to meeting the requirements and preferences of the user. Usability testing is indispensable for determining how well users interact with AIoT systems, identifying usability issues, and iteratively refining the design. Creating intuitive and user-friendly user interfaces is essential to considering human cognitive patterns and visual perception. Accessibility is also of the utmost importance, ensuring that AIoT systems are usable by individuals with disabilities via screen readers and voice control interfaces. Recognizing that users may desire varying degrees of control over AIoT systems, the design should account for these preferences while preserving the autonomy and intelligence of the system. Providing users with explicit explanations of how the AIoT system operates and makes decisions increases user trust and comprehension. Lastly, allowing users to override AI-generated decisions and provide feedback promotes a feeling of control and active participation in the AIoT system's operation. These factors contribute to an AIoT design that is user-centric, efficient, and user-friendly (Kübler et al., 2014).

Incorporating cognitive models and Human-Computer Interaction (HCI) principles into AIoT (Artificial Intelligence of Things) design is essential for creating systems that seamlessly align with human cognition, resulting in enhanced user experiences and reduced cognitive burden. It ultimately contributes to the widespread adoption of AIoT technologies by fostering positive interactions. In AIoT applications, user experience (UX) and emotional responses are paramount. An emotionally engaging design requires elements that evoke positive emotions, such as colours, visuals, and micro-interactions. Incorporating feedback mechanisms that recognize user interactions ensures that users feel heard and valued, which can significantly affect their emotional responses. In applications such as AIoT healthcare, where the dynamic conditions of patients and caregivers must be accounted for, an empathic approach to design is essential, considering users' emotional well-being. Prioritizing a user-centred design approach to customize AIoT applications to user requirements and preferences increases user satisfaction and emotional comfort. Usability testing is essential for identifying and resolving problem points in the user experience, increasing user satisfaction and decreasing user frustration. The harmonious integration of cognitive models and HCI principles into AIoT design, along with a focus on user experience and emotional responses, plays a crucial role in shaping AIoT systems that are intuitive, user-friendly, and conducive to positive interactions, thereby enhancing their adoption and success (Kurosu, 2020).

Ensuring that AIoT (Artificial Intelligence of Things) applications provide a gratifying and emotionally positive user experience necessitates a multifaceted strategy incorporating several crucial factors. Users are more likely to be satisfied if the system's performance is consistent and efficient. Exploring the potential of emotion recognition within AIoT applications will enhance the user experience by enabling the system to comprehend user emotions and tailor responses accordingly, resulting in more personalized and empathetic interaction. Valuable encourages users to actively provide feedback on their emotional responses and experiences with the system, as this feedback can inform ongoing system enhancements. Implementing effective computing techniques, in which AIoT systems detect and respond to user emotions in real time, ensures a more empathic and user-centric interaction, enhancing the user experience's emotional component.

Accessibility is essential, ensuring that AIoT applications are accessible to people with visual or auditory impairments. Simplified user interfaces reduce cognitive burden and simplify user interactions, enhancing the overall user experience. Context-aware design is essential for AIoT applications, allowing the system to adapt its behaviour based on user activities and preferences, improving its responsiveness to user requirements. By empowering users to customize the AIoT system to their preferences, customization promotes a more positive emotional response and a sense of control over their interactions. It is of the utmost importance to address user privacy and consent concerns, as transparency regarding data usage and secure data handling practices can positively influence emotional responses and establish Trust. In addition, it is essential to provide user support via chatbots, virtual assistants, or human assistance when necessary, as users who feel supported and resolve issues quickly are more likely to experience positive emotions. Effective resolution of user issues can convert negative emotional responses into positive ones, resulting in a more satisfying experience. To achieve a balance between user experience and emotions in AIoT applications, a comprehensive strategy incorporating design principles, user feedback, accessibility, and ethical considerations is required. Addressing these factors, AIoT applications can elicit positive affective responses, resulting in greater user satisfaction and engagement.

Incorporating AIoT (Artificial Intelligence of Things) into our existence has far-reaching effects on individuals' physical and mental health. Understanding these implications is essential for developing AIoT systems that foster positive outcomes and mitigate potential hazards (Murphy et al., 2021; De Kock et al., 2021). AIoT (Artificial Intelligence of Things) applications, particularly in healthcare, have the potential to significantly enhance overall health by providing individuals with valuable health insights and facilitating timely interventions. It is illustrated by wearable fitness and remote patient monitoring, which empower users to manage their health. As exemplified by smart home automation systems that enhance convenience, AIoT devices extend their benefits to daily life by streamlining tasks, reducing tension, and saving time. In addition, AIoT systems play a crucial role in enhancing physical and digital security, mitigating concerns regarding personal safety and data privacy. It is essential to recognize that the constant data collection by AIoT devices may raise privacy concerns, potentially causing individuals stress and anxiety if they believe their personal information is being misused. Security breaches and vulnerabilities in AIoT systems can negatively affect mental health by undermining confidence and causing distress. Excessive reliance on AIoT devices for decision-making and daily tasks may diminish an individual's self-efficacy and resiliency, negatively impacting mental health. It is essential to strike a balance between AI assistance and user control if users are to retain a sense of autonomy over their decisions and AIoT interactions. Transparency regarding data acquisition and utilization and obtaining informed consent can help individuals feel in control and reduce anxiety. There is the potential for IoT applications to provide mental health monitoring and interventions. Algorithms powered by artificial intelligence can detect signs of depression or anxiety and provide reasonable support. Stress management tools, mood monitoring, and guided meditation aim to improve mental health. In healthcare and mental health contexts, AIoT systems can detect emergencies, such as falls or sudden health deterioration, and trigger immediate assistance.

To ensure a positive and realistic experience with AIoT systems, educating users about their capabilities and limitations is crucial to prevent unrealistic expectations and reduce tension resulting from miscommunications. AIoT applications have the potential to have a significant impact on individual well-being, both in healthcare and daily life, with privacy, security, user autonomy, and mental health playing crucial roles in realizing their full potential (MilneIves et al., 2022; Dekker et al., 2020). Ensure that users know their rights and control over their data within AIoT systems, reducing their apprehension regarding data usage. In addition to AIoT interactions, encourage human-social interactions. Lack of social interaction can have detrimental effects on mental health. Encourage AIoT user communities in which people can share their experiences and find support. A multidisciplinary approach that considers technological, psychological, and ethical factors is required to comprehend the effects of AIoT on well-being and mental health. A balanced integration of AIoT into daily life can promote positive outcomes while mitigating potential risks and challenges, thereby contributing to individuals' well-being and mental health.

ORGANIZATIONAL THEORIES IN AIoT IMPLEMENTATION

Adopting AIoT (Artificial Intelligence of Things) technologies necessitates a well-structured, multi-component approach. Ahmetoglu et al. (2022) AIoT adoption begins with selecting a change management framework or model such as John Kotter's 8-Step Model, ADKAR, or Lewin's Change Management Model. It is crucial to define specific objectives and goals for AIoT integration and to ensure that these goals are effectively communicated throughout the organization (Adli et al., 2023; Hangl et al., 2022). To successfully execute this strategy, a dedicated team must spearhead the AIoT adoption process. This team should consist of champions and advocates of change who will play essential roles in facilitating communication and implementation. In addition, a comprehensive evaluation of the organization's current infrastructure is required to determine its compatibility with AIoT technologies and identify any voids requiring attention. The culture of the organization is another crucial factor. Potential change barriers can be identified by analyzing the prevalent culture, and the adoption strategy can be modified accordingly. Evaluating employees' skills and knowledge allows for identifying any AIoT-related skill and knowledge gaps, and training programs should be devised to address these deficiencies.

Potential risks and challenges associated with IoT adoption, such as cybersecurity threats, data privacy concerns, and regulatory compliance issues, should be transparently identified and addressed with employees and stakeholders. The participation of key stakeholders in the planning and decision-making processes is essential for a seamless transition, and employees are kept informed and actively involved through various means, including town hall meetings, feedback mechanisms, and regular updates. Focusing on AIoT technologies, data analytics, and related topics, training programs should be developed to improve the skills and knowledge of employees. Encouraging cross-functional teams and collaboration enriches the AIoT knowledge of employees. Identifying and empowering change champions to inspire and motivate their peers to embrace AIoT with the necessary tools and resources to advocate for its adoption and provide support is essential. Initiating small-scale AIoT prototype projects is also essential, as they enable testing in the real world and provide valuable feedback before full-scale deployment. Integral to the success of the AIoT adoption process are the lessons learned from the pilot projects, any necessary adjustments, and the application of the insights garnered to subsequent phases.

Adopting AIoT (Artificial Intelligence of Things) technologies necessitates a well-structured strategy that includes change management, ethical considerations, leadership, governance, and organizational learning. Strong leadership that defines a clear vision of how AIoT can transform the organization and aligns this vision with the larger strategic objectives forms the basis of this successful transition. A governance framework defines roles, responsibilities, and accountability for AIoT initiatives while incorporating ethical guidelines to ensure data integrity, privacy, and security.

Managing the enormous amount of data generated by AIoT (Artificial Intelligence of Things) devices requires effective data governance. Organizations implement exhaustive data governance policies covering various critical aspects to ensure responsible and effective management of this data. These policies address data quality by establishing standards to preserve the precision and dependability of data. Establishing access control mechanisms to determine who can access and modify data ensures the security and integrity of data. Data lifecycle management is essential for tracing data from creation to destruction, enabling organizations to make informed decisions regarding data retention and destruction.

Similarly, organizations must prioritize privacy and personal data protection. Privacy protection mechanisms ensure that personal information is handled with the utmost care and respect and that informed consent practices are upheld. Identifying potential vulnerabilities, such as data breaches and security dangers, relies heavily on risk assessments. Organizations can address these issues proactively and bolster their cybersecurity measures by conducting these assessments. Collaboration between various departments, such as IT, data science, operations, and business divisions, is essential for seamlessly integrating AIoT technologies into the organization's infrastructure. This collaboration facilitates improved decision-making and AIoT application optimization in addition to ensuring alignment with organizational objectives (Brous et al., 2020). Effective data governance is a multifaceted process encompassing data quality, access control, lifecycle management, privacy protection, informed consent, risk assessment, and collaboration, all necessary for successfully managing AIoT-generated data and leveraging it to achieve organizational objectives. Organizations can acclimate to the AIoT era more effectively and maximize its benefits by adopting these elements.

CASE STUDIES

Case Study 1: Economic Implications of AIoT Implementation in the Manufacturing Sector

A manufacturing company intended to revolutionize its operations by deploying AIoT (Artificial Intelligence of Things) technology throughout its production facilities. The company embarked on a transformational journey to increase efficiency, reduce costs, and obtain a competitive edge. By integrating AIoT sensors and predictive maintenance systems into its machinery, the company ensured continuous monitoring of equipment conditions, allowing for the prediction of maintenance requirements before the occurrence of critical problems. Concurrently, AI-powered quality control systems were implemented to reduce defects and optimize production processes.

This comprehensive AIoT integration produced remarkable results. The predictive maintenance system substantially reduced downtime, resulting in maintenance cost savings of millions of dollars. Automation facilitated by IoT-enabled AI ushered in an era of enhanced production line efficiency, resulting in decreased labour costs and increased output. Reduced defects resulted in less material waste and improved product quality, increasing customer satisfaction and repeat business. In the greater scheme of things, the company's enhanced productivity and product quality have positioned it to compete more effectively in the global market. The strategic AIoT implementation in the manufacturing sector had profound economic implications. The company realized significant cost reductions and improved efficiency and quality, strengthening its competitive position. Underscoring the immense impact of AIoT technology on manufacturing operations and economic outcomes, these transformational advances directly translated into increased profitability and robust growth.

Case Study 2: Behavioral Factors Influencing AIoT Adoption in the Healthcare Industry

A healthcare organization began implementing AIoT (Artificial Intelligence of Things) technologies to improve patient care and operational efficiency. The success of this implementation was contingent on healthcare professionals' acceptance of these novel tools. The company implemented IoT devices for remote patient monitoring, data analytics for comprehensive patient record management, and predictive analytics for disease management. Initially, healthcare professionals exhibited understandable reluctance to implement AIoT devices due to concerns that these technologies could replace their roles or jeopardize patient care. To surmount these apprehensions, the organization committed to extensive user training and embarked on a journey of clear communication, emphasizing that AIoT was intended as a supplement and not a replacement.

In addition to these factors, behavioural concerns arose, particularly regarding the ethical utilization of patient data. It was essential to address these concerns, necessitating the development of comprehensive data privacy and consent mechanisms. It was a crucial move in gaining the confidence of both healthcare professionals and patients. To completely integrate AIoT devices into healthcare workflows, healthcare professionals must adjust their routines and procedures. The organization provided support and administration of behavioural change to facilitate the integration process. Ultimately, the organization adopted AIoT technology by addressing these behavioural factors and ensuring healthcare professionals' acceptance. The results were tangible: improved patient care, more accurate diagnoses, and the streamlining of healthcare operations, ultimately demonstrating AIoT's central position in reshaping the healthcare landscape.

Case Study 3: Technological Challenges and Solutions in Deploying IoT in Smart Cities

The smart city program seeks to use AIoT (Artificial Intelligence of Things) technologies in different areas of urban life, including transportation, energy management, and public services, to establish a more inventive and efficient urban environment. The initiative entailed deploying thousands of IoT sensors and AI algorithms throughout the city to collect data on traffic patterns, energy consumption, waste management, and other topics. The overwhelming variety of data sources posed a difficulty, necessitating the establishment of a data hub and standardized data formats to ensure compatibility. In addition, the project's vast scope created scalability issues, which were effectively addressed by utilizing cloud-based solutions and edge computing for efficient data processing and storage. In this technologically advanced environment, protecting data and infrastructure from cyber threats was of the utmost importance, necessitating sophisticated cybersecurity measures, such as encryption and intrusion detection.

The project's complexity was exacerbated by using multiple vendors and technologies, necessitating the development of interoperability standards and application programming interfaces (APIs) to facilitate communication between devices and systems. Despite these technological obstacles, the successful implementation of AIoT in the smart city initiative resulted in significant enhancements to urban planning, traffic management, energy efficiency, and residents' overall quality of life. It transformed the city and served as a model for technological innovation and urban development in other smart cities worldwide.

CONCLUSION

In this investigation of AIoT (Artificial Intelligence of Things), we have investigated its theoretical foundations and underlying models. AIoT represents a transformational convergence of artificial intelligence and the Internet of Things, with enormous potential to reshape industries, boost productivity, and drive organizational transitions. The adoption of AIoT is guided by economic models and frameworks that evaluate costs and benefits, market dynamics, and competitive advantages. These models aid businesses in making informed decisions and comprehending the financial implications of AIoT. Understanding human decision-making and behaviour in AIoT systems is crucial for user acceptance and adoption. Ethical considerations and user trust are critical factors that influence interactions with AIoT technologies. Success in AIoT requires architectural frameworks, data management, security, privacy, and interoperability. These models provide the essential technical foundation for dependable and effective IoT systems.

User experience and cognitive models are integral components of AIoT design. They ensure that AIoT applications are aligned with human requirements and emotions, thereby improving the well-being of users. Models of change management, leadership, and governance allow organizations to navigate the complexities of AIoT implementation. Organizational learning and knowledge management are essential for nurturing an adaptive and innovative culture. IoT can reduce costs, increase productivity, and improve product quality in manufacturing, ultimately leading to a competitive advantage. The adoption of AIoT is dependent on addressing behavioural factors in healthcare. AIoT can enhance patient care and operational efficiency, improving healthcare outcomes. With the proper strategies, technological obstacles such as data integration and scalability can be overcome in smart cities. The Internet of Things has the potential to revolutionize urban living by enhancing transportation, energy efficiency, and overall quality of life. Encourage collaboration between economists, behavioural scientists, technologists, psychologists, and organizational experts to create a comprehensive understanding of AIoT. Focusing on data privacy, consent, and responsible use, develop comprehensive ethical frameworks to regulate AIoT applications. Prioritize user-centered design in AIoT application development to improve user experience and adoption. Change management effectiveness is vital. Leadership should provide AIoT adoption with support, resources, and a defined vision. To facilitate AIoT transitions and innovation, organizations should cultivate a culture of learning and knowledge sharing. Promote ongoing experimentation and failure-based learning to improve AIoT implementations and preserve competitiveness.

AIoT represents a convergence of various theories and models, presenting multiple opportunities. Its successful deployment is contingent on overcoming obstacles, comprehending user behaviour, nurturing a learning culture, and considering ethical ramifications. AIoT promises to reshape industries, enhance well-being, and drive the next wave of organizational transitions as it continues to evolve.

REFERENCES

Adi, E., Anwar, A., Baig, Z. A., & Zeadally, S. (2020). Machine learning and data analytics for the IoT. *Neural Computing & Applications*, *32*(20), 16205–16233. doi:10.1007/s00521-020-04874-y

Adli, H. K., Remli, M. A., Wong, K. N. S. W. S., Ismail, N. A., González-Briones, A., Corchado, J. M., & Mohamad, M. S. (2023). Recent Advancements and Challenges of AIoT Application in Smart Agriculture: A review. *Sensors (Basel)*, *23*(7), 3752. doi:10.3390/s23073752 PMID:37050812

Ahmetoglu, S., Cob, Z. C., & Ali, N. (2022). A Systematic Review of Internet of Things adoption in Organizations: Taxonomy, benefits, challenges and Critical factors. *Applied Sciences (Basel, Switzerland)*, *12*(9), 4117. doi:10.3390/app12094117

Archi, Y. E., & Benbba, B. (2023). The Applications of Technology Acceptance Models in Tourism and Hospitality Research: A Systematic Literature Review. *Journal of Environmental Management and Tourism*, *14*(2), 379. doi:10.14505/jemt.v14.2(66).08

Baskerville, R., Bunker, D., Olaisen, J., Pries-Heje, J., Larsen, T. J., & Swanson, E. B. (2014). Diffusion and Innovation Theory: past, present, and future contributions to academia and practice. In Springer eBooks (pp. 295–300). doi:10.1007/978-3-662-43459-8_18

Brous, P., Janssen, M., & Krans, R. (2020). Data governance as a success factor for data science. In Lecture Notes in Computer Science (pp. 431–442). doi:10.1007/978-3-030-44999-5_36

Cao, K., Liu, Y., Meng, G., & Sun, Q. (2020). An overview of edge computing research. *IEEE Access : Practical Innovations, Open Solutions*, *8*, 85714–85728. doi:10.1109/ACCESS.2020.2991734

Chen, M., & Wu, H. (2021). Real-time intelligent image processing for the Internet of Things. *Journal of Real-Time Image Processing*, *18*(4), 997–998. doi:10.1007/s11554-021-01149-0

Choung, H., David, P., & Ross, A. (2022). Trust and ethics in AI. *AI & Society*, *38*(2), 733–745. doi:10.1007/s00146-022-01473-4

De Kock, J. H., Latham, H. A., Leslie, S. J., Grindle, M., Munoz, S. A., Ellis, L., Polson, R., & O'Malley, C. M. (2021). A rapid review of the impact of COVID-19 on healthcare workers' mental health: Implications for supporting psychological well-being. *BMC Public Health*, *21*(1). Advance online publication. doi:10.1186/s12889-020-10070-3 PMID:33422039

Dekker, I., De Jong, E. M., Schippers, M. C., De Bruijn-Smolders, M., Alexiou, A., & Giesbers, B. (2020). Optimizing students' mental health and academic performance: AI-Enhanced Life Crafting. *Frontiers in Psychology*, *11*, 1063. Advance online publication. doi:10.3389/fpsyg.2020.01063 PMID:32581935

Dizdarević, J., Carpio, F., Jukan, A., & Masip-Bruin, X. (2019). A survey of communication protocols for the Internet of Things and related challenges of FOG and cloud computing integration. *ACM Computing Surveys*, *51*(6), 1–29. doi:10.1145/3292674

Dwivedi, Y. K., Rana, N. P., Jeyaraj, A., Clement, M., & Williams, M. D. (2017). Re-examining the Unified Theory of Acceptance and Use of Technology (UTAUT): Towards a revised theoretical model. *Information Systems Frontiers*, *21*(3), 719–734. doi:10.1007/s10796-017-9774-y

Frei-Landau, R., Muchnik-Rozanov, Y., & Avidov-Ungar, O. (2022). Using Rogers' diffusion of innovation theory to conceptualize the mobile-learning adoption process in teacher education in the COVID-19 era. *Education and Information Technologies*, *27*(9), 12811–12838. doi:10.1007/s10639-022-11148-8 PMID:35702319

Gadatsch, A. (2023). IT Investment Calculation and Total Cost of Ownership Analysis: IT Standards as a tool for IT Controlling. In Springer eBooks (pp. 75–93). doi:10.1007/978-3-658-39270-3_6

Gardašević, G., Veletić, M., Maletić, N., Vasiljević, D., Radusinović, I., Tomović, S., & Radonjić, M. (2016). The IoT architectural framework, design issues and application domains. *Wireless Personal Communications*, *92*(1), 127–148. doi:10.1007/s11277-016-3842-3

Gaspars-Wieloch, H. (2017). Project Net Present Value estimation under uncertainty. *Central European Journal of Operations Research*, *27*(1), 179–197. doi:10.1007/s10100-017-0500-0

Николов, Н. (2020). Research of MQTT, CoAP, HTTP and XMPP IoT Communication protocols for Embedded Systems. *2020 XXIX International Scientific Conference Electronics (ET)*. 10.1109/ET50336.2020.9238208

Hagendorff, T. (2020). The Ethics of AI Ethics: An Evaluation of Guidelines. *Minds and Machines*, *30*(1), 99–120. doi:10.1007/s11023-020-09517-8

Hangl, J., Behrens, V. J., & Krause, S. (2022). Barriers, Drivers, and Social Considerations for AI adoption in Supply Chain Management: A Tertiary study. *Logistics*, *6*(3), 63. doi:10.3390/logistics6030063

Hay, L., Cash, P., & McKilligan, S. (2020). The future of design cognition analysis. Design Science, 6. doi:10.1017/dsj.2020.20

Howes, A. (1995). Cognitive Modelling: Experiences in Human-Computer Interaction. In Springer eBooks (pp. 97–112). doi:10.1007/978-94-011-0103-5_8

Hung, L. (2021). Adaptive devices for AIoT systems. *2021 International Symposium on Intelligent Signal Processing and Communication Systems (ISPACS)*. 10.1109/ISPACS51563.2021.9651095

Jung, J., Gohar, M., & Koh, S. J. (2020). COAP-Based streaming control for IoT applications. *Electronics (Basel)*, 9(8), 1320. doi:10.3390/electronics9081320

Kübler, A., Holz, E. M., Riccio, A., Zickler, C., Kaufmann, T., Kleih, S. C., Staiger-SälZer, P., Desideri, L., Hoogerwerf, E. J., & Mattia, D. (2014). The User-Centered Design as Novel perspective for evaluating the usability of BCI-Controlled Applications. *PLoS One*, 9(12), e112392. doi:10.1371/journal.pone.0112392 PMID:25469774

Kurosu, M. (2020). Human-Computer Interaction. Design and user experience. Lecture Notes in Computer Science. doi:10.1007/978-3-030-49059-1

Kuzlu, M., Fair, C., & Güler, Ö. (2021b). Role of Artificial Intelligence in the Internet of Things (IoT) cybersecurity. *Discover the Internet of Things*, 1(1), 7. Advance online publication. doi:10.1007/s43926-020-00001-4

Lemus, L., Félix, J. M., Fides-Valero, Á., Benlloch-Dualde, J., & Martínez-Millana, A. (2022). A Proof-of-Concept IoT system for remote healthcare based on interoperability standards. *Sensors (Basel)*, 22(4), 1646. doi:10.3390/s22041646 PMID:35214548

Lin, Y., Chen, S., Tsai, C., & Lai, Y. (2021). Exploring computational thinking skills training through augmented reality and AIoT learning. *Frontiers in Psychology*, 12, 640115. Advance online publication. doi:10.3389/fpsyg.2021.640115 PMID:33708166

Marangunić, N., & Granić, A. (2014). Technology acceptance model: A literature review from 1986 to 2013. *Universal Access in the Information Society*, 14(1), 81–95. doi:10.1007/s10209-014-0348-1

Marikyan, D., Papagiannidis, S., & Alamanos, E. (2020). Cognitive dissonance in technology adoption: A study of smart home users. *Information Systems Frontiers*, 25(3), 1101–1123. doi:10.1007/s10796-020-10042-3 PMID:32837263

Milne-Ives, M., Selby, E., Inkster, B., Lam, C. S., & Meinert, E. (2022). Artificial intelligence and machine learning in mobile apps for mental health: A scoping review. *PLOS Digital Health*, 1(8), e0000079. doi:10.1371/journal.pdig.0000079 PMID:36812623

Murphy, K., Di Ruggiero, E., Upshur, R., Willison, D. J., Malhotra, N., Cai, J., Malhotra, N., Lui, V., & Gibson, J. L. (2021). Artificial intelligence for good health: A scoping review of the ethics literature. *BMC Medical Ethics*, 22(1), 14. Advance online publication. doi:10.1186/s12910-021-00577-8 PMID:33588803

Nagano, H. (2020). The growth of knowledge through the resource-based view. *Management Decision, 58*(1), 98–111. doi:10.1108/MD-11-2016-0798

Ortega, B. (2020). Cost-Benefit analysis. In Springer eBooks (pp. 1–5). doi:10.1007/978-3-319-69909-7_600-2

Paris, I. L. B. M., Habaebi, M. H., & Zyoud, A. (2023). Implementation of SSL/TLS Security with MQTT Protocol in IoT Environment. *Wireless Personal Communications, 132*(1), 163–182. doi:10.1007/s11277-023-10605-y

Prabha, C., Mittal, P., Gahlot, K., & Phul, V. (2023). Smart Healthcare System Based on AIoT Emerging Technologies: A Brief Review. In *Proceedings of International Conference on Recent Innovations in Computing* (pp. 299–311). 10.1007/978-981-99-0601-7_23

Razzaq, A. (2020). A Systematic review of software architectures for IoT systems and future direction to adopting a microservices architecture. *SN Computer Science, 1*(6), 350. Advance online publication. doi:10.1007/s42979-020-00359-w

Reinhardt, K. (2022). Trust and trustworthiness in AI ethics. *AI and Ethics, 3*(3), 735–744. doi:10.1007/s43681-022-00200-5

Ryan, M. (2020). In AI, we trust ethics, artificial intelligence, and reliability. *Science and Engineering Ethics, 26*(5), 2749–2767. doi:10.1007/s11948-020-00228-y PMID:32524425

Schürmann, T., & Beckerle, P. (2020). Personalizing Human-Agent interaction through cognitive models. *Frontiers in Psychology, 11*, 561510. Advance online publication. doi:10.3389/fpsyg.2020.561510 PMID:33071887

Seriki, O. (2020). Resource-Based view. In Springer eBooks (pp. 1–4). doi:10.1007/978-3-030-02006-4_469-1

Shen, M. (2023, August 15). Blockchains for Artificial Intelligence of Things: A Comprehensive survey. *IEEE Journals & Magazine.* https://ieeexplore.ieee.org/abstract/document/10105989/authors#authors

Taher, M. (2011). Resource-Based view theory. In Springer eBooks (pp. 151–163). doi:10.1007/978-1-4419-6108-2_8

Talebkhah, M., Sali, A., Marjani, M., Gordan, M., Hashim, S. J., & Rokhani, F. Z. (2020). Edge computing: Architecture, Applications and Future Perspectives. *2020 IEEE 2nd International Conference on Artificial Intelligence in Engineering and Technology (IICAIET).* 10.1109/IICAIET49801.2020.9257824

Tellioğlu, H. (2021). User-Centered design. In Springer eBooks (pp. 1–19). doi:10.1007/978-3-030-05324-6_122-1

Ursavaş, Ö. F. (2022). Technology Acceptance Model: history, theory, and application. In Springer eBooks (pp. 57–91). doi:10.1007/978-3-031-10846-4_4

Vagdatli, T., & Petroutsatou, K. (2022). Modelling Approaches of Life Cycle Cost–Benefit Analysis of Road Infrastructure: A Critical Review and Future Directions. *Buildings*, *13*(1), 94. doi:10.3390/buildings13010094

Van Wynsberghe, A. (2021). Sustainable AI: AI for sustainability and the sustainability of AI. *AI and Ethics*, *1*(3), 213–218. doi:10.1007/s43681-021-00043-6

Vermesan, O., Bahr, R., Ottella, M., Serrano, M. M., Karlsen, T., Wahlstrøm, T., Sand, H. E., Ashwathnarayan, M., & Gamba, M. T. (2020). Internet of Robotic Things intelligent connectivity and platforms. *Frontiers in Robotics and AI*, *7*, 104. Advance online publication. doi:10.3389/frobt.2020.00104 PMID:33501271

Yang, R., & Wibowo, S. (2022). User trust in artificial intelligence: A comprehensive conceptual framework. *Electronic Markets*, *32*(4), 2053–2077. doi:10.1007/s12525-022-00592-6

Ye, L., Wang, Z., Jia, T., Ma, Y., Shen, L., Zhang, Y., Li, H., Chen, P., Wu, M., Liu, Y., Jing, Y. P., Zhang, H., & Huang, R. (2023). Research progress on the low-power artificial intelligence of things (AIoT) chip design. *Science China. Information Sciences*, *66*(10), 200407. Advance online publication. doi:10.1007/s11432-023-3813-8

KEY TERMS AND DEFINITIONS

Artificial Intelligence of Things (AIoT): Integrating artificial intelligence algorithms and techniques into IoT systems to enhance data analysis, decision-making, and automation processes.

Behavioral Science: Interdisciplinary field examining human behaviour, encompassing psychology, sociology, anthropology, and economics, to understand actions, decisions, and interactions.

Economics: Study how societies allocate scarce resources to satisfy unlimited wants and needs, encompassing production, distribution, consumption, and exchange of goods and services.

IoT (Internet of Things): Network of interconnected devices capable of collecting and exchanging data, enabling automation, monitoring, and control of physical objects or environments.

Models: Simplified representations of real-world systems or processes used for analysis, prediction, and problem-solving, aiding comprehension and decision-making.

Organizational Studies: Examination of organizations and their functioning, focusing on structures, processes, behaviours, and strategies to enhance effectiveness, efficiency, and adaptability.

Psychology: Study of the human mind and behaviour, encompassing thoughts, feelings, perceptions, and actions, aiming to understand and explain individual and group dynamics.

Technology: Application of scientific knowledge, tools, and techniques to solve practical problems, improve processes, and enhance human capabilities, driving innovation and progress.

Chapter 11

Trends and Challenges in AIoT Implementation for Smart Home, Smart Buildings, and Smart Cities in Cloud Platforms

V. Santhi

https://orcid.org/0000-0002-4274-7474
Vellore Institute of Technology, India

Yamala N. V. Sai Sabareesh
Vellore Institute of Technology, India

Ponnada Prem Sudheer
Vellore Institute of Technology, India

Villuri Poorna Sai Krishna
Vellore Institute of Technology, India

ABSTRACT

Smart homes, smart buildings, and smart cities increase the quality of living by creating a world that is better and more secure, reducing dependency on human needs and efforts. This chapter investigates the complex landscape of security threats resulting from the integration of AIoT (artificial intelligence of things) and cloud platforms within the domains of smart homes, smart buildings, and smart cities. It examines the distinctive challenges and vulnerabilities that arise when these technologies converge. These technologies hold the potential to significantly enhance the quality of life, reinforce home security, and facilitate elderly care.

DOI: 10.4018/979-8-3693-0993-3.ch011

However, their effectiveness hinges on access to extensive data about users' homes and private lives. This chapter underscores the critical need to address the ensuing security and privacy concerns, which present substantial barriers to the widespread adoption of these promising technologies.

1. INTRODUCTION

In the previous days when home automation was considered as an add-on facility, it was limited to high end residences. It has now become an aspect of living owing to the growing significance of technology, in our day-to-day routines. Thanks to the embrace of innovations such as the Internet of Things (IoT) and Artificial Intelligence (AI). The smart homes have transformed into practical aids that streamline our everyday chores.

Smart homes, characterized by the remote control of appliances through interconnected devices, whether via the Internet of Things (IoT) or AI capabilities, offer an innovative way to enhance daily living (Weingärtner, 2019). These connected appliances, operating autonomously while adapting to user preferences through wireless and wired connections (Pentikousis, 2000), define the essence of smart homes a network of interconnected devices and services that respond automatically to predefined rules, provide remote access through mobile apps or web browsers (Lin et al., 2015) and deliver alerts and notifications to users. Leveraging AI and IoT technologies, these homes incorporate sensors, lighting, meters, and other connected devices to gather and analyse data, optimizing household functions and utilities, ultimately enhancing daily life's convenience and efficiency (Jalal et al., 2018). This chapter delves into the transformative potential of AI and the Internet of Things in augmenting the intelligence of smart homes by cloud service.

In the world of contemporary urban development, it becomes apparent that shifting societal preferences and lifestyles, driven by technological advancements, necessitate a re-evaluation of urban planning and construction practices. This shift emphasizes the construction of mobile residences characterized by heightened openness and robust interconnections among their constituent parts. These modern dwellings embody qualities such as adaptability, responsiveness, dynamism, and resilience, evolving in tandem with occupant needs and technological innovations. Smart buildings epitomize this evolution, embodying core principles of flexibility, accommodation, and multi-functionality, designed to provide the needs to resident's requirements in efficiency, security, and comfort (Koo et al., 2016). Their intelligent design places paramount importance on optimizing living spaces to enhance inhabitant well-being as mentioned in Braun et al. (2018). While smart technology aims to improve internal space efficiency and user-centric design concepts, certain studies in the world of smart building researchers have tended to prioritize technological implementation over broader aspects of habitability and spatial functionality. Despite efforts to introduce ambient intelligence into homes on a smaller scale, the widespread adoption of smart building concepts in practice remains a work in progress. Understanding the factors contributing to user satisfaction becomes pivotal for designers and engineers striving to align products with user preferences. It is essential to recognize mere availability of sophisticated features does not guarantee their acceptance or improved quality of life (Liu et al., 2019). Smart buildings, the fusion of intelligence (Barik et al., 2017; Batra et al., 2021) such as ML (machine learning), operational efficiency, material science, and construction practices, prioritize adaptability in achieving development goals, including energy efficiency (Martins et al., 2021), longevity, occupant comfort, and overall satisfaction. With the capacity to gather and interpret diverse data sources over time, smart buildings evolve and respond intelligently to changing circumstances, seamlessly integrating comfort, safety, security, design excellence, and long-term sustainability.

So, when we talk about "Smart City" which emerged in the 1990s, focusing on the impact of modern Information and Communication Innovations on urban systems. Organizations like the California Organized for Smart Communities played a crucial role in investigating how cities could leverage information innovations to become "smart". Smart cities encompass a wide range of applications that often involve socio-technical interaction between citizens and pervasive devices, known as Artificial Intelligence and the Internet of Things (AIoT). These cities aim not only for innovation but also to facilitate economic, social, and environmental well-being, make monitoring more aware, interactive, and efficient (Javed et al., 2023). The smartness of a city is driven by the emergent AIoT Cloud platform, integrating Artificial Intelligence and the Internet of Things as a robust technological foundation for smart city initiatives. Two crucial components of this pervasive network are the Internet of Things (IoT) and cloud computing. In smart cities, various electronic devices interconnect through applications, such as cameras in monitoring systems and sensors in transportation networks (Ozer et al., 2023). These devices can be organized by geographical location and analysed using sophisticated analytical systems, enhancing the city's efficiency and data-driven decision-making (Olaniyi et al., 2023). Cloud computing plays a pivotal role in advancing internet-based computing within smart cities, leveraging information technology capabilities as a resource. As smart devices extend beyond the cloud infrastructure, the IoT significantly improves efficiency, performance, and data throughput. Within the cloud infrastructure, the AIoT benefits from increased productivity, enhanced execution, and greater payload capacity. The integration of IoT and cloud technologies mentioned by Sergi, I and Montanaro in Sergi et al. (2021) represents the future of secure logistics and web-based advancements that align seamlessly with IoT systems in smart cities.

1.1 AI and IoT in Smart Homes, Buildings, and Cities, With Its Impact on Smart Homes

The emergence of Artificial Intelligence (AI), initially coined as a term during the historic Dartmouth conference in 1956, represents a pivotal milestone in the world of computer science. AI serves as a fundamental catalyst reshaping the landscape of various industries, notably construction projects, by endeavoring to replicate human-like logic, learning, cognition, reasoning, problem-solving, planning, and data storage capabilities (Kharb, 2023). Its profound impact on everyday life is propelled by the relentless advancement of Information and Communication Technology (ICT) and Robotic Technology (RT). AI demonstrates the capacity to comprehend its surroundings and systematically pursue objectives through a blend of analytical and intuitive actions. The exponential surge in research interest and investment in AI since the mid-twentieth century is attributable to its ability to provide intentional, intelligent, and adaptable solutions to intricate engineering and scientific challenges is very much cleared by Julia Musto in recent conference. Additionally, AI systems have proven instrumental in immersive learning phases, facilitating the classification of extensive databases, and offering valuable insights through visualizations, explanations, and interpretations of intricate models. The amalgamation of scientific data analysis, machine learning, cognitive science, and data theory enables automated programming, optimization, information compression, and modeling. Machine learning, a cornerstone of AI, plays a pivotal role in acquiring and synthesizing diverse data sources (Battina, 2017), culminating in intelligently adaptive decision-making, harnessing insights from structured and unstructured data. We have a better future if we can use AI properly in this world.

Artificial Intelligence (AI) serves as a transformative force in modern building management and operations. AI systems, capable of emulating human traits such as perception, reasoning, interaction, and learning, offer unparalleled potential to revolutionize the way buildings are managed and optimized. These systems adapt seamlessly to evolving information, eliminating the needs for constant reprogramming when faced with unforeseen scenarios. Beyond traditional applications, AI's integration into smart buildings contributes significantly to energy efficiency, automation, control, and safety enhancement (Sergi et al., 2021). It is noteworthy that AI's role extends to addressing critical aspects of smart building research, encompassing expert systems, fuzzy logic, genetic algorithms, machine learning, machine vision, natural language processing, neural networks, and pattern recognition, with emerging focus areas in deep learning and natural language processing.

The continued evolution of technology, particularly the Internet of Things (IoT), is ushering in a widespread adoption of AI in the construction of smart buildings. These innovative structure's seamlessly merge data collection, intelligent sensing, and responsible decision-making, with a keen emphasis on the intricate interplay between humans, machinery, and the environment. Smart buildings, characterized by extensive IoT device integration, are becoming increasingly prevalent, and the industry is actively transitioning towards simulating these intelligent structures in virtual environments Tiago Andrade, Daniel Bastos mentioned in his paper that how the IoT is being used in this present scenario (Andrade et al., 2019). The concept of the digital twin, while not limited to smart buildings, finds relevance in this context. A digital twin represents a comprehensive and dynamic depiction of a physical asset or system, ranging from its molecular to geometric characteristics. It encompasses a probabilistic simulation that incorporates various physical mechanisms and scales. More crucially, it serves as an operationally responsive and forward-looking representation, capturing the essence of a building's physical structure and its evolution over time, enriched by sensor data. In the world of smart buildings, the digital twin which explains about the physical structure of the text or data as said in Singh et al. (2021) in which the concept extends beyond mere structural and functional design, encompassing the ongoing evolution and real-time insights derived from sensor data, paving the way for smarter, more efficient, and responsive built environments.

In the smart cities, artificial intelligence (AI) is emerging as an effective tool that can transform urban living article (Ortega-Fernández et al., 2022) explains how AI does shape shifting into a safe human in urban environment. These AI-powered cities are becoming adept at utilizing energy and resources more efficiently, safeguarding the environment, enhancing the quality of life for citizens, and accelerating the adoption of cutting-edge Information and Communication Technologies (ICT). Explainable AI is playing a pivotal role in the development of smart cities, ensuring transparency and accountability. The prominence of AI in smart cities is already evident, as it has risen to the forefront of research and development in this field. These cities require large-scale decision-making capabilities to manage daily activities effectively.

As AI concepts in smart cities evolve alongside complementary innovations, a paradigm shift is underway to drive and transform AI-driven ventures into a pervasive AI-driven mode. To facilitate researchers in embarking on their work in the domain of AI for smart cities there is an analysis on how badly this Artificial intelligence is cropped all over the world (Bughin et al., 2018), there's a growing need for a rigorous and informative study that offers current, projected, and future insights. Such studies provide comprehensive insights and analyses of various AI innovations custom fitted for smart cities.

The advantages of AI in enabling smart city solutions are multifaceted. They encompass improved water supply, effective energy management, effective waste management, and the moderation of issues like activity clog, clamour contamination, and contamination. Generally, smart city initiatives have focused on generating data and securing modern insights into a city's complexity and dynamics. AI is presently lifting these cities to another level by empowering decision-makers to saddle this wealth of information and data for informed decision-making. Whereas smart city AI applications offer numerous benefits such as automation and efficiency, they moreover raise important regulatory issues. These include concerns related to the equitable distribution of benefits, privacy, legal frameworks, and ethical contemplations. Given the assorted points of view displayed by various analysts, it is imperative to dive into the strengths, weaknesses, and overall impact of AI calculations on smart cities.

China stands out as a pioneer within the worldwide landscape of smart cities Pu Liu and Zhenghong Peng, brought some report making on how China is grown in this field (Liu et al., 2013) with the biggest number of such cities within the world. These cities utilize a wide cluster of innovations, including metering gadgets, cameras, inserted sensors, data mining, processing, and AI-based analysis techniques to manage urban spaces and public resources effectively. AI holds the promise of significantly enhancing effectiveness and the quality of life in future smart cities. It achieves this by tapping into tremendous pools of assets and joining advanced advances such as machine vision, natural language processing (NLP), machine learning (ML), robotics, and more. AI-based applications are being deployed across various domains within smart cities to create more connected, sustainable, and responsive urban environments.

The rapid evolution of information and communication technologies (ICT) (Joseph, 2002) is driving the scope and scale of smart city development. This transformation is largely facilitated by the advent of the Internet of Things (IoT). Smart cities are reaping substantial benefits from information and computing technologies, particularly through IoT-enabled solutions. These solutions, utilizing gadgets like smart energy meters, security devices, and appliances for domestic well-being, enhance convenience and quality of life for various communities. The Internet of Things (IoT) represents a ground breaking communication paradigm that aims to establish an innovative framework for connecting a multitude of advanced devices to the Internet. This concept is instrumental in making the Internet more immersive and pervasive, and it is gaining momentum as operators, vendors, manufacturers, and enterprises recognize its potential. IoT-based smart city applications can be categorized based on various factors like network type, coverage, flexibility, and end-user inclusions. A smart city in the IoT context comprises billions of interconnected devices. Its success relies on the ability to provide network connectivity to every IoT device equipped with sensing capabilities. These devices can utilize various communication systems such as open Wi-Fi, Bluetooth, cellular networks, and satellites to interact with cloud-based application centres. However, ensuring seamless connectivity in smart cities presents challenges like providing network access to highly mobile devices, managing network transitions, and ensuring connectivity for massively deployed devices in areas with limited communication infrastructure.

The IoT paradigm centres around smart and self-configuring devices interconnected through global grid infrastructures. IoT devices are typically characterized as distributed objects with limited processing capabilities, aiming to enhance the reliability, performance, and security of smart cities and their infrastructure. Various Internet technologies, including WIFI, 2G, 3G, 4G, and Power Line Communication (PLC) K.C. Abraham mentioned in his re-search, facilitate data transmission over long distances, particularly in applications related to smart homes, smart cities, power system monitoring, energy management, and renewable energy integration. The application layer is where data is received and processed, enabling the design of efficient control, distribution, and management strategies. Devices can be grouped based on their geographical positions and evaluated using analytical systems. Sensor services play a crucial role in gathering specific data and have been applied in projects ranging from cyclist and vehicle monitoring to parking lot management. IoT infrastructure has found applications in air and noise pollution control, vehicle development, and surveillance systems. When we talk about Leading tech giants like Apple, Google, and Amazon have made significant investments in their smart assistants and AI capabilities for home automation. Siri, Google Assistant, and Alexa are well-known examples of these AI-driven platforms. Artificial Intelligence, empowers computers to perform tasks automatically, often relying on data collected and processed using various machine learning and deep learning techniques. It's the intelligence that allows these systems to execute specific functions without human intervention.at the same time internet of Things (IoT) refers to the technology that enables internet-connected devices and appliances to communicate, operate remotely, share real-time data, and respond to voice commands. These devices span from everyday household items like lighting, fans, air conditioners and washing machines to more substantial equipment in larger structures, such as pumps and fire alarm systems.

In smart homes and buildings, IoT frameworks supply valuable data, while AI leverages this data to execute tasks that reduce human effort. Smart devices equipped with AI and IoT integration can respond to user voice commands or pre-programmed AI instructions from a distance. For instance, Google's Nest thermostat learns user behaviour over time, automatically adjusting the temperature for comfort when occupants are home and optimizing energy efficiency when they're away. AI and IoT holds significant promise for smart homes. Raw sensor data collected from interconnected devices can be transformed into behavioural patterns that enhance daily life by encrypting the data (Al Rikabi et al., 2021). AI-enabled devices can learn residents' routines and predict optimal experiences. For instance, if no one is home, the system won't activate heating, fans, or lights and will automatically secure the doors.

Imagine a scenario where a user prepares a meal in a smart oven or stove, while AI monitors the meal's internal temperature. If the food reaches the ideal temperature, AI can adjust the cooking temperature to prevent overcooking. The AI can then notify the user when the meal is ready to be taken out of the oven or burner. AI's capacity to learn and anticipate user preferences extends to other areas, such as a smart kitchen being prepped before a user arrives home to cook.

The potential of IoT and AI isn't confined to new homes; there are retrofit options that can convert existing devices, such as switches, into smart ones. Even old air conditioners can be upgraded to offer remote access through smart apps or AI-based cloud servers (Sadeeq et al., 2021). These wireless solutions simplify deployment and reduce the need for extensive electrical work, making smart living accessible to a broader audience. For tech-savvy households, the fusion of AI and IoT in smart homes offers personalized automation that adapts to your daily routine, surpassing traditional usage patterns. As technology advances and more devices integrate seamlessly, the gains in smart home automation will be substantial.

1.2 The Current State of AIoT in Smart Environments: Trends and Developments

In this present living scenario, the integration of Artificial Intelligence and the Internet of Things (AIoT) within smart environments, powered by cloud platforms, represents a dynamic landscape characterized by ongoing trends and significant developments as of 2023. This burgeoning field is at the forefront of technological innovation, shaping the way we interact with and optimize our surroundings. The synergy of AI and IoT technologies is driving transformative changes across various domains, from urban planning to industrial automation, healthcare, and beyond. In this ever-evolving ecosystem, cloud platforms play a pivotal role, serving as the backbone that enables seamless connectivity, data analysis, and scalability. As we delve into the current state of AIoT in smart environments, it becomes evident that key trends and developments are poised to shape the future of this exciting field. landscape of AIoT implementation, smart homes, buildings, and cities are experiencing transformative advancements.

In smart homes, AI-driven voice assistants like Amazon Alexa, Google Assistant, and the most popular Ai Siri such as home pod have made seamless device control through natural language commands a reality, while enhanced security and surveillance systems provide facial recognition and real-time alerts. Additionally, energy management solutions harness AI to optimize resource consumption. Smart buildings utilize occupancy sensors and predictive maintenance powered by machine learning, streamlining space utilization, and reducing operational downtime. Meanwhile, in smart cities, AIoT is revolutionizing traffic management, environmental monitoring, waste collection, public safety, and infrastructure maintenance, thereby contributing to enhanced quality of life and sustainable urban development where we can them as smart cities. As technology continues to evolve, the integration of AI and IoT is poised to further redefine how we live, work, and interact within these interconnected smart environments.

When we talk about this domain of smart homes and buildings, the integration of IoT (Internet of Things) and Artificial Intelligence (AI) is of paramount importance, facilitating the optimization of processes and the augmentation of user experiences. These technologies have emerged as foundational elements in the functionality of mobile applications designed to oversee and manage smart environments. Notably, the convergence of IoT and AI in these contexts engenders the comprehensive collection and analysis of voluminous datasets, often referred to as "thing data." This invaluable data corpus is effectively harnessed by AI algorithms, enabling real-time insights and process automation as they collect information they are programmed to share with AI (Gaur et al., 2019) in such scenario they can harvest better results. For instance, within the purview of smart homes, data gleaned from temperature sensors can be employed by AI-driven thermostats to discern user preferences and dynamically adjust heating or cooling systems, ensuring both comfort and energy efficiency. Moreover, the fusion of IoT and AI bears significant implications for enhancing security within smart environments. Utilizing data from surveillance cameras, motion detectors, and door sensors, AI algorithms vigilantly scrutinize activities to detect anomalies and subsequently activate security protocols, fortifying the protective measures in place. Beyond security enhancements, these technologies assume a pivotal role in optimizing energy utilization and resource management. Smart lighting systems, as an exemplar, adeptly adapt brightness levels based on ambient light conditions and occupancy patterns, thereby curbing energy consumption. In the ambit of smart buildings, the continuous monitoring and analysis of thing data afford the prospect of predictive maintenance for equipment, thus extending their operational lifespan and diminishing downtimes. Furthermore, IoT and AI afford users and building administrators centralized command and profound insights through mobile applications this can be much better in future with the future AI (Gill et al., 2022), delivering the convenience of remote environment management, real-time adjustments, and informative notifications grounded in empirical data. In summation, the confluence of IoT and AI, drawing upon the wellspring of thing data, holds transformative potential within smart homes and buildings, ushering in tangible advancements in comfort, security, energy efficiency, and the holistic user experience. These technologies persist as pivotal catalysts in the evolutionary trajectory of smart environments, progressively enhancing their efficiency, security, and user-friendliness.

2. SMART HOMES/BUILDINGS

In 2023, a "smart home" is a technologically advanced residential space where various internet-connected devices such as appliances, lighting, locks, cameras, and more are seamlessly integrated, as shown in figures 1 and 2. This integration allows homeowners to remotely control and share data, significantly enhancing convenience and security. Smart homes utilize the Internet of Things (IoT) to maximize energy efficiency and offer centralized control through integrated digital platforms, often accessible via smartphones or other user-friendly interfaces.

When we talk about smart homes, artificial intelligence (AI) plays a pivotal role in transforming raw sensor data from connected devices into practical daily life enhancements. These AI-powered devices exhibit a keen understanding of residents' patterns and can anticipate and optimize their experiences. For instance, they intelligently refrain from activating heating, fans, or lights when no one is at home, and they automatically secure doors in the absence of occupants.

Imagine a scenario where a user is cooking with a smart oven or stove and AI is actively monitoring the meal's internal temperature. When the meal reaches its ideal temperature, AI intervenes by adjusting the cooking settings to prevent burning. The user is then promptly notified when the meal is perfectly prepared. AI holds the potential to learn and pre-emptively cater to users' preferences. In a smart kitchen, for instance, AI can preconfigure the environment before the user arrives home to start cooking.

2.1 Security Concerns in Smart Homes

In the world of smart home applications, a myriad of security threats has been identified, ranging from internal to external in nature. Often, internal threats stem from a lack of resident proficiency in securing their homes effectively. We are sure that certain companies involved in the manufacturing or provision of services for smart home applications do not consistently prioritize the implementation of requisite security measures to mitigate potential vulnerabilities (Prachi Tiwari, 2022). In the context of this research endeavour, we undertook an in-depth analysis of prevalent security challenges that are best to be used in smart home applications.

The perceptions of elderly individuals regarding smart homes (SHs) were studied, revealing that privacy concerns were more prominent among participants than ease of use. There are many articles that say the security threats possessed by IoT and AI In the smart environment, it is much more clearly explained by in (Fabrègue et al., 2023). While participants were less concerned about the nature of the data itself, they expressed strong reservations about how companies handled their data, particularly fearing increased advertising and data profiteering. An investigation into user's knowledge and mental models of the internet indicated that those with more comprehensive technical models perceived greater privacy threats. In a field study analysing 14 smart homes, concerns arose about security-critical devices. Remote access was viewed as a double-edged sword, providing additional control but also increasing security concerns. Study participants also voiced worries about the individual privacy of household members when using smart home technologies. For example, children might feel uncomfortable about their parents being able to monitor their whereabouts.

An online survey asked participants about their perceived privacy consequences of using smart home technologies. Approximately one-third of the time, others discussed concerns unrelated to privacy. Users were presented with randomized internet of things (IoT) scenarios with varying privacy implications. The results revealed that users were more comfortable with data collection in public settings but less comfortable with biometrics or data sharing with third parties. They expressed a preference for being notified about data collection in scenarios perceived as uncomfortable.

Despite extensive research on technical security concerns related to SH technologies, relatively little attention has been given to end-user security and privacy concerns among smart homeowners, indicating that convenience often outweighs personal privacy concerns for adopters. However, many participants expressed limited concern about potential threats.

Recognizing this gap, our study aims to analyse the specific concerns of a substantial interview sample of end users related to the security and privacy of smart home technologies. Our focus is primarily on users with little or no prior experience with smart homes, aiming to identify potential barriers to adoption. The interview study includes various smart home concepts with differing levels of security and privacy, allowing us to explore a range of concerns.

2.1.1 Vulnerabilities in Software Applications Within Smart Homes

In the world of modern smart homes, a plethora of interconnected devices, collectively known as smart devices, rely heavily on the intricate software applications that underpin their functionalities. These software applications, painstakingly crafted by human programmers, stand as the cornerstone of seamless smart home operation. Yet, paradoxically, it is this human element that engenders a potent factor—software vulnerabilities. These vulnerabilities, stemming from programming oversights or errors, constitute proverbial chinks in the Armor, creating potential entry points for nefarious actors' intent on infiltrating smart homes, thereby undermining their security and integrity.

The susceptibility of software vulnerabilities to exploitation is a dynamic facet, intrinsically tethered to the level of awareness that malicious hackers possess regarding existing flaws within specific software applications. In contexts where data confidentiality stands as a paramount concern, such as within the healthcare sector, vulnerabilities assume a gravity that is hard to overstate. The ramifications of successful exploitation of these vulnerabilities can be dire, as hackers gain access to sensitive patient data, potentially besmirching the reputation of healthcare institutions. Furthermore, the educational arena, represented by schools and universities, does not remain immune to the ominous spectre of software attacks. In such environments, the consequences may manifest as data losses or unauthorized alterations, with certain parties, particularly students, standing to benefit.

It is of paramount importance to underscore the profound implications of residents' proclivity to customize codes for an array of household appliances, ranging from doors and alarms to gate locks, alongside comprehensive audio and video surveillance systems that vigilantly monitor internal and external activities (Komninos et al., 2014), where it is clearly mentioned how security measures take place in the smart environment. In the grim eventuality of hackers breaching the fortress of gateway software, an unsettling vulnerability emerges—vast troves of data become susceptible to theft or manipulation, rendering the sanctum of home security porous.

When the dark underbelly of software vulnerabilities is laid bare and subsequently exploited, the ramifications are manifold:

1. Eavesdropping: Malicious actors can surreptitiously intercept and meticulously monitor all network activities, as clearly shown in a paper by Federico Maggi and Alberto Volpato (Maggi et al, 2011), granting them unfettered access to a treasure trove of sensitive data. This intrusion into the private domain of smart home residents represents a grievous violation of privacy, as attackers could potentially amass comprehensive information spanning years.

2. Data Modification: With the capability to manipulate data by altering passwords or tampering with transmitter-receiver configurations, malicious actors enable unauthorized data extraction. In the context of smart homes and public service facilities, such as hospitals, the spectre of data manipulation or deletion looms ominously. The paper (Kodada et al., 2012) stated clearly how this data modification takes place.

3. Password-Based Attacks: Armed with stolen or compromised passwords, hackers gain the power to modify network listings, manipulate network addresses, delete historical network data, or even recalibrate device settings. Particularly within the healthcare domain, as we can find in (Xu et al., 2023), with its best mitigations, where physicians rely on sensors to diligently monitor patients' medication regimens and health parameters, hackers with access to

the database wield the power to manipulate critical data, thereby dispatching falsified information to both medical professionals and patients alike.

The vulnerabilities deeply entrenched within the tapestry of smart home software applications present profound security risks. The ramifications of these vulnerabilities range from the dire spectre of data breaches to the insidious manipulation of data. The imperative to address and mitigate these vulnerabilities is paramount to safeguarding the sanctity, security, and privacy of smart home environments.

When examining each class of technologies individually, such as AI, IoT, and cloud services, it becomes evident that they each harbour distinct security concerns. Consequently, our research aims to comprehensively investigate and delineate the specific security concerns associated with each of these domains. Furthermore, we delved into the corresponding mitigation strategies and implemented models within each respective context. This approach enables us to gain a holistic understanding of the nuanced security landscapes inherent to AI, IoT, and cloud services, and we can find the best model and type that should be used in the smart environment to be on the safe side, as shown in Figure 1.

2.1.2 Ensuring Patch Delivery Resilience in Smart Home Networks

The process of patching vulnerabilities in smart home devices is inherently tied to the regular updating of software. However, when this process encounters difficulties, distinguishing whether it is due to natural system failures or the covert actions of malicious hackers becomes a complex task. In such instances, the integrity of smart homes may be compromised, leading to potential data modifications.

The reception and application of patches present a formidable challenge, particularly for companies directly engaged with the smart home ecosystem. These companies often leverage wireless sensor technologies to disseminate patches to residents. Paradoxically, the urgency of promptly installing these patches is juxtaposed with the risk of user rejection. When residents are prompted to update their devices, they may be engaged in critical activities or have their smart home appliances in active use (Tawalbeh et al., 2020). This paper shows how good the system updates can turn the security challenges from down to top, leading to hesitation or refusal to execute the updates.

It is important to note that these patches encompass a spectrum from software-level updates to hardware-level enhancements. The transition from software to hardware updates introduces an inherent risk, underscoring the imperative for seamless automation. Failure in the delivery and application of patches within smart homes is a recurring issue, particularly in establishments such as hospitals, schools, banks, industries, and government buildings where the security and integrity of critical data are paramount.

Even the occurrence of patch delivery failures can be attributed to a variety of factors, including, but not limited to, connectivity issues with the internet, the continuous interconnection of smart homes with the Internet of Things (IoT), hardware malfunctions, or even power outages. These multifaceted challenges underscore the complexity of ensuring consistent and effective patch management in the dynamic landscape of smart homes.

2.1.3 Access Control in Smart Homes: Challenges and Solutions

In smart homes, the proliferation of interconnected devices has reached staggering proportions. Smart devices, ranging from thermostats to cameras, offer unparalleled convenience and automation but concurrently introduce a complex web of security challenges. For security-conscious residents, a widespread practice involves setting distinct passwords for each device, a strategy considered paramount in ensuring the sanctity of their dwelling (Xu et al., 2023), as shown in this paper by Ming Xu and Jitao Yu. However, this practice, while effective in principle, is not devoid of pitfalls. One prominent concern lies in the human propensity to forget passwords, a vulnerability that, if exploited, could potentially compromise the very security these measures are designed to enhance.

Moreover, the act of sharing access credentials, albeit often well-intentioned, carries inherent dangers. Sharing passwords with trusted individuals may inadvertently extend access privileges to those who should not possess them, creating a paradox where the quest for security inadvertently fosters vulnerabilities. This underscores the importance of not only password management but also the need for sophisticated authentication mechanisms and stringent access controls within smart home ecosystems.

Further complicating matters is the intricate network of communication among devices within and outside the smart home environment. In an era where seamless connectivity reigns supreme, homeowners find themselves in situations where their mobile applications serve as conduits for notifications, seeking their approval for various actions. These actions may include granting access via door cameras or adjusting thermostat settings remotely. However, this prompts a critical question: How can homeowners, in this era of digital interconnectedness, reliably distinguish between legitimate requests from authorized users and those originating from malicious actors seeking surreptitious entry?

Mitigating these multifaceted security risks demands a holistic approach to smart home security (Apthorpe et al., 2017). Among the foremost strategies is the deployment of robust access control mechanisms that extend beyond traditional passwords. Biometric authentication, two-factor authentication (2FA), and multifactor authentication (MFA) stand as formidable tools for bolstering user identification and authorization. These mechanisms elevate the security threshold, reducing the likelihood of unauthorized access and fortifying the integrity of smart home environments.

Furthermore, encryption emerges as an indispensable shield against eavesdropping and data interception within the smart home network. Leveraging encryption protocols, such as WPA3 for Wi-Fi communication and Transport Layer Security (TLS) for data transmission, safeguards sensitive information from prying eyes. This cryptographic defence ensures that even if unauthorized parties intercept data traffic, deciphering the encrypted information remains an insurmountable challenge.

Additionally, adopting a stringent policy of connecting only trusted devices to the home network serves as a critical defensive measure. This prudent approach precludes the integration of potentially compromised or unverified devices that could serve as vectors for cyberattacks.

The security of smart homes is undeniably complex. The extensive proliferation of smart devices and their interconnected nature calls for a multifaceted approach to access control and security. Robust authentication mechanisms, encryption protocols, and device trustworthiness assessments are instrumental in fortifying the smart home's security posture. This comprehensive perspective underscores the importance of staying vigilant in safeguarding the privacy and security of residents within the evolving ecosystem of smart homes.

2.2 Security Challenges in AI-Powered Worlds

When examining the prevalent challenges confronting AI, it becomes evident that several issues persist. To address these concerns effectively, it is imperative to identify and implement the most robust mitigation systems. Our primary objective is to discern and select the foremost mitigation strategies currently employed by certified companies within the AI landscape. Subsequently, we aim to integrate these chosen systems seamlessly into the contemporary AI framework, which is intertwined with various embedded devices.

2.2.1 Data Privacy and Protection

Concern: AI systems often require large volumes of data for training. Ensuring the privacy and protection of this data is crucial. Unauthorized access or data breaches can lead to privacy violations and the misuse of personal information.

Mitigation:

1. Data loss prevention (DLP) (Daubner et al., 2023) is a basic component of information assurance, planned to anticipate unauthorized access, spillage, or robbery of touchy data. DLP innovations consist of different apparatuses and forms that help organizations keep control over their information.
2. Storage with Built-in Data Protection Choosing the proper capacity arrangement is essential for guaranteeing the security of your data. Cutting-edge capacity advances presently come equipped with built-in information security highlights, advertising extra layers of security.

firewalls play a significant part in information assurance by acting as an obstruction between your inside frameworks and the exterior world (Yang et al., 2023). They can help anticipate unauthorized access and protect your information from different dangers.

Identity and access management (IAM) (Glöckler et al., 2023) systems are designed to manage user identities and access rights across your organization. By centralizing authentication and authorization processes, IAM can help streamline information security endeavours and move forward in security.

Endpoint protection devices such as laptops (Siji et al., 2023), smartphones, and other mobile devices are often defenceless targets for cyberattacks. Endpoint protection technologies are planned to secure these gadgets and the data they contain.

2.2.2 Adversarial Attacks

Concern: Adversarial attacks involve manipulating input data to deceive AI models. Attackers can change pixels in an image or words in a text to fool the system into making incorrect predictions.

Mitigation:

Limited-memory BFGS (L-BFGS) The Limited-memory Brayden-Fletcher-Goldfarb-Shanno (L-BFGS) (Sahu et al., 2023) strategy may be a non-linear gradient-based numerical optimization calculation to play down the number of irritations included in pictures.

1. Fast Gradient Sign Method (FGSM) (Wibawa et al., 2023) A simple and fast gradient-based strategy is utilized to create antagonistic cases to play down the most extreme sum of annoyance included in any pixel of the picture to cause misclassification.
2. Jacobian-based Saliency Map Attack (JSMA) Unlike FGSM in this paper, whose keen and educated paper experience is mixed (Wiyatno et al., 2018), the strategy employs highlight choice to play down the number of highlights,

thereby causing misclassification. Level annoyances are included to highlight iteratively concurring to saliency esteem by diminishing.

2.2.3. Model Vulnerabilities

Concern: AI models themselves can be vulnerable to attacks. Attackers may exploit the model's architecture or parameters to manipulate the model's output or extract sensitive information.
  Mitigation:

1. Employ random, universally unique identifiers (UUIDs).
2. Establish strong authorization protocols.
3. Adopt the zero-trust security framework.
4. Explain ability and Interpretability:

Concern: Some AI models, like deep neural networks, are often viewed as black boxes, making it challenging to understand their decision-making processes. This lack of transparency can be a security concern in critical applications.
  Mitigation: Developing and using more interpretable AI models, like decision trees or rule-based systems, can enhance transparency. Explainable AI (XAI) techniques can also provide insights into model decisions. We can find one of the hybrid detection systems in (Dias et al., 2021).

2.2.4. Model Poisoning

Concern: Model poisoning involves manipulating the training data to compromise the model's performance. Attackers can introduce malicious data during training.
  Mitigation: Careful data validation and cleansing processes can help mitigate model poisoning. Additionally, using secure data sources and regularly monitoring unusual data patterns are important.

1. SVM Resistance Enhancement: This approach, called K-LID-SVM, counters label-flipping attacks, which are a vulnerability in SVMs, by introducing a weighted SVM with K-LID computation. K-LID is a novel metric for identifying outliers in high-dimensional spaces. It achieves higher stability against various label-flipping attack variants and reduces misclassification errors by an average of 10% across real-world datasets (Ramirez et al., 2022), making it superior to traditional LID-SVM.
2. Bagging Classifier: Utilized as an ensemble method to detect poisoning samples and counteract the influence of outliers on training data.

3. Kernel-Based SVM: Improves robustness against DP attacks by reducing the SVM slackness penalty and replacing the dual SVM objective function with a probability-based approach for mislabelling.
4. Antidote: A statistical defence scheme that treats anomaly detection to mitigate the effects of poisoning attacks by discarding outliers in training data.

All these methods are explained clearly in (Younes et al., 2020).

2.2.5. Transfer Learning Risks

Concern: Transfer learning relies on pre-trained models, which may introduce security risks if the source model is compromised or if the fine-tuning process is not secure.

Mitigation: Secure transfer learning involves ensuring the source models are trusted and adopting secure fine-tuning practices (Tripuraneni et al., 2020). Regular model validation can help detect security issues.

2.2.6. Data Integrity

Concern: Data integrity is critical for AI systems. Data manipulation or tampering can lead to incorrect decisions, as shown in Article (Vuković et al., 2014), and compromised security.

Mitigation: Implementing data integrity checks and secure data storage practices can safeguard against data tampering.

2.2.7. Bias and Fairness

Concern: AI models can inherit biases from training data, resulting in discriminatory outcomes. Bias can be both an ethical and a security concern.

Mitigation: Addressing bias in AI models involves carefully curating training data and employing fairness-aware algorithms. Regular audits for bias and fairness are important. Analysts and professionals have proposed different approaches to relieve predisposition in AI, as we can find in (Mehrabi et al., 2021). These approaches incorporate pre-processing information, show choice, and post-processing choices. In any case, each approach has its confinements and challenges, such as the need for differing and agent-prepared information, the trouble of recognizing and measuring diverse sorts of inclination, and the potential trade-offs between reasonableness and precision. Moreover, there are moral contemplations around how to prioritize distinctive sorts of predisposition and which bunches to prioritize within the relief of predisposition. Despite these challenges, relieving predisposition in AI is fundamental for making reasonable and even-handed frameworks that advantage all people and

257

society. Progressing inquiries about and improvements to moderation approaches are fundamental to overcoming these challenges and guaranteeing that AI frameworks are utilized to the advantage of all.

2.2.8. Robustness

Concern: AI systems must be robust against unexpected inputs or environmental conditions to prevent vulnerabilities.

Mitigation: robustness testing, diverse training data, and scenario-specific training can enhance AI system robustness.

1. After detection, mitigation strategies are employed to address the identified robustness risks.
2. Procedures may incorporate show fine-tuning, including or supplanting demonstrate modules, or re-assessment utilizing vigoro checklists. For example, backdoors can be evacuated by fine-tuning show parameters.
3. Anonymous inputs can be sifted, and models can be reinforced to handle them way better.

2.2.9. Deployment and Operational Security

Concern: The deployment environment of AI systems must be secure to prevent unauthorized access or manipulation.

Mitigation: Implementing strong access controls, monitoring for anomalies (Otuoze et al., 2018), and timely security updates are essential for operational security.

2.2.10. Malicious Use

Concern: AI technology can be exploited for malicious purposes, such as generating fake content or automating cyberattacks (Brundage et al., 2018).

Mitigation:

1. Educate employees: Employees are frequently the weakest interface within the information security chain. It is critical to teach representatives information judgment best practices, such as making solid passwords, being cautious about almost everything they press on, and being mindful of phishing assaults.
2. Keep software up to date: Software merchants frequently discharge overhauls that fix security vulnerabilities. It is vital to keep the program up-to-date to decrease the risk of being misused by assailants.

Have a plan in place: On the occasion of an information breach or other security incident, it is critical to be able to reply rapidly and minimize the disturbance to your trade. Make it beyond any doubt to have an organized input and test it frequently.

2.2.11. Legal and Ethical Concerns

Concern: Compliance with laws and ethical guidelines is essential for responsible AI use. Violations can lead to legal (Tippins et al., 2021) and reputational risks.

Mitigation: Strict adherence to AI-related regulations, ethical guidelines, and industry standards is crucial. We can find the recent debates on the ethical ton of AI in the press meetings and the government call taken in this recent event .

2.2.12. Supply Chain Security

Concern: The AI supply chain, including hardware and software components, must be secure to prevent tampering or the insertion of malicious components.

Mitigation:

1. Vendor Assessment and Due Diligence: Conduct intensive security evaluations and due diligence on sellers and providers. Guarantee they follow the best cybersecurity hones and measures.
2. Supplier Security Agreements: Implement legally binding understandings with providers that incorporate security clauses and prerequisites, counting occurrences, announcements, and security reviews.
3. Supply Chain Visibility: Improve perceivability in the supply chain to screen and track products and information. This incorporates the use of advances like blockchain, IoT, and RFID for real-time tracking.
4. Data Encryption: Scramble touchy information in travel and at rest to ensure it from unauthorized access. This is often the basis for securing information because it moves through the supply chain.
5. Secure Communication: Execute secure communication channels and conventions to ensure information is traded between supply chain accomplices and frameworks.

2.2.13. Authentication and Authorization

Concern: Proper authentication and authorization mechanisms are essential to controlling access to AI systems.

Mitigation:

1. Strong Authentication Methods: Organizations ought to empower or require the use of solid confirmation strategies, such as biometrics, keen cards, and equipment tokens, in addition to or rather than passwords.

2. MFA Implementation: Executing MFA ought to be a need. MFA includes an extra layer of security by requiring clients to supply two or more verification components some time ago.

3. Security Awareness Training: Customary preparation and mindfulness programs for representatives can help them recognize phishing endeavours and social design strategies, diminishing the probability of falling casualties to these assaults.

4. Identity and Access Management (IAM) Solutions: Use IAM arrangements that give centralized control and administration of client get-to-rights, counting role-based get-to-control (RBAC) and authorizations.

5. Single Sign-On (SSO): Implementing SSO solutions can streamline the login process for clients while keeping up security. Clients verify once and pick up numerous frameworks and applications.

2.3 Security Challenges in the IoT-Powered World

When discussing security threats within the IoT ecosystem, it is essential to segment IoT devices into categories such as wearables, RFID, sensors, and actuators. Attacks can potentially occur through the networking layer, and even if hackers breach the firewall at a certain point, it is crucial to explore methods for mitigating these risks. Let's begin by examining potential attack vectors and their associated solutions.

One primary avenue for attacks is through the networking layer. IoT devices often communicate over networks, making them susceptible to various threats, including eavesdropping, man-in-the-middle attacks, and data interception. To address these concerns, implementing robust encryption protocols and secure communication channels can significantly enhance IoT security. Additionally, regular network monitoring and intrusion detection systems can help identify suspicious activities in real-time, allowing for prompt responses.

Another vulnerability arises when attackers attempt to breach the firewall. Although firewalls are essential for protecting IoT networks, they may not be fool proof. In such cases, multi-factor authentication (MFA) and strong access controls can bolster security by limiting unauthorized access. Regularly updating firewall rules and firmware can also help close potential security gaps.

Furthermore, it is essential to consider the security of individual IoT device categories like wearables, RFID, sensors, and actuators. Each of these devices can have unique vulnerabilities. To mitigate these risks, manufacturers should prioritize secure-by-design principles, ensuring that devices are resistant to common attacks.

Regular firmware updates should be made available to patch known vulnerabilities, and end-users should be educated on best practices for device security, such as changing default passwords and keeping devices up-to-date.

We see that addressing security threats within the IoT ecosystem involves a multi-faceted approach. By implementing robust encryption, secure communication channels, access controls, and device-specific security measures, we can significantly reduce the risks associated with IoT deployments. Furthermore, ongoing research and development in IoT security are essential to stay ahead of emerging threats and vulnerabilities in this rapidly evolving landscape. KH.

2.3.1 Sleep Deprivation Attacks in the IoT

In the Internet of Things (IoT) landscape, where efficiency and battery life are paramount, sleep deprivation attacks emerge as a pressing security concern. These attacks manipulate IoT devices, often powered by replaceable batteries, by keeping them perpetually awake, thereby accelerating power consumption. This relentless drain on resources can have detrimental effects on IoT devices, leading to a denial of service (DoS) scenario as batteries become depleted. The motivations behind sleep deprivation attacks may range from simple battery exhaustion (Alotaibi et al., 2022) to more sinister purposes, such as espionage or disruption.

2.3.1.1 Mitigation Strategies

To safeguard against sleep deprivation attacks in IoT environments, a multifaceted approach is imperative. Beyond the strategies outlined earlier, here are additional mitigation measures:

Device Authentication: Implement stringent device authentication protocols to ensure that only authorized devices can communicate with IoT networks (Malviya et al., 2022).

Network Segmentation: Divide IoT networks into segments to limit the attack surface. This way, even if one segment is compromised, it doesn't necessarily compromise the entire network (Abdel-Rahman, 2023).

Behaviour Analysis: Employ machine learning and behavioural analysis to detect anomalies in device behaviour that may indicate a sleep deprivation attack (Bhanpurawala et al., 2022). This proactive approach can help identify threats in real-time.

2.3.2 Battery Health Monitoring

Implement battery health monitoring systems to keep tabs on the status of device batteries (Bhanpurawala et al., 2022). This enables timely battery replacement or recharging, minimizing the impact of prolonged attacks.

2.3.3 Security Updates

Maintain a proactive stance on security by ensuring that devices receive regular security updates and patches (Albalawi et al., 2022). This helps in fortifying devices against emerging threats.

We can most probably safeguard our IoT devices, by which smart environments will not be affected, so we need to adopt these comprehensive mitigation strategies. IoT ecosystems can better defend against sleep deprivation attacks by preserving the integrity of the network, extending device lifespans, and ensuring consistent and reliable service.

2.3.4 Code Tampering Threats in IoT Security

2.3.4.1 Malicious Code Injection and Mitigation Strategies

Malicious code injection attacks in the context of IoT security present a formidable challenge. These attacks involve the surreptitious insertion of unauthorized code into IoT devices, opening the door to a range of detrimental outcomes (Shrivastava et al., 2022). This malicious code can coerce IoT nodes into executing unintended functions, potentially leading to unauthorized data access or manipulation. In more severe cases, attackers can gain complete control over compromised devices, posing significant security risks. To counteract these threats, it is imperative to adopt comprehensive security measures. Manufacturers must adhere to secure coding practices and conduct regular vulnerability assessments (Karie et al., 2021). Network segmentation and access control mechanisms should be implemented to restrict unauthorized access, while intrusion detection and anomaly detection systems can help identify suspicious activities. Emphasizing secure coding practices, timely updates, and end-user education can collectively fortify IoT ecosystems against malicious code injection attacks, preserving the integrity and security of these interconnected systems. In the rapidly evolving landscape of IoT security, safeguarding against malicious code injection attacks is paramount. These attacks, which exploit vulnerabilities in IoT devices to insert unauthorized code, can have far-reaching consequences. To address this threat effectively, a multi-faceted approach is imperative. Below, we outline key mitigation methods:

1. Secure Coding Practices: Manufacturers bear a significant responsibility in ensuring IoT security. Employing secure-by-design principles during device development, including rigorous code reviews, vulnerability assessments (Sanjana, et al., 2022), and adherence to best coding practices, can minimize vulnerabilities from the outset.

2. Regular Firmware Updates: Timely patching through regular security updates and firmware upgrades is essential. These updates frequently contain fixes for known vulnerabilities, thus closing potential entry points for malicious actors (Zhu., 2019).

3. Network Segmentation: IoT devices should be isolated on separate network segments from critical infrastructure (Abdollahi et al., 2021). This isolation limits the potential impact of an attack, confining it to a designated network area even if code injection occurs.

4. Access Control and Authentication: Implementing multi-factor authentication (MFA) and robust access controls is crucial for preventing unauthorized access to IoT devices. Restricting access to authorized users and devices enhances security (Thomaz et al., 2023).

5. Intrusion Detection and Anomaly Detection: Employ real-time monitoring using intrusion detection systems (IDS) and anomaly detection systems (ADS) to continuously scrutinize IoT networks for unusual behaviour patterns (Pathak et al., 2021). Rapid detection enables swift responses to potential threats.

6. Code Validation and Signing: To ensure the authenticity of firmware and software updates, require digital signatures. This practice guarantees that only trusted updates are applied to IoT devices (Ioannidis et al., 2023).

7. Security Education: Educate both end-users and administrators on IoT security best practices. This includes guidance on changing default passwords, recognizing signs of compromise, and maintaining up-to-date devices and software (Chen et al., 2023).

8. Behaviour Analysis: Implement behavioural analysis to understand the expected conduct of IoT devices (Yang et al., 2022). Any deviations from established norms should trigger alerts, prompting further investigation.

9. Zero Trust Architecture: Adopt the Zero Trust model, where trust is never assumed, and continuous authentication and validation of devices and users occur, even within the network perimeter (Wang et al., 2023).

10. Security Standards and Regulations: Stay informed about IoT security standards and regulations relevant to your industry or region. Compliance with these standards helps establish a baseline for security practices (Shaaban et al., 2022).

When securing IoT ecosystems against malicious code injection attacks necessitates can be like a proactive, multi-layered security strategy. Manufacturers, administrators, and end-users all play vital roles in protecting these interconnected systems. By implementing the outlined mitigation methods, organizations can substantially reduce the risks associated with malicious code injection, thereby ensuring the enduring reliability and security of their IoT environments.

2.3.5 Eavesdropping Threats in IoT Security

In the domain of IoT (Internet of Things) security, eavesdropping attacks present a formidable challenge. These attacks occur when malicious actors clandestinely intercept and monitor the communication between IoT devices or between an IoT device and a central network. The attack sequence unfolds as follows: first, the attacker gains unauthorized access to the communication channel or network (Wang et al., 2022), exploiting vulnerabilities or weak security measures. Subsequently, data packets exchanged among IoT devices are captured and analysed for valuable information, including sensitive sensor data, user information, or proprietary business data. Notably, the attacker may also manipulate data, inject false commands, or tamper with data integrity, potentially leading to erroneous actions within the IoT system. Moreover, eavesdropping attacks can expose authentication credentials, enabling unauthorized access, and pose a severe threat to privacy if personal information is compromised (James et al., 2023). To mitigate these risks, IoT systems must employ robust security measures, such as device authentication, IPsec security channels, cryptography, error detection, and rigorous risk assessments. These collective efforts serve to protect IoT environments, ensuring data confidentiality, integrity, and privacy while thwarting the insidious threat of eavesdropping attacks.

2.3.5.1 Mitigation Methods Against Eavesdropping Attacks in IoT Environments

Eavesdropping attacks pose a significant threat to the security of IoT (Internet of Things) environments, particularly in open and unsecured settings. To counteract this threat and protect sensitive data, a series of effective mitigation methods can be employed:

1.   Device Authentication:

One fundamental strategy to enhance IoT security involves the implementation of robust device authentication procedures. This ensures that IoT devices are required to verify their identity before sending or receiving data within the network (Malviya et al., 2022). By enforcing stringent authentication, the network can prevent unauthorized devices from gaining access, thereby fortifying overall security.

2.   IPsec Security Channels:

Employing Internet Protocol Security (IPsec) channels proves to be a highly effective countermeasure against eavesdropping attacks. IPsec provides crucial functions such as authentication and encryption, ensuring the legitimacy of the parties engaged in communication (Mei et al., 2021) while safeguarding the confidentiality of data being transmitted. This security framework empowers the network to differentiate between legitimate data senders and fraudulent ones, effectively thwarting eavesdropping, and node tampering attempts.

3.   Cryptography:

The utilization of cryptographic techniques plays a pivotal role in securing IoT data (Rana et al., 2022). Cryptography involves encoding information in a manner that renders it unreadable without the corresponding decryption key. By employing cryptographic protocols, data is protected from unauthorized access and prying eyes.

4.   Error Detection:

The implementation of error detection mechanisms is another essential facet of IoT security (Swessi et al., 2022). These mechanisms serve to identify and rectify data transmission errors that malicious actors may seek to exploit. By swiftly detecting and addressing tampering attempts, these error detection processes enhance data integrity.

5.   Risk Assessment:

Comprehensive risk assessments are integral to identifying and mitigating potential vulnerabilities within IoT systems (Elmarady et al., (2021). Through proactive evaluation of security weaknesses, organizations can bolster their overall security posture and reduce the likelihood of falling victim to eavesdropping attacks.
In synergy, these mitigation methods work to uphold the security and integrity of IoT-based smart environments (Kornaros, G. 2022). effectively safeguarding against data security breaches and ensuring the confidentiality and privacy of sensitive information.
When we talk about the networking related attacks the most popular attack, we can name it as which is:

2.3.6 DDoS Attack

In the landscape of IoT (Internet of Things) networks, Distributed Denial of Service (DDoS) attacks represent a formidable threat characterized by a well-orchestrated assault on the availability of target servers or networks (Kumari et al., 2023). The attack unfolds in a systematic manner, commencing with the compromise of a multitude of IoT devices that are transformed into a botnet under the control of malicious actors. These compromised devices serve as the means to execute the attack by inundating the target with an overwhelming deluge of requests or traffic. This orchestrated onslaught is designed to saturate the target's resources, such as bandwidth (Tushir et al., 2020) and processing power, rendering it incapable of responding to legitimate user requests effectively. The distributed and coordinated nature of DDoS attacks makes them particularly challenging to thwart, underscoring the critical importance of robust security measures within IoT ecosystems to mitigate their disruptive potential.

2.3.6.1 Mitigation Methods Against DDoS Attacks in IoT Networks

Safeguarding against DDoS (Distributed Denial of Service) attacks stands as an imperative. These attacks, characterized by orchestrated efforts to inundate a target server or network with an overwhelming volume of requests, pose a formidable threat (Chaganti et al., 2023). To counteract the disruptive potential of DDoS attacks within IoT ecosystems, a set of robust mitigation strategies comes into play:

1. Traffic Filtering: Implementing comprehensive traffic filtering mechanisms represents a frontline defence against DDoS attacks. These mechanisms are adept at identifying and isolating malicious (Gupta et al., 2022) traffic patterns, thereby curbing the impact of an ongoing attack. By distinguishing legitimate requests from malicious ones, network resources can be preserved, ensuring continued availability and reliability.
2. Device Hardening: The foundation of IoT security lies in device hardening. Ensuring that all IoT devices are securely configured is pivotal. This involves maintaining up-to-date firmware, robust authentication mechanisms, and stringent access controls. Such measures fortify the devices (Kiran et al., 2022) against compromise and their potential enlistment into botnets orchestrated by attackers.
3. Network Monitoring: Continuous vigilance in the form of network monitoring plays a pivotal role. By scrutinizing network traffic for unusual patterns and anomalies, security professionals can detect the early signs of an impending DDoS attack (Jmal et al., 2023).Timely identification equips organizations

with the knowledge required to mount a rapid response and mitigate the threat effectively.

4. Rate Limiting: The judicious application of rate limiting mechanisms to incoming traffic is a prudent strategy. By imposing rate limits on requests, the network can thwart the rapid influx of traffic during an attack (Feraudo et al., 2024), thereby preventing resource exhaustion and preserving service availability for legitimate users.

5. Cloud-Based DDoS Protection: Leveraging cloud-based DDoS protection services is increasingly crucial. These services function as robust shields that can absorb and dissipate the impact of malicious traffic in the cloud, safeguarding the on-premises infrastructure from being overwhelmed. This approach ensures that legitimate traffic flows (Salah et al., 2022). unimpeded to its intended destination, even in the face of a DDoS onslaught.

these mitigation strategies form a comprehensive defence against the disruptive forces of DDoS attacks within IoT networks (Ibrahim et al., 2022). By diligently implementing these measures, organizations can fortify their IoT ecosystems, ensuring the uninterrupted availability, reliability, and security of critical services.

2.3.7 Routing Attacks in IoT Networks

Routing attacks present a formidable challenge within IoT (Internet of Things) networks. These attacks, often instigated by malicious nodes, can disrupt the normal flow of data by manipulating routing paths. Common tactics employed include route redirection, the creation of routing loops, interference with device movement, and the dissemination of false error messages. These actions can convolute the IoT system, causing data leakage, unauthorized access, network congestion, and degradation of overall network performance. The ramifications of routing attacks are significant, necessitating a robust defence. Mitigation strategies encompass secure routing protocols (Patel et al., 2023), intrusion detection systems, access control, data validation, and continuous network monitoring, all of which work collectively to bolster the resilience of IoT networks against routing attacks. By understanding the nature of these attacks and implementing comprehensive security measures, IoT systems can better safeguard their data and ensure the integrity and efficiency of routing processes.

The mitigation methods that need to be applied to be safe form this attack is.

2.3.7.1 Secure Routing Protocols

Implementing robust and secure routing protocols specifically designed for IoT networks is paramount. These protocols incorporate advanced encryption and authentication mechanisms to ensure the integrity and confidentiality of routing information (Ganesh, D. E. 2022). By adopting these protocols, IoT systems can fortify their defences against routing attacks and maintain secure routing pathways.

2.3.7.2 Intrusion Detection Systems (IDS)

Employing Intrusion Detection Systems within the IoT network is crucial for identifying and responding to anomalous routing behaviour indicative of attacks. These systems continuously monitor network traffic (Heidari et al., 2023), flag suspicious routing changes, and initiate corrective actions to restore normal network operation. IDS serves as an essential layer of defence against routing attacks.

2.3.7.3 Access Control Measures

Stringent access control measures must be enforced to restrict access to critical routing components and configurations. Only authorized personnel should possess the privileges to modify routing tables or protocols (Khalid et al., 2023). Access control ensures that routing information remains in the hands of trusted individuals, mitigating the risk of unauthorized routing manipulation.

2.3.7.4 Data Validation Mechanisms

To bolster security, IoT networks should incorporate robust data validation mechanisms that verify the authenticity and integrity of routing information before it is accepted and acted upon. By rigorously validating routing updates, networks can detect and reject false or malicious (Navarro et al., 2022) routing data, preventing it from influencing routing decisions.

2.3.7.5 Continuous Network Monitoring

Real-time network monitoring plays a pivotal role in detecting routing attacks promptly. It involves the continuous observation of network behaviour to identify deviations or suspicious routing activities (Sundberg et al., 2023). With vigilant monitoring in place, IoT systems can swiftly detect and respond to routing attacks, mitigating potential disruptions, and safeguarding network integrity.

Incorporating these five mitigation methods as part of a comprehensive security strategy empowers IoT networks to defend against routing attacks effectively (Agiollo et al., 2021). By prioritizing secure protocols, intrusion detection, access control, data validation, and vigilant monitoring, IoT systems can maintain the reliability and security of their routing processes, ensuring uninterrupted and secure data transmission.

2.3.8 Mitigation Strategies Against Traffic Analysis Attacks in IoT Security

2.3.8.1 Traffic Analysis Attacks in IoT Security

Traffic analysis attacks represent a critical concern within the domain of IoT (Internet of Things) security. These attacks are characterized by the adversary's initial reconnaissance phase, where network information is surreptitiously gathered. Common methods for such reconnaissance include port scanning and packet sniffers. Subsequently, attackers engage in the analysis of network traffic originating from user devices (Hafeez et al., 2019), with the goal of extracting sensitive user information. A notable characteristic of traffic analysis attacks is their passive nature, rendering them challenging to detect due to their non-invasive approach .

2.3.8.2 Mitigation Strategies at the Network Layer

To effectively mitigate the substantial risks posed by traffic analysis attacks, a series of countermeasures can be implemented at the network layer:

1.   Authentication Mechanisms:

Robust authentication mechanisms play a pivotal role in preventing unauthorized access to data residing on sensor nodes (Ortega-Fernandez et al., 2023). These mechanisms necessitate that only authenticated and authorized entities, whether users or devices, are granted access to the IoT (Ali et al., 2019) network. This layer of security serves as a fundamental gatekeeper against potential intrusions.

2.   Encryption Processes:

Encryption emerges as a cornerstone in IoT security. It involves the transformation of data into an unreadable format without the corresponding decryption key. By employing encryption during data transmission, (Albalawi et al., 2022) IoT systems fortify their defences, rendering sensitive information impervious to prying eyes during transit. This formidable safeguard significantly increases the complexity of any potential data extraction attempts through traffic analysis (Jangjou et al., 2022).

3.   Secure Routing Algorithms:

Preserving the integrity and confidentiality of data exchanges among sensor nodes (Gudla et al., 2023) hinges on ensuring routing security. The adoption of secure routing algorithms is instrumental in achieving this objective. For instance, the strategic utilization of multiple paths for secure routing not only enhances the network's resilience to faults but also bolsters overall system performance by rectifying errors. Noteworthy among these protocols is the Ad hoc On-demand Multipath Distance Vector (AOMDV) routing protocol (Patra et al., 2022), which exhibits the capability to effectively counter various attack vectors, including the intricate wormhole attacks.

By integrating these network layer countermeasures, including robust authentication mechanisms, encryption processes, and secure routing algorithms, IoT ecosystems can fortify their defences against traffic analysis attacks (Roberts et al., 2023). In doing so, these environments are poised to uphold the confidentiality and integrity of their data, thereby shielding sensitive information from potential malevolent actors, and preserving the sanctity of IoT security.

2.4 Security Challenges in Cloud Platform in This Smart World

2.4.1 Cloud Malware Injection in Cybersecurity

Cloud malware injection is a highly concerning cybersecurity threat in the world of cloud computing. This attack vector encompasses various techniques employed by malicious (Madhvan et al., 2023) actors to gain unauthorized control, inject malicious code, or introduce rogue virtual machines into cloud environments. These attackers often disguise themselves as legitimate services, making detection challenging. Once inside the cloud, they access, capture, and potentially modify sensitive data. Cloud malware injection (WaqasAhmad1 et al., 2022). not only compromises data integrity but also poses risks of identity theft, financial fraud, and system compromise. To mitigate these risks, robust security measures are imperative, including stringent access controls, regular security scanning, encryption, continuous monitoring, and user education. Understanding the intricacies of cloud malware injection and implementing comprehensive security strategies are vital for safeguarding cloud resources and maintaining the security and integrity of cloud-based services and data in research paper.

We need to take this Mitigation Strategies such as:

Robust Access Controls: Implement stringent access controls and authentication mechanisms to prevent unauthorized access to cloud resources (Aljumah et al., 2020).

- Security Scanning: Regularly scan cloud environments for vulnerabilities and malware (Prabhavathy et al., 2022), employing intrusion detection and prevention systems.
- Encrypted Communication: Encrypt data transmissions (Tanveer et al., 2023) and storage within the cloud to protect sensitive information from interception.
- Continuous Monitoring: Employ real-time monitoring and analysis tools to detect (Sundberg et al., 2023) and respond to suspicious activities promptly.
- User Education: Educate cloud users and administrators about security best practices and the potential risks associated with cloud malware injection.

The main thing about this cloud malware injection represents a sophisticated and evolving threat to cloud computing (Arul et al., 2022).Understanding the attack techniques and implementing comprehensive security measures are essential for safeguarding cloud resources, maintaining data integrity, and ensuring the overall security of cloud-based services and data.

2.4.2 Storage Security Challenges in IoT Environments

Storage attacks in IoT-based smart environments represent a formidable security challenge. These attacks encompass a range of threats, from data breaches to unauthorized access and data corruption, all of which can result in compromised or falsified information. Whether data is stored in the cloud or on local storage devices, vulnerabilities exist in both domains (Tanveer et al., 2023). Data replication, a common practice for redundancy and accessibility, inadvertently expands the attack surface. Furthermore, diverse user access levels within smart environments add complexity to securing data storage. Mitigating these threats requires robust security measures, including encryption, access control, security audits, intrusion detection systems (Sharadqh et al., 2023), and compliance with data protection regulations. The role of secure backups is pivotal in restoring data integrity in the event of compromise. As smart environments continue to evolve, emerging technologies like blockchain are being explored to enhance the security of data storage. Addressing storage attacks is paramount in maintaining the confidentiality and integrity of the vast amount of user data at stake in IoT-based ecosystems.

2.4.3 Mitigation Strategies

Effective mitigation strategies against storage attacks involve comprehensive encryption of stored data, robust access control mechanisms, regular security audits, intrusion detection systems (Sharadqh et al., 2023), and continuous monitoring of data access and usage. Additionally, organizations must adhere to data protection regulations and compliance standards to ensure data security and privacy.

2.4.4 The Role of Secure Backups

Secure and regularly updated data backups are essential in mitigating the impact of storage attacks. In the event of data compromise, (Kumar et al., 2022) backups enable the restoration of vital information to minimise disruption and data loss.

2.4.5 Emerging Technologies

As IoT ecosystems evolve, emerging technologies such as blockchain are being explored to enhance the security of data storage. Blockchain's decentralized and tamper-resistant nature offers potential solutions for ensuring data integrity in storage. Storage attacks in IoT-based smart environments are a significant concern due to the sheer volume of data involved and the potential consequences of data compromise or manipulation. Robust security measures, vigilant monitoring, and compliance with data protection (Whaiduzzaman et al., 2022) regulations are imperative to safeguard sensitive information stored in both cloud and local storage solutions.

2.4.6 Side-Channel Attack

Side-channel attacks represent a potent class of security threats, wherein attackers exploit subtle, unintended information leakage from a system's implementation (Nassiri Abrishamchi et al., 2022). These attacks capitalize on exposed side properties, such as power consumption patterns, processing timing variations, or even acoustic emissions. The vulnerability arises from the absence of robust security measures in handling IoT data, such as the unsafe practice of storing unencrypted data on cloud platforms or IoT devices themselves.

1.   Web Application Scanners and Web Firewall Applications: These tools serve as the first line of defence against web-based attacks. Web application scanners identify vulnerabilities in web applications, while web firewall applications (Nasrullayev et al., 2023) act as a protective barrier, detecting and blocking malicious traffic and requests. Together, they safeguard against a myriad of web-based threats.
2.   Cryptography Techniques: Employing advanced cryptographic techniques is essential to protect sensitive data within the platform layer. Encryption ensures that data remains confidential and secure, even if an attacker gains access to

the underlying infrastructure (Rana et al., 2022). Robust encryption methods are vital for safeguarding IoT systems.

3.   Memory Protection (e.g., Helper Safe): Memory protection mechanisms, such as Helper Safe, guard against tampering attempts. They ensure the integrity of memory pages, preventing unauthorized modifications. This is crucial for maintaining data consistency and security within the platform layer.

2.4.7 Fragmentation Redundancy for Cloud Security

To enhance data security and thwart confidential information leakage in cloud environments, the use of fragmentation redundancy is paramount. This strategy involves dividing and distributing data across multiple fragments or parts stored on different servers (Youssef, A. E. 2020). By dispersing data in this manner, even if one fragment is compromised, the attacker gains access to only a portion of the data, limiting the extent of potential damage and maintaining data confidentiality and integrity. Incorporating these platform layer security concepts and fragmentation redundancy measures is critical for IoT systems to safeguard sensitive data, protect against various attacks, and ensure robust security throughout the platform layer and cloud storage.

2.4.8 Reprogram Attacks in IoT Security

IoT (Internet of Things) security, the spectre of "Reprogram Attacks" looms large, particularly when network programming systems are utilized for the remote reprogramming of IoT devices. This form of cyber threat involves the exploitation of vulnerabilities within the programming process, effectively affording malicious actors the means to assert control over a significant portion of an IoT network (Bekri et al., 2022). In this discourse, we embark upon a more profound exploration of reprogram attacks, delving into their intricacies, and shedding light on pivotal aspects concerning mitigation strategies:

2.4.8.1 Vulnerabilities in Network Programming Systems

At the heart of reprogram attacks lies a predilection for targeting vulnerabilities inherent in network programming systems. These systems, often responsible for remotely updating or altering IoT device firmware (Mahbub, 2022), can serve as potential entry points for malevolent entities should they remain inadequately secured.

2.4.8.2 Unauthorized Code Execution

Within the framework of reprogram attacks, assailants adeptly exploit programming frailties to introduce unauthorized code into IoT devices during the reprogramming process (Zhu., 2019). This subterfuge is a direct path to device compromise, potentially bestowing upon attackers a degree of control over these devices and unfettered access to the overarching network.

2.4.8.3 Scale and Impact

Distinguishing reprogram attacks from other cyber threats is their capacity for significant scale and impact. Given that IoT networks (Neshenko et al., 2019) frequently encompass an extensive array of interconnected devices, the successful compromise of a single device can precipitate a chain reaction, endowing perpetrators with control over a substantial portion of the network.

2.4.8.4 Mitigation Strategies

Effectively countering reprogram attacks necessitates the deployment of a multifaceted strategy, which encompasses:

a. Secure Programming Systems: The bastion of defence lies in the implementation of robust security measures within network programming systems (Steingartner et al., 2023). This includes, but is not confined to, the encryption of communication channels, the adoption of secure authentication methods, and the periodic updating and patching of these systems to address emerging vulnerabilities.

b. Code Signing and Validation: The employment of code signing, and validation mechanisms assumes a pivotal role in ensuring that only trusted and authenticated firmware updates gain acceptance by IoT devices (Zheng et al., 2022). This serves as an effective bulwark against unauthorized code execution.

c. Access Control: Stringent access controls must be enforced, limiting access to network programming systems (Steingartner et al., 2023) solely to authorized personnel. Such restrictions serve as a barrier against potential attackers.

d. Network Segmentation: IoT networks should be subjected to prudent segmentation, limiting the spread of reprogram attacks. This involves the isolation of critical devices (Abdel-Rahman, 2023) from those of lesser import, averting the rapid dissemination of compromises.

e. Regular Device Updates: The unerring application of security updates and patches to IoT devices assumes paramount importance. Ensuring that devices receive and incorporate timely updates acts as a robust defence against the exploitation of known vulnerabilities.

When I say the spectre of reprogram attacks casts a formidable shadow over IoT networks, given their potential for extensive ramifications. To safeguard IoT systems (Mishra et al., 2021), it is imperative that network programming systems be fortified, code signing, and validation mechanisms be employed, access controls be implemented, network segmentation be practiced, and a proactive stance towards device security updates be embraced. Through the assiduous adoption of these measures, IoT networks can appreciably diminish the perils posed by reprogram attacks, concurrently reinforcing the overall security of their systems.

2.4.9 Sniffing Attacks in IoT Environments

A sniffing attack in the context of IoT (Internet of Things) systems unfolds when a malicious actor deploys a specialized sniffer application with the intent of exploiting system vulnerabilities. This covert makeover carries substantial consequences for IoT networks. Within a sniffing attack, the attacker surreptitiously infiltrates the IoT system by deploying a tailored sniffer application. This application is specifically crafted to eavesdrop on network communications, intercepting and capturing data packets discreetly as they traverse the network.

2.4.9.1 Unauthorized Access to Network Information

Once embedded within the system, the sniffer application quietly intercepts and scrutinizes network traffic, gaining access to critical network information, including the content of data packets, origin and destination addresses, and communication protocols (Saritepeci et al., 2023). Through the interception of unencrypted data, malicious actors may potentially extract sensitive information or gain insights into network operations.

2.4.9.2 Corrupted System and Access to Confidential Data

Sniffing attacks possess the potential to compromise the integrity of the IoT ecosystem, leading to a corrupted environment where data privacy, confidentiality (Xia et al., 2019), and system reliability are gravely imperilled. Successful attacks may provide attackers unauthorized access to confidential user data, including personally identifiable information (PII), authentication credentials, sensitive sensor readings or proprietary business data, with repercussions ranging from privacy violations to financial losses.

To thwart the perils posed by sniffing attacks in IoT systems, a comprehensive security strategy is imperative(Goel et al., 2023). The robust implementation of encryption, network segmentation, intrusion detection systems, regular software maintenance, and informed user practices are instrumental in safeguarding IoT networks against these surreptitious cyber threats.

1. Encryption: Implementing robust encryption protocols for data in transit renders intercepted data indecipherable to would-be attackers. This ensures that even if data packets are captured, their content remains confidential (Jangjou et al., 2022).
2. Network Segmentation: Dividing IoT networks into distinct zones with stringent access controls constrains the reach of sniffing attacks (Kamruzzaman et al., 2023). Unauthorized access to sensitive network segments becomes substantially more challenging.
3. Intrusion Detection Systems (IDS): The deployment of IDS within the IoT infrastructure aids in identifying anomalous network activity that may signal a sniffing attack. Real-time monitoring and alerts facilitate swift responses to security breaches (Heidari et al., 2023).
4. Regular Software Updates: Maintaining up-to-date devices and software within the IoT ecosystem, complete with essential security patches, serves to close potential vulnerabilities that attackers could exploit (Zhu., 2019).
5. User Education: Raising awareness among end-users and administrators regarding the risks associated with sniffing attacks and the imperative of robust security practices is instrumental in thwarting these surreptitious cyber threats (Tanveer et al., 2023).

Sniffing attacks constitute a formidable menace to IoT systems, clandestinely intercepting network communications. Their potential for unauthorized access to sensitive information underscores the critical need for steadfast security measures, including encryption, network segmentation, intrusion detection, software maintenance and informed user practices, to shield against these insidious cyber perils.

3. ROLE OF AI IN SMART HOMES/BUILDINGS

Artificial Intelligence (AI) has assumed a pivotal role in the evolution of smart homes and buildings, fundamentally transforming these spaces into highly efficient, secure, and user-centric environments (Sleem et al., 2023). Its multifaceted impact extends across various domains, including automation, energy efficiency, security, and personalized experiences, ushering in a new era of intelligent living.

Revolutionizing Automation:

AI is at the forefront of redefining automation within smart homes and buildings. It orchestrates a seamless symphony of smart devices and appliances, regulating lighting, heating, cooling, and security systems with unparalleled precision. AI systems adapt dynamically to user preferences and real-time conditions (Zakzouk et al., 2023), optimizing energy consumption and delivering tailor-made comfort. Through continuous learning from user behaviour and environmental cues, AI ensures that homes operate with peak efficiency, enhancing convenience and peace of mind.

3.1 Energizing Efficiency

AI's profound impact on energy efficiency is undeniable. Advanced algorithms continuously scrutinize energy consumption patterns, making real-time adjustments to optimize energy usage (Guo et al., 2019). For example, AI can dynamically fine-tune heating and cooling systems based on occupancy and weather forecasts, significantly reducing energy wastage and utility costs. This eco-conscious approach aligns seamlessly with sustainability goals, reducing the carbon footprint of smart homes and buildings.

3.2 Security Beyond Compare

AI introduces a paradigm shift in security and surveillance within smart homes and buildings. Leveraging AI-powered technologies such as facial recognition, object detection, and anomaly detection, sophisticated security systems can recognize authorized residents and visitors while remaining vigilant for signs of suspicious activities or potential intruders (Taiwo et al., 2022). This elevated level of security fortifies the safety of occupants and enhances overall peace of mind.

3.3 Voice Assistants and Beyond

AI's integration with voice assistants and virtual concierges has revolutionized user interaction with smart devices. Prominent examples include Amazon's Alexa, Google Home, Ivee Sleek, Jibo, Athom Homey, Apple HomePod, and Josh Micro (Edu et al., 2020). These voice-driven AI systems empower users to effortlessly manage various functions, from checking the weather to making online orders or summoning transportation services. Natural language interaction redefines convenience in daily life.

3.4 Proactive Maintenance

Predictive maintenance emerges as a critical application of AI in smart buildings. AI meticulously analyses data streams from sensors and systems, pre-emptively identifying when equipment, such as elevators, HVAC or lighting, requires maintenance or repair (Aguilar et al., 2021). This proactive approach minimizes downtime, extends equipment lifespan, and reduces maintenance costs.

3.5 Illuminating Intelligence

AI extends its brilliance to lighting control, seamlessly adjusting lighting levels and colour temperatures in response to natural light, time of day, and user preferences (Chinchero et al., 2020). This not only enhances energy efficiency but also creates inviting and comfortable living environments that adapt to occupant's needs and moods.

3.6 Personalized Living

AI's power is most evident in its capacity to personalize living spaces. By analysing user behaviour and preferences, AI enables homes to anticipate and cater to individual needs (Yigitcanlar et al., 2021). For example, when occupants enter specific rooms, a smart home can dynamically adjust room temperature, select music playlists or fine-tune lighting, crafting an immersive and enjoyable living experience.

3.7 Guardian of Health

Health and wellness monitoring represent an emerging frontier for AI in smart homes (Mukherjee et al., 2020). A multitude of sensors and devices enable AI to monitor vital signs, activity levels, and sleep patterns of occupants. This data is leveraged to provide recommendations for healthier living and to trigger alerts for medical attention when needed.

3.8 Intelligent Appliances

AI-driven optimization reigns supreme in smart appliances. From refrigerators and ovens to washing machines, AI learns usage patterns and adeptly adjusts settings to maximize efficiency (Dhanalakshmi et al., 2023). These appliances streamline daily routines while conserving valuable resources.

3.9 Safety First

AI seamlessly integrates into safety and emergency response systems. Collaborating with fire and smoke detectors, carbon monoxide detectors, and security cameras, AI provides early warnings and rapid emergency responses (Khan et al., 2022). In the event of threats, these AI-enhanced systems swiftly notify occupants and emergency services, bolstering overall safety.

To provide further clarity, we have defined six core clusters of AI functions:

1. Activity Recognition:

Activity recognition is a fundamental AI function in smart environments. AI systems analyse data from various sensors to discern and understand human actions and movements within a space (Franco et al., 2021). This capability allows smart homes and buildings to recognize activities such as walking, sitting, sleeping or cooking. It's particularly useful for enhancing security and safety. For instance, if an AI system detects an abnormal or potentially dangerous activity, it can raise alerts or trigger specific responses, ensuring the well-being of occupants. This technology finds applications in products like Hive Link and Essence Care@Home, where it provides valuable insights into daily routines and can even detect emergencies.

2. Data Processing:

Data processing is the backbone of AI in smart homes and buildings. AI algorithms excel at extracting meaningful insights from vast and diverse datasets. They identify patterns, trends and correlations within data, enabling smart systems to make informed decisions. For example, in devices like the August Smart Lock + Connect and Nest Protect, AI processes data to optimize security settings and provide real-time information on environmental conditions (Dong et al., 2021). Data processing is crucial for making smart homes more efficient, secure, and responsive to user needs.

3. Voice Recognition:

Voice recognition is a transformative AI function that allows users to interact with smart devices and systems through natural language (Alexakis et al., 2019). AI-driven voice assistants like Amazon's Alexa, Google Home, and others leverage sophisticated voice recognition technology. These systems understand spoken commands and queries, enabling users to control various aspects of their homes by simply speaking. Voice recognition extends beyond basic commands to facilitate conversations, answer questions, and even provide entertainment. This function makes smart homes more accessible and user-friendly, enhancing the overall living experience.

4.   Image Recognition:

Image recognition is a powerful AI capability used extensively in security and surveillance applications within smart homes and buildings (Jaihar et al., 2020). AI systems can analyse images and videos from cameras to identify and categorize objects, faces, emotions, and scenes. This technology is employed in products like Lighthouse, Nest Cam, and Honeywell Smart Home Security System. It enables features such as facial recognition, motion detection and anomaly alerts, enhancing security and providing valuable insights into the environment. Image recognition is a critical component for creating safer and smarter living spaces.

5.   Decision-Making:

Incoming data and autonomously determine the appropriate actions. For instance, in products like Arlo Ultra and Ecobee4, AI evaluates data from sensors and devices to make decisions like triggering alarms, adjusting settings or initiating specific responses. Rapid and effective decision-making is essential for maintaining security, energy efficiency, and overall functionality within smart homes and buildings(Guo et al., 2019). It ensures that the systems can respond swiftly to changing conditions and user preferences.

6.   Prediction-Making:

Prediction-making is a forward-looking AI function that relies on historical data and real-time sensor inputs. AI systems analyse this data to generate predictions, patterns, and trends (Kaya et al., 2021). In smart homes, prediction-making is used in products like Nest Thermostat and Viaroom home. For example, by analysing past temperature trends and current weather data, AI can predict the optimal thermostat settings to maintain comfort while conserving energy. Prediction-making enhances automation and allows smart homes to anticipate user needs and adapt proactively. AI functions encompass a wide range of capabilities that collectively enable smart homes and buildings to deliver efficiency, security, convenience, and personalization. By harnessing these AI functions, these environments become more responsive, adaptive, and user-centric, ultimately enhancing the quality of life for occupants.

4. ROLE OF IOT DEVICES IN SMART ENVIRONMENT

In various application domains, the deployment of IoT technology has demonstrated significant potential for enhancing productivity. One prominent area where IoT technology is making a profound impact is in factory and plant automation as its much better proved by Iqbal H. Sarker in (Sarker et al., 2023), often categorized under the Industrial Internet of Things (IIoT). As elucidated by research, IIoT applications have become instrumental in streamlining industrial processes. Key concerns in this context encompass aspects such as reliability, which is bolstered by redundancy measures, robust security protocols and the adaptability of IoT solutions to endure challenging environmental conditions.

Furthermore, IoT's transformative capabilities extend to the healthcare sector in paper (Krishnamoorthy et al., 2023) shows us how better Wireless body area networks (WBAN) can be used in particularly in the networking of biomedical instruments and databases within hospitals. This integration not only holds the promise of significantly augmenting the quantity and availability of diagnostic and treatment decisions, but it also carries profound implications for rural and remote clinics. These clinics stand to benefit greatly from IoT-enabled access to specialist opinions, thereby overcoming geographical barriers and enhancing healthcare services. Extension of medical instrumentation to the home environment has yielded remarkable improvements in the quality of life for patients while simultaneously reducing the frequency of hospital readmissions. Consequently, IoT technology continues to carve a path towards more efficient and effective solutions in both industrial and healthcare settings, underpinning a promising future for enhanced productivity and improved quality of life. Over the past two decades, the automotive industry has witnessed a remarkable transformation driven by the proliferation of electronics, characterized by the integration of numerous networked microprocessors. The next phase of evolution in this sector is poised to usher in a new era of vehicle-to-vehicle and vehicle-to-infrastructure communication. The critical drivers shaping this transition encompass standardization, security, and cost-efficiency.

The world of Transport and Logistics has embraced the widespread use of RFID tags for the tracking the measures to use is clearly explained in (Ali et al., 2023) and monitoring of shipments, pallets, and even individual items, thereby optimizing supply chain operations. Ongoing research endeavours are directing their focus towards the development of smart tags endowed with the ability to record and report essential transport conditions, ranging from shock and tilt to temperature, humidity and pressure. The primary impetus behind this pursuit lies in achieving cost-effective solutions and enabling seamless communication with the multitude of tags simultaneously deployed. The influence of Internet of Things (IoT) technology transcends the automotive and logistics sectors, extending its disruptive reach across a diverse spectrum of industries such as entertainment, dining, public transport, sport and fitness, telecommunications, manufacturing, hotels, education, environmental science, robotics, and retail. In these varied domains, IoT stands as a pivotal catalyst for innovation and success, prompting industries to make substantial investments in cutting-edge technologies. To ensure the robust security and uninterrupted availability of their systems, businesses in these sectors are increasingly availing themselves of specialized IT support, either through in-house expertise or external providers, aligning their technological infrastructure with their evolving operational needs.

The intricate landscape of the Smart Environment, focusing on the nuances of professional system design, installation, and configuration. Notably, professional support is readily available when incorporating smart electronics as an integral part of a new construction project. The majority of Smart Environment IoT (Internet of Things) technology implementations occur incrementally, as needs evolve and necessitate the retrofitting of existing spaces. In this context, there is often a lack of ongoing professional guidance, both in the initial design phase and during the operation of IoT systems within the Smart Environment.

While some established Smart Environment standards, such as X.10 powerline-carrier communications exist . It is essential to highlight that these standards were conceived before the integration of environmental control networks with the Internet and therefore it lacks vital security measures. In the present landscape, a plethora of networking standards is available for the Smart Environment, including Zwave, Insteon, Bluetooth, Zigbee, Ethernet, WIFI, RS232, RS485, C-bus, UPB, KNX, EnOcean, and Thread, each with its own set of advantages and drawbacks. Efficiently and securely managing a heterogeneous network featuring various protocols (Ramesh et al., 2023) remains a formidable challenge, especially when entrusted to non-experts. Consequently, this paper underscores the significance of addressing these challenges to ensure the seamless operation and security of Smart Environment IoT systems in an increasingly interconnected world. And mainly when we see in the Smart Buildings potentially provides additional comfort and security, as well as enhanced ecological sustainability. For example, a smart air conditioning system can use a wide variety of household sensors and web-based data sources to make intelligent operating decisions, rather than simple manual or fixed-schedule control schemes (Jha, 2023). The smart air conditioning system can predict the expected building occupancy by tracking location data to ensure the air conditioner achieves the desired comfort level when the building is occupied and saves energy when it is not. In addition to enhanced comfort, the Smart Building can assist with independent living for the aging. The Smart Building can assist with daily tasks such as cleaning, cooking, shopping, and laundry. Low-level cognitive decline can be supported with intelligent building systems to provide timely reminders for medication. Home health monitoring can signal caregivers to respond before expensive and disruptive hospitalization is needed. None of these benefits is likely to be taken up if the Smart Building system is not secure and trusted.

5. ROLE OF CLOUD PLATFORM IN THE SMART ENVIRONMENT

Evolution of Cloud Computing and Its Contemporary Impact over the course of several decades, computer technology has continuously evolved, facilitated by the proliferation of Internet connectivity. Notably, in 1961, Professor John McCarthy of the Massachusetts Institute of Technology (MIT) introduced the concept of cloud computing (Zawaideh et al., 2022). McCarthy's visionary insights included the concept of pay-as-you-go models for contemporary cloud platforms. He anticipated a future where computing infrastructure would be organized as public utilities, allowing subscribers to pay for the capacity they utilize while granting access to a wide array of programming languages on large-scale systems.

5.1 The Ubiquity of Cloud Computing

Cloud computing has emerged as an omnipresent and convenient network access technology, simplifying the provisioning of services and the sharing of resources with minimal administrative effort (Thyagaturu et al., 2022). This technological paradigm is defined by several key characteristics and service models, providing a standardized approach to their utilization.

Key Characteristics of Cloud Computing Include:

1. On-demand self-service
2. Scalability to connect multiple devices
3. Resource pooling for efficient resource utilization
4. High levels of resiliency for uninterrupted operations
5. Measurement services for performance assessment
6. Diverse Cloud Deployment Models

Cloud computing services are typically categorized into three primary deployment models to meet diverse business needs.

1. Private Cloud: Privately managed and operated by a single organization or group, the private cloud, also known as an internal or in-house cloud (Kaur et al., 2023), offers controlled access to resources and services. It emphasizes security, privacy, and governance.
2. Public Cloud: The public cloud, provided by large-scale infrastructures, is accessible to a wide range of users, including large industry groups. This model is maintained by organizations such as public institutions (Dzulhikam, 2022).

3. Hybrid Cloud: The hybrid cloud(Abbes et al., 2023) configuration combines two or more unique clouds, such as a private cloud and a community cloud or public cloud. It facilitates the integration of standardized technology for data and applications, allowing companies to maintain a stable environment by merging the benefits of private and public cloud solutions Maniatis mentioned in detailed in his work.Service Models Based on Purpose

5.2 Deployment Models Align With Specific Service Models Tailored to User Objectives

These service models encompass:

Software as a Service (SaaS): A comprehensive software solution that delivers cloud applications, basic IT infrastructure (Marian et al., 2022), and platforms. Typically, accessible via web browsers(Rahman, 2023), it offers a multi-tenant environment and usage-based pricing. Notable examples include Gmail.

Infrastructure as a Service (IaaS): This service encompasses infrastructure-related components such as server resources, IP, network, storage, and power for server operation(Petrovska et al., 2022). IaaS streamlines infrastructure management, reducing cost and complexity by eliminating the need to own and maintain physical servers and data centres(George, 2023).

Platform as a Service (PaaS): PaaS provides a complete development and deployment environment within the cloud, facilitating the development of a wide range of cloud-based applications(Kalp Chobisai et al., 2022), from basic apps to complex enterprise solutions.

5.2.1 The Democratization of Technology: Low-Code and No-Code Solutions

In parallel with the evolution of cloud computing, a notable trend has emerged in the form of low-code and no-code solutions. These tools and platforms empower individuals and organizations to create applications and leverage data-driven solutions without extensive coding experience(Venkata Subhadu, 2023).

Low-code and no-code solutions extend to various domains, including website development, web applications, and even the creation of AI-powered applications. These advancements substantially lower the barriers to entry for businesses seeking to harness the power of artificial intelligence (AI) and machine learning (ML) technologies(Rokis et al., 2022). Many of these innovative solutions are hosted in the cloud, providing users with accessible as-a-service options. Tools like Figma, Air table, and Zoho exemplify this trend, enabling users to perform tasks that once required coding expertise(Abidin, 2021). This trend underscores the ever-growing significance of cloud technology, especially in the year 2023(Cheah et al., 2023) and beyond, by enhancing accessibility and user-friendliness, thus democratizing access to technology and innovation.

5.3 Empowering Smart Homes With Cloud Computing

1. Remote Monitoring and Control:
   - Cloud computing enables seamless remote monitoring and control of smart home devices. Homeowners can access(Chataut et al., 2023) and manage their devices, such as thermostats, security cameras, and lights, from anywhere with an internet connection.
   - This functionality enhances convenience and security, allowing homeowners to adjust settings, view real-time footage, and receive alerts on their smartphones or computers(Motevalli et al., 2023).
2. Data Analytics and Machine Learning:
   - Cloud platforms facilitate advanced data analytics and machine learning algorithms within smart homes. These technologies analyse data(Rehman et al., 2022) collected from sensors and devices to optimize various aspects of daily living.
   - For example, machine learning algorithms can learn household preferences and adjust heating or cooling schedules to maximize energy efficiency while maintaining comfort.
3. Voice Assistants and Automation:
   - Cloud-based voice assistants, such as Amazon Alexa and Google Assistant, play a pivotal role in smart homes. These assistants are integrated with cloud(Kumar,2020) services to interpret natural language commands.
   - Residents can use voice commands to control devices(Mao et al., 2023), access information, set reminders, and even automate routines. This integration enhances user experience and simplifies daily tasks(Duong, 2023).
4. Firmware Updates and Security:
   - Manufacturers leverage cloud services to deliver firmware updates and security patches to smart home devices.
   - Regular updates ensure that devices remain current and secure(Yang et al., 2023), addressing vulnerabilities and improving overall system reliability. This cloud-driven approach enhances the long-term performance and safety of smart home ecosystems(Saad et al., 2023).

5.4 Enhancing Smart Buildings With Cloud Technology

1. Energy Management:
   ○ Cloud platforms collect data from sensors and devices deployed throughout smart buildings. This data includes information on temperature, occupancy, lighting usage, and more.
   ○ Advanced analytics and machine learning algorithms process this data to optimize energy consumption(Kumar et al., 2021). For instance, heating and cooling systems can be adjusted in real-time based on occupancy patterns, leading to significant energy savings.
2. Predictive Maintenance:
   ○ Cloud-hosted machine learning models analyses data from building sensors to predict equipment failures.
   ○ Predictive maintenance schedules enable proactive servicing of building systems(Gholamzadehmir et al., 2020), preventing costly downtime and enhancing the operational lifespan of critical equipment.
3. Occupant Experience:
   ○ Cloud-based applications and user interfaces empower building occupants with control over their environment.
   ○ These interfaces may include room booking systems, visitor management tools, and parking reservations(Aliero et al., 2021). Occupants can conveniently access these services, enhancing their overall experience within the smart building.
4. Security and Access Control:
   ○ Cloud-based access control and video surveillance systems provide robust security solutions for smart buildings.

Access permissions, granting or revoking access in real-time. Additionally, cloud-connected security cameras enable continuous monitoring(Lokshina et al., 2019), with alerts and video feeds accessible from anywhere.

5.5 Transforming Smart Cities With Cloud Services

1. Traffic Management:
   ○ (Kumar et al., 2020) Cloud platforms process data from an array of sources, including traffic sensors, cameras, and GPS devices, to optimize traffic flow and reduce congestion in smart cities(Kim et al., 2021).
   ○ Dynamic traffic management systems use real-time data to adjust traffic signals, reroute vehicles, and provide commuters with accurate

information, improving overall transportation efficiency(Swain et al., 2023).

2. Waste Management:
    ◦   Smart waste bins equipped with sensors transmit data to cloud platforms, revolutionizing waste management in urban areas(Aroba et al., 2023).
    ◦   The cloud facilitates efficient waste collection scheduling based on fill levels(Ijemaru et al., 2023), reducing unnecessary pickups and associated costs while contributing to cleaner, more sustainable cities.

3. Emergency Response:
    ◦   Cloud-based systems are essential for real-time communication and coordination among emergency services, public safety agencies(Shiny et al., 2023), and city officials.
    ◦   These systems streamline emergency response processes, allowing for rapid deployment of resources during critical situations and enhancing overall disaster preparedness and public safety.

4. Data Analytics for Urban Planning:
    ◦   Cloud services aggregate and analyses diverse urban data sources, including air quality sensors, energy consumption data, and citizen behaviour patterns(Ali et al., 2023).
    ◦   The insights gained from data analytics support informed urban planning decisions, sustainable development initiatives, and evidence-based policies for building smarter and more resilient cities.

5. Citizen Engagement:
    ◦   Cloud-based applications and digital platforms connect citizens with city services and resources.
    ◦   Through these platforms, citizens can access information, report issues, participate in civic activities, and engage with their communities(Lepore et al., 2023), fostering a sense of belonging and involvement in shaping the future of their cities.

The integration of cloud computing technologies plays a pivotal role in revolutionizing the functionalities and efficiencies of smart homes, smart buildings, and smart cities(Belli et al., 2020). Cloud platforms enable remote access, data analytics, automation, and enhanced security, ultimately contributing to the creation of more sustainable, interconnected, and liveable urban environments.

6. The set of device types that we need to use to make the smart environment much smarter and private (Table 1).

Table 1. Device types that make the smart environment smarter and more private

Item	Image Quality/ Video Quality	Area Covered	Network Access Capability	Connectivity	AI and Computer Vision Capabilities	Power Source	Privacy and Security	User Interface and Ease of Use	Message Transfer Speed between IOT Devices
CCTV CAMERA	2 Megapixel 1/2.8" Progressive CMOS, 1920(H)x1080(V)	56.2° (wide), 2.6° (tele)	100Base-TX/10BaseT (RJ-45)	Wi-Fi 2.4 & 5 GHz	YES	48 W	YES	Auto & Manual, 2 Inputs; 1 Output	30 frames per second for smooth video, 60fps
	1/2.8" CMOS sensor, 3840X2160 (8.0 Megapixels) at 25 frames/sec	Horizontal: 105°-31, Vertical: 55°-17°, Pan: 360°, Tilt: 0°~90°	One 10M/100Mbps; RJ45	Wi-Fi 2.4 & 5 GHz	YES	DC 12.0V~2A, <24W	YES	Auto & Manual	15fps, 25fps video
	5 MP CMOS sensor	360 DEGREES	GbE RJ45 port	Wi-Fi 2.4 & 5 GHz	YES	37 to 57VDC	YES	Auto & Manual	30fps & 60fps
	4 megapixels, 2560X1440 resolution	160 DEGREES	RJ45 port	Wi-Fi 2.4 & 5 GHz	YES	100V-240V	YES	Auto & Manual	2-4Mbps
	Security Features	System Programs/ Security	Speed	LAN/WAN Ports	Processor & Memory	AI Integration	MU/MIMO	Beam Forming	Additional features
SMART ROUTER	YES	WEP, WPA-PSK, WPA-Enterprise (WPA/WPA2, TKIP/AES), 802.11w/PMF	6.5 Mbps to 1.7 Gbps (MCS0 - MCS9 NSS1/2/3/4, VHT 20/40/80/160)	5 GbE Ports: 1 WAN and 4 LAN, GbE RJ45 port	ARM Cortex-A57 Quad-Core at 1.7 GHz. 2 GB DDR RAM	YES	2.4 GHz 2x2, 5 GHz 4x4	Dual-band, quad-polarity antenna	License-free SD-WAN, Wire Guard, L2TP and OpenVPN server, OpenVPN client, OpenVPN and IPsec site-to-site VPN, One-click Teleport and UID VPN, Policy-based WAN and VPN routing, DHCP relay, Customizable DHCP server, IGMP proxy, IPv6 ISP support
	YES	WEP, WPA, WPA2, WPA/WPA2-Enterprise (802.1 x)	5 GHz: 1300 Mbps (802.11 ac), 2.4 GHz: 600 Mbps (802.11 n)	1 x Gigabit WAN Port, 4 x Gigabit LAN Ports	Qualcomm QCM502 Quad-Core processor	YES	3x3 MI-MIMO	Dual band	Wi-Fi Router Archer A9, Power Adapter, RJ45 Ethernet Cable
	YES	WPA3, WPA2, TLS vl.2+, VPN passthrough, IPv6, NAT, UPnP, port forwarding, DHCP, static IP, and cloud connectivity	AX3000	2 x WAN, 2 x LAN	1 GHz dual-core processor, 512MB RAM, 4GB flash storage	YES	2x2 MU-MIMO	Dual band	Intelligently routes traffic to reduce drop-offs and dead spots.
	YES	WPA3	BE27000 (11,530 + 8,647 + 5,765 +	One (1) 10 Gig Internet WAN port	Quad-core 2.2Ghz processor 4GB Flash and 2GB RAM	YES	2x2 MU-MIMO, 3x3 MU-MIMO	Quad-Band	unleashes speeds up to 27Gbpst for unparalleled performance and coverage for your whole home, from the

Table 1. Continued

	Fire Safety	Sensor Type	Connectivity	Power Source	CO Detection	Area to be Placed	Maintenance/Building Codes	AI Integrated Service	Additional Features
				One (1) 10Gbps Ethernet LAN port Four (4) 10/100/1000/2500Mbps Multi-Gig Ethernet LAN ports		4x4 MU-MIMO		front door to the back yard and the basement to the rooftop
SMOKE/HOME SENSOR	YES	Photoelectric	WiFi, Bluetooth	Battery	YES	Wall	OS requirements: Windows 7 or later; Mac OS 10.9 or later; Chrome OS 38 or later; Browser requirements: Firefox 33 or later; Google Chrome 38 or later; Internet Explorer 11 or later; Safari 7 or later	YES	Safe & Sound is a smart smoke and carbon monoxide alarm that has a built-in speaker, microphone, and Alexa voice assistant. Sensors: Photoelectric and electrochemical
	YES	Photoelectric	WiFi, Bluetooth	Battery	YES	Wall		YES	This mini smoke alarm packs reliable performance into a sleek and modern design with its advanced photoelectric sensor that provides fast and accurate smoke detection.

	Technology Type	Detection Range	Field of View	Sensitivity adjustment	Response time	Environmental Conditions	Coverage Pattern	Connectivity	Integration with Smart home Echo System
MOTION SENSOR	Aeotec's Gen5 technology	5m	120°	5m	1sec	Temperature, Humidity, UV range.	5m	App	Home Assistant
	passive infrared (PIR) technology	12m	160°	5m	1sec	Temperature, Humidity, IP rating.	12m	AI assistant	Send you a notification when someone is at your door.
	Connectivity	Special Features	Mobile App	Voice Control	Key less Entry Methods	Security Features	Remote Access/Guest Access	Integration with Echo System	Power Source

289

Table 1. Continued

SMART LOCK

Control Options			Unlocking Methods	Security Features	Remote	Power
NFC	With the equipped IR sensor, the lock automatically wakes up once it detects you.	YES	fingerprint, number pad, RFID tag, RFID CARD	Push Pull Two Way Latch Mortise. Fingerprint, RF card, password, lock. Anti-hacking encryption protocol. Anti-theft mode. Automatic wake-up on approaching.	YES	4 x AA batteries

SMART BULB/LAMP

BULB type	Brightness Level	Colour Temperature	Colour Options	Connectivity	App Control	Voice Control/Special Features	Dimness	Energy Efficiency/ Life Span
smart LED light bulbs	1100 lumens	6500 kelvins	Multicolour	Wi-Fi, Bluetooth	YES	control and dim your lights using your phone, using voice commands, or using automations	YES	9 W, 22 years
smart LED light bulbs	1200 lumens	6500 kelvins	Multicolour	Wi-Fi, Bluetooth	YES	With the built-in mic in the control box, your strip lights can pick up on the beats of your favourite music. You can also use the Govee Home App to access various music mode to customize the colours and lighting effects.	YES	9 W, 20,000 hours

SMART FAN

Control Options	Speed & Air Flow	Integration with Smart home Echo System	Power Source	Smart Sensors	Noise Levels	Remote Control	Lighting Integration	Special Features
Smart home voice control Mobile App Control Remote control.	6 speeds	Smart Assistants	20 Watts	YES	20-35db	YES	YES	Conveniently quiet. The LED lights ceiling fan had setting the timer function
Pull chain. Remote control	5 speeds	Smart home platforms	30-70 watts.	YES	40-50 decibels (dB)	YES	YES	Energy efficiency, Quiet operation, Smart home integration

SMART REFRIGERATOR

Touch Screen Display	Connectivity	Power Source	Camera	Voice Control	App Integration	Food Management	Temperature Zone	Special Features
YES	IFTTT, WIFI	120v	YES	YES	YES	YES	37 degrees Fahrenheit in the fresh food and 0 degrees Fahrenheit in the freezer sections.	LED Back Lit Wall Fingerprint Resistant Stainless. Easily wipe away smudges and fingerprints for a look that's always sparkling clean. Adjustable Temperature Drawer, Built-in Wi-Fi, Soft-Close Vegetable Drawer.

Table 1. Continued

Touch Screen Display	Connectivity	Power Source	Voice Control	Cycle Options	Water Efficiency/ Power Efficiency	Noise Level	Mobile Apps	Child Lock/ Self Cleaning/ Delay Lock
YES	WIFI, BLUETOOTH	115v	YES	YES	YES	YES	37 degrees Fahrenheit in the fresh food and 0 degrees Fahrenheit in the freezer sections.	-
SMART WASHING MACHINE YES	WIFI, BLUETOOTH	120v	YES	10	4248 Gallons/year	50db	YES	YES
YES	WIFI	220v	NO	11	19 Gallons/cycle	49db	YES	YES

Automation Mechanism	Control Options	Integration with Smart home Echo System	Connectivity	Sensor Support	Security	Remote Access	Power Source/Energy Efficiency	Special Features
SMART CURTAINS YES	Remote control Smartphone app Voice commands Smart home automations	Smart Home Control Assistants	WIFI	motion sensors, door and window sensors, and light sensors	Ring and ADT	YES	12V	allows you to schedule your curtains to open and close at specific times, create custom scenes, and group your curtains together for easy control.
YES	Voice control, remote control, smart phone app, wall switch.	Smart Home Control Assistants	WIFI	Temperature sensors Door and window sensors Motion sensors Light sensors	Ring and ADT	YES	12V	Use the remote-control feature to close your curtains from the comfort of your couch, Use the smartphone app to schedule your curtains to open and close at specific times, so you can wake up to natural light or come home to a well-lit home.

Brewing Capacity	Brewing Options/Speed	Coffee Options/Grinding Option	Programmability	Connectivity	Voice Assistant	Temperature Control	Security	Special Features
SMART COFFEE MAKER capacity of 12 cups	10-15 min	You can choose from three different brew strengths: mild, medium, and strong. The built-in burr grinder offers eight different grind settings, from coarse to fine.	iOS 8 or higher, Android 4.1x or higher	WIFI	YES	YES	We don't have remote direct access to system permission without the owner	Brew temperature: You can choose from three different brew temperatures: low, medium, and high.
capacity of 10 cups	10-15 min	Choose your strength with Classic, Rich, Over Ice, Cold Brew, or Specialty	iOS 8 or higher, Android 4.1x or higher	WIFI	YES	YES	causing a permanent denial of service on the device	This brewing system gives you the ability to brew hot, flavourful cups of coffee and tea or over-ice beverages

Table 1. Continued

	Key Features	Connectivity	Voice Control	Device Support	Security & Privacy	Support & Updates	User Interface/ Energy Efficiency	Connectivity through Cloud Platform	Special Features
SMART HOME AUTOMATED ARTIFICIAL INTELLIGENCE	4-mac Siri, Sound Recognition, multi-room audio	802.11n Wi-Fi, Peer-to-peer discovery for easy guest access, Bluetooth 5.0, Thread, Ultra Wide band chip for device proximity	YES	5 horn-loaded tweeters, Neodymium magnets, bass calibration mic, computational audio, room sensing, Dolby Atoms, Air-Play, stereo pair	Encrypted and anonymous IDs protect your identity	Home-Pod with the latest version of software, 802.11 Wi-Fi Internet Access	Line voltage: 100V to 240V AC, Frequency: 50Hz to 60Hz, Certified by energy star	YES	Highly secured with self-updating and healing features in there security deficiency skills
	Sensor Type	**Accuracy & Precision**	**Data Sampling Rate**	**Connectivity**	**Mobile App**	**Alerts & Notifications**	**Integration with Smart home Echo System**	**Power Source**	**Special Features**
SMART AIR QUALITY SENSORS	Radon, particulate matter (PM2.5)	highly accurate.	Every 10 minutes Hourly	WiFi, Bluetooth	YES	YES	IFTTT, Voice Assistant	6AA Batteries	
	particulate matter (PM) sensor PM1, PM2.5, and PM10 particles	highly accurate.	PM1, PM2.5, and PM10 particles in the air every 1 minute.	WiFi, Bluetooth	YES	YES	IFTTT, Voice Assistant	5V	
	Detection Technology	**Radiation Types**	**Energy Range**	**Accuracy & Precision**	**Response Time**	**Altering/Alarming/Security**	**Data Altering & Storage**	**Connectivity**	**Special Features**
SMART RADIOACTIVE SENSORS	NaI(Tl)-detector with high-quality micro photo multiplier, software switch for R or 5y energy response and calibration	Gamma rays	60 keV to 1.3 MeV	accurate high dose rate radiation measurement	less than 10 seconds	YES	NO	WIFI	detecting radiation by measuring the light flashes produced when radiation interacts with a scintillator material.
	Sensor Type	**Detection Range**	**Adjustable Sensitivity**	**Fiel of View**	**Response Time**	**Light Sensing**	**Mounting Options**	**Power Source**	**Special Features**
SMART BRIGHTNESS LEVEL SENSORS	photodiodes, photoresistors, phototransistors, and photovoltaic light sensors	0-1000lux	YES	120°	respond to changes in the amount of light received	YES	WALL	Corded Electric	Detect light instantly when there is a brightness change and shows on APP as "Notice the light is detected"
	Detection Method	**Accuracy**	**Alert Mechanism**	**Connectivity**	**Battery Life**	**Coverage Area**	**Integration with Smart home Echo System**	**Power Source**	**Special Features**
SMART WATER LEAK DETECTOR	Water detection sensor, Temperature sensor, Humidity sensor.	highly accurate.	YES	WiFi	2 years	10-foot radius of its location.	Smart Assistants	12VDC	Smart Leak Sensing, Remote Monitoring, Automatic Shut-Off, Freeze Warning, Low Battery Alert.

6. PROPOSED METHODOLOGY FOR TRANSFORMING HOME, BUILDING, AND CITIES INTO SMART AND SECURE ENVIRONMENTS

6.1 To Overcome the Challenges in Smart Home We Need To

When we talk about the smart home the world is craving for such a place where everyone purely depends on the devices that can come handy when needed so when the discussion comes to smart home it has special gadgets to control things like automated security systems from a distance. This kind of home has special gadgets and sensors to keep the person safe and monitor them. It is connected to a smart unit that analyses the sensor data and looks out for any dangerous or suspicious situations. It also connects to a control centre that is far away. The RCC (remote control centre) makes sure that there is help available in case of an emergency. It acts as a connection between the person at home and the different people involved in providing care, like those who respond to alarms, emergency experts, social workers, family members or volunteers who help with caregiving. Technicians are people who fix and maintain machines and equipment, while people who manage and control computer system administrators. When comes to the sensors we use automatic devices like motion sensors to control lights, locks that move on their own for the entrance door, machines to open doors and windows and curtains and blinds that move automatically which helps the facilitator to be comfortable when using such gadgets we can find such controlled much more fluidly when we use those then comes to lights that turn on when they sense movement, locks that can be controlled with a motor and doors and windows that can open and close by themselves. They also have blinds and curtains that can open and close with a motor. It could be helpful to have some automatic systems that can detect smoke, gas that can catch fire easily, when a microwave oven is being used incorrectly, or if the water in the bath is too hot or too cold such iot devices can be controlled by a single connected voice assistants when can completely balance the home environment it helps people use it simply by their voice to control the lights, curtains, gates, and door locks in their home. They can also start complex actions by saying a simple phrase like "I am at home" or "I left home" as they mentioned (Tavares, et al.,2021).

To control the light using your voice, you need three things: Internet, a voice assistant, and a connected smart home system. A voice assistant is a software agent that uses voice recognition technology to do things or help you with tasks. Voice-activated home systems that can do tasks automatically have the power to make life easier and more enjoyable, while also making daily activities run more smoothly. Friendly cars have been gaining popularity in recent years. By using voice commands, drivers can control various functions in their car without having to physically touch any buttons or knobs. This not only enhances convenience but also reduces distractions on the road, making driving safer. Additionally, voice control technology in cars contributes to the overall goal of sustainability by minimizing the need for manual operation and promoting the use of renewable energy sources. Friendly houses are good for people with disabilities because they help them live a better life that they couldn't have in the past. Many advantages help and assistance provided in the place where one works. As we placed in the dig1 we can see how the devices are placed in a unique manner by which the homes turn secured and smart which even helps use save energy by Artificial Intelligence (AI) and the Internet of Things (IoT) which combined brought this transformation in a way we can conserve energy in homes. By using analytics and machine learning AIoT systems analyses consumption patterns. Optimize the operation of heating, cooling, lighting, and appliances. Smart thermostats equipped with AIoT capabilities adjust temperature settings based on occupancy and real time data. Homeowners can make decisions, about energy consumption through monitoring and feedback (Syamala et al.,2023). Additionally remote control and automation of devices help reduce power usage. Moreover, AIoT technology enables load balancing, integration of energy sources and provides valuable insights for behavioural change. This continuous learning technology does not improve energy efficiency. Also empowers homeowners to actively contribute to environmental sustainability making it an essential part of smart and eco-friendly living. We brought all the best devices together with proper specifications in the table1 in which we even mentioned the proper specifications.

In this figure 1 we have all the IOT (internet of things) which are connected to a single source which is router and that is controlled by the AI (artificial intelligence) which are basically called voice assistance these devices are like smart bulb where it is used in smart home to modify the LED light's settings to correspond with the ambient brightness, pledges to decrease the amount of electricity the building uses. In this smart bulb, the main tasks that need to be focused on are controlling, monitoring, and maintaining. The aim of this smart bulb design is to keep the LED light bright by managing how strong it is inside the structure. In this smart bulb, it's important to control how bright the LED lights . This is because it can help save electricity by using the LDR sensor. The LDR sensor can tell how bright or dark it is in the surroundings [LIGHT]. We have smart sensors to detect the movement and capture the result and it works accordingly and then comes water quality sensor [WATER], smoke sensor detects smoke in the air, including both particles and gases. It can also measure the amount of smoke present. They have been here for a long time. Now, with the invention of IoT Figure 1, these devices are even better at their job. They are connected to a system that quickly tells the user if anything goes wrong in different businesses. Smoke detectors are used a lot in factories, offices, and places where people stay overnight to find things that might be dangerous for safety. This system helps protect people who work in dangerous areas. It is better at keeping them safe than older systems [smoke detector].

6.2 To Overcome the Challenges in Smart Buildings We Need To

smart building is building which is having with special gadgets to control things like automated security system from a distance. These smart sensors are now necessary part of everyday life. Touch sensors can tell when someone touches a Smart-phone screen. Sensors can be found in different shapes and sizes. There are many kinds of sensors in a typical car. Tire pressure sensors check if a tire is flat or needs air.

Figure 1. Smart home with well-secured devices placed properly in the home

Self-driving cars have special sensors called ultrasonic sensors. These sensors use sound waves to figure out how far away the car is from things around it. There are small detectors in everything we interact with every day. For example, water quality sensor is used for keeping track of the quality of water which is used for many different purposes. They are such a device that are used to check the condition of water in places where water is delivered. Non-drinkable water and dirty water can cause contamination in the distribution system. Leaking pipes in areas with low water pressure can also let polluted water enter the system. Microbes can also grow in the pipes. To avoid water damage from appliances like washing machines and dishwashers, these sensors can be used to solve these problems. Now, with the invention of IoT, these devices are even better at their job. Smoke detectors are used a lot in factories, offices, and places where people stay overnight to find things that might be dangerous for safety as we talk about this, we came up with a design in Figure 2. This system helps protect people who work in dangerous areas. It is better at keeping them safe than older systems these humidity sensors are used in factories and homes to control temperature and air circulation systems. They are also used to keep medicines and other important things safe in cars, museums, factories, gardening places, weather stations, paint and coatings companies, hospitals, and the medicine industry.

Smart buildings are at the forefront of modern architectural and infrastructural innovation, embodying a multifaceted set of attributes that redefine our built environment. At the heart of this transformation is automation, a fundamental concept that underpins smart building's ability to seamlessly integrate and manage automatic devices and functions. This automation transcends mere convenience, it streamlines processes, enhances energy efficiency, and augments security protocols, creating an environment that anticipates and responds to occupants needs. Hallmark of smart buildings is their remarkable multi-functionality. Within a single structure, diverse roles converge, optimizing space utilization and resource allocation. For instance, a conference room can quickly transform into a presentation space, ensuring that space is used to its full potential, minimizing waste and cost.

However, the true power of smart buildings lies in their adaptability. By harnessing the data generated within and around them, these structures continuously learn, predict, and respond to evolving user preferences and external environmental stressors. This adaptability extends beyond the mere physical aspects of the building it encompasses the optimization of energy consumption, climate control, and resource allocation, ultimately enhancing occupant comfort and well-being. Interactivity further distinguishes smart buildings as hubs of connectivity. They foster engagement among occupants and provide platforms for seamless interaction. Whether it's through touchless interfaces, IoT (Internet of Things) devices, or collaborative spaces, smart buildings promote communication, collaboration, and an enhanced user experience.

Figure 2. In smart buildings, comprehensive network coverage seamlessly interconnects IoT devices, all routed directly to the cloud for efficient communication and data management

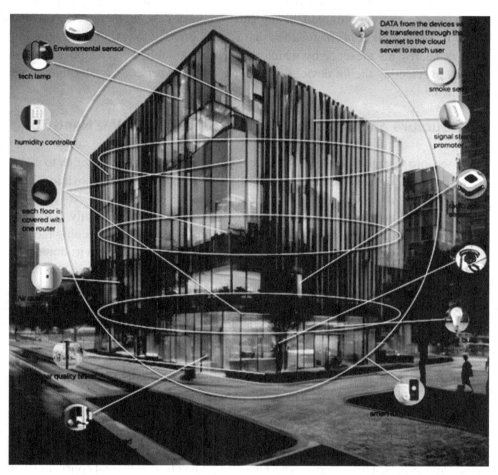

Perhaps most notably, smart buildings prioritize efficiency across the board. They are designed to provide not only energy efficiency but also substantial time and cost savings. Through the implementation of advanced technologies, such as automated lighting and climate control, predictive maintenance, and energy management systems, these buildings reduce resource wastage and operational expenses while optimizing overall performance from Figure 2. In essence, smart buildings represent an architectural revolution that converges innovation, sustainability, and user-centric design. They are the embodiment of a future where our built environments not only serve our immediate needs but also adapt and evolve to meet the demands of a rapidly changing world. The main theme of this chapter is to provide the best way to secure the place in which the information that is gathered or generated by the iot devices in the building will be sent to the cloud by the best networking strategies. In which the cloud server will read the information generated and then it encrypts the data if it seems to be confidential then that is transferred to the AI which is guiding the user nearby his presence by which this can be possible we can find such best options in selecting the iot devices we made a table for better understating we can refer the table1.

6.3 Structure and the Challenges in Smart Cities We Need To

As we see into smart cities, where AI, IoT and cloud services comes together to make a new chapter, in development unfolds, brimming with potential and innovation. This journey is about improving the lives of those who reside in these cities. Just imagine a city where AI assists in predicting traffic patterns making your daily commute smoother and safer. Picture IoT sensors spread throughout the city keeping tabs on air quality and energy consumption to ensure a more sustainable environment for everyone. All of this becomes possible due to the computing power of the cloud – not only storing vast amounts of data but also providing access whenever and wherever its needed. As we venture down this thrilling path smart cities emerge on the horizon promising a future where technology adds value to our lives strengthens our communities and constructs a world for us all. "Smart cities" are cities that utilize innovation to make strides different things like lighting, traffic, emergency administrations, tourism, and more. Imaginative modern Jobs in cities are likely to progressively utilize and centre on innovation. Our focus is on the necessities of specific circumstances or purposes.

Light sources are commonly used in smart cities and are an important part of the Internet of Things (IoT). Many governments are using IoT technology to save money and energy. The system has a router that connects devices and confirms their identity. Moving many device connections to a smart pole variety of purposes. It can be used to control the brightness and colour of the lights in a room, as well as schedule when the lights turn on and off. Additionally, it can be integrated with other smart devices, such as motion sensors or voice assistants, to create a more convenient and efficient lighting system. Overall, smart lighting offers many benefits and options for personalized lighting experiences.

The smart lighting can be used for a variety of purposes a variety of tasks:

1.   Controls for lighting.
2.   Cameras for surveillance.
3.   Electronic billboards.

4. Access to wireless technology.

Using AIoT (Artificial Intelligence of Things) in smart houses makes the maintenance and management of streetlights easier. It is useful and doesn't cost too much. We can make the streetlights work together by installing a system that makes them all light up at the same time. Using sensors and connecting them to a cloud management

T-service. Smart lighting systems are advanced technologies that allow you to control your lights in a convenient and efficient way. Monitor light, people, and vehicle movement, and then combine it with previous information and the current situation. Gather data and analyse it. To make the lighting schedule better road change to red. This indicates to drivers that they should stop their vehicles and allow the pedestrians to cross safely. The crossing can be activated when a bus is getting close to the bus stop. This will also make the streetlights turn on. It could be improved by making it brighter, when we talk about the transportation system in smart cities is growing quickly. Applications are programs or software that are designed to perform specific tasks or functions on a computer or other electronic device. They can be used for a wide range of purposes, such as communication, entertainment, productivity, or information retrieval. In simpler terms, applications are like tools or instruments that help you do different things on your computer or device savings and efficiency by incorporating smart technology into their operations and infrastructure. By using innovative tools and systems, transportation companies and cities can save money and improve their overall transportation processes, making it easier and more efficient for people and goods to move around. This can lead to reduced expenses and better service for everyone involved. Saving money, feeling safe, planning the best way to travel, and having a great experience as a passenger people believe that trains, which have made shipping less necessary. Trains and wireless passenger connections are getting better with more improvements. They assist the efficient public transportation organization that serves busy cities and is run by government officials.

Power buses are vehicles or structures that carry electrical power from a power source to other devices or locations. The smart transportation system's IoT response has many different functions.

All these devices are operated using the router:

1. The vehicle circuit system and a remote association are between the motors and the smarts expedite centre. This increment concerns with respect to the move from the existing analogy network to IP-based voice-to-voice communication.
2. Collection of comfortable fares and mobile ticketing are by means of firewall, IPsec VPN, social isolation, and confirmation.

3. With online service, we can keep an eye on and manage our groups and devices, such as getting updates from large gatherings and monitoring vehicles. These improvements help with transportation interaction. The app provides an extra layer of security and protection for their messages and calls. It also enables location tracking, which can help in emergency situations. Overall, it gives peace of mind to all the users.

The process of making progress or improving something. when more green cars are used, smart cities will have a cleaner and healthier air. areas, there are other changes happening in the environment. For people who drive cars, the aim is to lessen traffic jams and pollution in the air. To make a proper smart city we need to take smart water management system, smart meters, smart remote and social security where IoT-based smart city technology can help improve social security by providing real-time tracking, analysis, and other options. The public security systems can guess how likely a crime is to happen by looking at information from sensors and CCTV cameras in the city, as well as data from social media and readings Figure 3. The police can stop or punish bad people with the help of these social security applications. The answer is to utilize connected devices in the intelligent city. For instance, when a crime occurs, the device details are sent to a cloud.

Figure 3. Connected smart city: Cloud-powered IoT integration for seamless information access

In simple terms, the platform examines the data and determines who the criminal is. The platform figures out how far away a gunshot is from a phone by looking at the time it takes for the sound to reach the phone. The software in the cloud can notify the police using a mobile app, air quality checker, smart infrastructure, and many more we can find the process that happen when the data is getting encrypted and how it is being transported to the cloud server in Figure 3.

CONCLUSION

This paper particularly shows how smart homes, smart buildings, and smart cities, the rapid integration of interconnected devices through AIoT and cloud platform which has undeniably by which they can be mitigated through the methods proposed transformed our lives, offering unprecedented convenience and efficiency. However, this digital revolution has ushered in a host of security challenges that should not be underestimated. These challenges encompass a wide array of threats, from the possibility of unauthorized access to critical systems and data breaches to even the potential for physical security risks. To effectively address these multifaceted concerns, we came up with a theoretical method and proper technological architectures in our paper in which cloud server encryption and authentication mechanisms must be in place to protect both data and access to these smart systems and privacy of the user. Implementing secure communication protocols should be a standard practice. Ensuring that software updates are not only frequent but also essential. These updates patch known vulnerabilities and ensured that systems are running with the latest security measures. Network segmentation, which involves isolating critical systems from non-critical ones, can be an effective strategy to contain potential breaches and limit unauthorized access. User education plays a pivotal role in mitigating security risks. Raising awareness about the importance of strong, regularly changed passwords, and educating users about the risks associated with connecting to unsecured networks can significantly reduce vulnerabilities. Collaboration between the public and private sectors is paramount. Establishing industry standards and regulations can help address security concerns collectively, fostering a safer environment for smart technology adoption.

In an ever-evolving landscape, it's essential for organizations to develop and regularly update incident response plans. These plans help in reacting swiftly and effectively to security breaches when they occur. As technology continues to evolve, security measures and practices must adapt accordingly to stay ahead of potential security challenges. Being proactive in monitoring and adapting to new threats and staying vigilant in adopting emerging technologies and security practices, is crucial in safeguarding the potential benefits of smart environments while minimizing the associated risks. This comprehensive approach is vital to harness the full potential of smart technologies while ensuring the safety and privacy of individuals and the integrity of critical infrastructure in our increasingly interconnected world.

REFERENCES

Abbes, W., Kechaou, Z., Hussain, A., Qahtani, A. M., Almutiry, O., Dhahri, H., & Alimi, A. M. (2023). An Enhanced Binary Particle Swarm Optimization (E-BPSO) algorithm for service placement in hybrid cloud platforms. *Neural Computing & Applications*, *35*(2), 1343–1361. doi:10.1007/s00521-022-07839-5

Abdel-Rahman, M. (2023). Advanced cybersecurity measures in IT service operations and their crucial role in safeguarding enterprise data in a connected world. *Eigenpub Review of Science and Technology*, *7*(1), 138–158.

Abdollahi, A., Pradhan, B., & Alamri, A. (2021). RoadVecNet: A new approach for simultaneous road network segmentation and vectorization from aerial and google earth imagery in a complex urban set-up. *GIScience & Remote Sensing*, *58*(7), 1151–1174. doi:10.1080/15481603.2021.1972713

Abidin, A., Senin, N., & Manaf, A. A. A. (2021). A preliminary study of low-code/no-code ecosystem practices: translating design student views on crafting interactive design. *NVEO-Natural Volatiles & Essential Oils Journal*, 10244-10258.

Agiollo, A., Conti, M., Kaliyar, P., Lin, T. N., & Pajola, L. (2021). DETONAR: Detection of routing attacks in RPL-based IoT. *IEEE Transactions on Network and Service Management*, *18*(2), 1178–1190. doi:10.1109/TNSM.2021.3075496

Aguilar, J., Garces-Jimenez, A., R-Moreno, M. D., & García, R. (2021). A systematic literature review on the use of artificial intelligence in energy self-management in smart buildings. *Renewable & Sustainable Energy Reviews*, *151*, 111530. doi:10.1016/j.rser.2021.111530

Albalawi, A. M., & Almaiah, M. A. (2022). Assessing and reviewing of cyber-security threats, attacks, mitigation techniques in IoT environment. *Journal of Theoretical and Applied Information Technology*, *100*, 2988–3011.

Alexakis, G., Panagiotakis, S., Fragkakis, A., Markakis, E., & Vassilakis, K. (2019). Control of smart home operations using natural language processing, voice recognition and IoT technologies in a multi-tier architecture. *Designs*, *3*(3), 32. doi:10.3390/designs3030032

Ali, A., Smartt, C., Lester, E., Williams, O., & Greedy, S. (2023). High capacity chipless RFID tags for biomass tracking application. *International Journal of Microwave and Wireless Technologies*, *15*(5), 742–752. doi:10.1017/S1759078722000745

Ali, I., Sabir, S., & Ullah, Z. (2019). Internet of things security, device authentication and access control: a review. *arXiv preprint arXiv:1901.07309.*

Ali, M., Naeem, F., Adam, N., Kaddoum, G., Adnan, M., & Tariq, M. (2023). Integration of Data Driven Technologies in Smart Grids for Resilient and Sustainable Smart Cities: A Comprehensive Review. *arXiv preprint arXiv:2301.08814.*

Ali, M. L., Ismat, S., Thakur, K., Kamruzzaman, A., Lue, Z., & Thakur, H. N. (2023, March). Network Packet Sniffing and Defense. In *2023 IEEE 13th Annual Computing and Communication Workshop and Conference (CCWC)* (pp. 499-503). IEEE. 10.1109/CCWC57344.2023.10099148

Aliero, M. S., Qureshi, K. N., Pasha, M. F., Ghani, I., & Yauri, R. A. (2021). Systematic mapping study on energy optimization solutions in smart building structure: Opportunities and challenges. *Wireless Personal Communications*, *119*(3), 2017–2053. doi:10.1007/s11277-021-08316-3

Aljumah, A., & Ahanger, T. A. (2020). Cyber security threats, challenges and defence mechanisms in cloud computing. *IET Communications*, *14*(7), 1185–1191. doi:10.1049/iet-com.2019.0040

Alotaibi, Y., & Ilyas, M. (2022). Security risks in internet of things (IoT): a brief survey. In *Proceedings of the 26th World Multi-Conference on Systemics, Cybernetics and Informatics (WMSCI 2022)* (pp. 1-5). 10.54808/WMSCI2022.01.6

AlRikabi, H. T. S., & Hazim, H. T. (2021). Enhanced data security of communication system using combined encryption and steganography. *iJIM, 15*(16), 145.

Andrade, T., & Bastos, D. (2019, June). Extended reality in iot scenarios: Concepts, applications and future trends. In *2019 5th Experiment International Conference (exp. at'19)* (pp. 107-112). IEEE.

Apthorpe, N., Reisman, D., & Feamster, N. (2017). Closing the blinds: Four strategies for protecting smart home privacy from network observers. *arXiv preprint arXiv:1705.06809.*

Aroba, O. J., Xulu, T., Msani, N. N., Mohlakoana, T. T., Ndlovu, E. E., & Mthethwa, S. M. (2023, March). The adoption of an intelligent waste collection system in a smart city. In *2023 Conference on Information Communications Technology and Society (ICTAS)* (pp. 1-6). IEEE. 10.1109/ICTAS56421.2023.10082750

Arul, E., & Punidha, A. (2022, February). Artificial Intelligence to Protect Cyber Security Attack on Cloud E-Learning Tools (AIPCE). In *International Conference on Computing, Communication, Electrical and Biomedical Systems* (pp. 29-37). Cham: Springer International Publishing. 10.1007/978-3-030-86165-0_3

Barik, R. K., Gudey, S. K., Reddy, G. G., Pant, M., Dubey, H., Mankodiya, K., & Kumar, V. (2017, December). FogGrid: Leveraging fog computing for enhanced smart grid network. In *2017 14th IEEE India Council International Conference (INDICON)* (pp. 1-6). IEEE.

Batra, R., Song, L., & Ramprasad, R. (2021). Emerging materials intelligence ecosystems propelled by machine learning. *Nature Reviews. Materials*, *6*(8), 655–678. doi:10.1038/s41578-020-00255-y

Battina, D. S. (2017). The Role of Machine Learning in Clinical Research: Transforming the Future of Evidence Generation. *Future*, *4*(12).

Bekri, W., Layeb, T., Rihab, J. M. A. L., & Fourati, L. C. (2022, May). Intelligent IoT Systems: Security issues, attacks, and countermeasures. In 2022 International Wireless Communications and Mobile Computing (IWCMC) (pp. 231-236). IEEE.

Belli, L., Cilfone, A., Davoli, L., Ferrari, G., Adorni, P., Di Nocera, F., Dall'Olio, A., Pellegrini, C., Mordacci, M., & Bertolotti, E. (2020). IoT-enabled smart sustainable cities: Challenges and approaches. *Smart Cities*, *3*(3), 1039–1071. doi:10.3390/smartcities3030052

Bhanpurawala, A., El-Fakih, K., & Zualkernan, I. (2022). A Formal Assisted Approach for Modeling and Testing Security Attacks in IoT Edge Devices. *arXiv preprint arXiv:2210.05623*.

Braun, T., Fung, B. C., Iqbal, F., & Shah, B. (2018). Security and privacy challenges in smart cities. *Sustainable Cities and Society*, *39*, 499–507. doi:10.1016/j.scs.2018.02.039

Brundage, M., Avin, S., Clark, J., Toner, H., Eckersley, P., Garfinkel, B., . . . Amodei, D. (2018). The malicious use of artificial intelligence: Forecasting, prevention, and mitigation. *arXiv preprint arXiv:1802.07228*.

Bughin, J., Seong, J., Manyika, J., Chui, M., & Joshi, R. (2018). Notes from the AI frontier: Modeling the impact of AI on the world economy. *McKinsey Global Institute, 4*.

Chaganti, R., Bhushan, B., & Ravi, V. (2023). A survey on Blockchain solutions in DDoS attacks mitigation: Techniques, open challenges and future directions. *Computer Communications*, *197*, 96–112. doi:10.1016/j.comcom.2022.10.026

Chataut, R., & Phoummalayvane, A. (2023). *Unleashing the Power of IoT: A Comprehensive Review of IoT Applications, Advancements, and Future Prospects in Healthcare, Agriculture, Smart Homes, Smart Cities, and Industry 4.0.* Academic Press.

Chen, R., Wu, X., & Liu, X. (2023). RSETP: A Reliable Security Education and Training Platform Based on the Alliance Blockchain. *Electronics (Basel)*, *12*(6), 1427. doi:10.3390/electronics12061427

Chinchero, H. F., & Alonso, J. M. (2020, June). A review on energy management methodologies for LED lighting systems in smart buildings. In *2020 IEEE International Conference on Environment and Electrical Engineering and 2020 IEEE Industrial and Commercial Power Systems Europe (EEEIC/I&CPS Europe)* (pp. 1-6). IEEE. 10.1109/EEEIC/ICPSEurope49358.2020.9160796

Chobisai, K., Kaul, V., Kumar, D., & Gupta, K. P. A. V. (2022). *Performance Analysis Of Platform As A Service (Paas).* For Storage In Multi Cloud Environment.

Cp, V., Kalaivanan, S., Karthik, R., & Sanjana, A. (2022, February). Blockchain-based IoT Device Security. In *2022 2nd International Conference on Artificial Intelligence and Signal Processing (AISP)* (pp. 1-6). IEEE. 10.1109/AISP53593.2022.9760674

Daubner, L., & Považanec, A. (2023, August). Data Loss Prevention Solution for Linux Endpoint Devices. In *Proceedings of the 18th International Conference on Availability, Reliability and Security* (pp. 1-10). 10.1145/3600160.3605036

Dhanalakshmi, R., Benjamin, M., Sivaraman, A., Sood, K., & Sreedeep, S. S. (2023). Machine Learning-Based Smart Appliances for Everyday Life. In *Smart Analytics, Artificial Intelligence and Sustainable Performance Management in a Global Digitalised Economy* (pp. 289–301). Emerald Publishing Limited. doi:10.1108/S1569-37592023000110A015

Dias, T., Oliveira, N., Sousa, N., Praça, I., & Sousa, O. (2021, December). A hybrid approach for an interpretable and explainable intrusion detection system. In *International Conference on Intelligent Systems Design and Applications* (pp. 1035-1045). Cham: Springer International Publishing.

Dong, B., Shi, Q., Yang, Y., Wen, F., Zhang, Z., & Lee, C. (2021). Technology evolution from self-powered sensors to AIoT enabled smart homes. *Nano Energy*, *79*, 105414. doi:10.1016/j.nanoen.2020.105414

Duong, L. (2023). *The techniques of IoT and it's applications for smart homes: internet of things techniques and standards for building smart homes*. Academic Press.

Dzulhikam, D., & Rana, M. E. (2022, March). A critical review of cloud computing environment for big data analytics. In *2022 International Conference on Decision Aid Sciences and Applications (DASA)* (pp. 76-81). IEEE. 10.1109/DASA54658.2022.9765168

Edu, J. S., Such, J. M., & Suarez-Tangil, G. (2020). Smart home personal assistants: A security and privacy review. *ACM Computing Surveys*, *53*(6), 1–36. doi:10.1145/3412383

Elmarady, A. A., & Rahouma, K. (2021). Studying cybersecurity in civil aviation, including developing and applying aviation cybersecurity risk assessment. *IEEE Access : Practical Innovations, Open Solutions*, *9*, 143997–144016. doi:10.1109/ACCESS.2021.3121230

Fabrègue, B. F., & Bogoni, A. (2023). Privacy and Security Concerns in the Smart City. *Smart Cities*, *6*(1), 586–613. doi:10.3390/smartcities6010027

Feraudo, A., Popescu, D. A., Yadav, P., Mortier, R., & Bellavista, P. (2024, January). Mitigating IoT Botnet DDoS Attacks through MUD and eBPF based Traffic Filtering. In *Proceedings of the 25th International Conference on Distributed Computing and Networking* (pp. 164-173). 10.1145/3631461.3631549

Franco, P., Martinez, J. M., Kim, Y. C., & Ahmed, M. A. (2021). IoT based approach for load monitoring and activity recognition in smart homes. *IEEE Access : Practical Innovations, Open Solutions*, *9*, 45325–45339. doi:10.1109/ACCESS.2021.3067029

Ganesh, D. E. (2022). Analysis of wireless sensor networks through secure routing protocols using directed diffusion methods. *International Journal of Wireless Network Security*, *7*(1), 28–35.

Gaur, B., Shukla, V. K., & Verma, A. (2019, April). Strengthening people analytics through wearable IOT device for real-time data collection. In *2019 international conference on automation, computational and technology management (ICACTM)* (pp. 555-560). IEEE. doi:10.1109/ICACTM.2019.8776776

George, A. S., & Sagayarajan, S. (2023). Securing Cloud Application Infrastructure: Understanding the Penetration Testing Challenges of IaaS, PaaS, and SaaS Environments. *Partners Universal International Research Journal*, *2*(1), 24–34.

Gholamzadehmir, M., Del Pero, C., Buffa, S., Fedrizzi, R., & Aste, N. (2020). Adaptive-predictive control strategy for HVAC systems in smart buildings–A review. *Sustainable Cities and Society, 63*, 102480. doi:10.1016/j.scs.2020.102480

Gill, S. S., Xu, M., Ottaviani, C., Patros, P., Bahsoon, R., Shaghaghi, A., Golec, M., Stankovski, V., Wu, H., Abraham, A., Singh, M., Mehta, H., Ghosh, S. K., Baker, T., Parlikad, A. K., Lutfiyya, H., Kanhere, S. S., Sakellariou, R., Dustdar, S., ... Uhlig, S. (2022). AI for next generation computing: Emerging trends and future directions. *Internet of Things : Engineering Cyber Physical Human Systems, 19*, 100514. doi:10.1016/j.iot.2022.100514

Glöckler, J., Sedlmeir, J., Frank, M., & Fridgen, G. (2023). A systematic review of identity and access management requirements in enterprises and potential contributions of self-sovereign identity. *Business & Information Systems Engineering*, 1–20.

Goel, A., & Sahil, G. (2023). Implementing privacy and data confidentiality within the framework of the Internet of Things. *Journal of Data Protection & Privacy, 5*(4), 374–387.

Gudla, S., & Kuda, N. (2023). A reliable routing mechanism with energy-efficient node selection for data transmission using a genetic algorithm in wireless sensor network. *Facta Universitatis, Series. Electronics and Energetics, 36*(2), 209–226. doi:10.2298/FUEE2302209G

Guo, X., Shen, Z., Zhang, Y., & Wu, T. (2019). Review on the application of artificial intelligence in smart homes. *Smart Cities, 2*(3), 402–420. doi:10.3390/smartcities2030025

Gupta, B. B., Chaudhary, P., Chang, X., & Nedjah, N. (2022). Smart defense against distributed Denial of service attack in IoT networks using supervised learning classifiers. *Computers & Electrical Engineering, 98*, 107726. doi:10.1016/j.compeleceng.2022.107726

Hafeez, I., Antikainen, M., & Tarkoma, S. (2019, March). Protecting IoT-environments against traffic analysis attacks with traffic morphing. In *2019 IEEE International Conference on Pervasive Computing and Communications Workshops (PerCom Workshops)* (pp. 196-201). IEEE. 10.1109/PERCOMW.2019.8730787

He, S., Ren, W., Zhu, T., & Choo, K. K. R. (2019). BoSMoS: A blockchain-based status monitoring system for defending against unauthorized software updating in industrial Internet of Things. *IEEE Internet of Things Journal, 7*(2), 948–959. doi:10.1109/JIOT.2019.2947339

Heidari, A., & Jabraeil Jamali, M. A. (2023). Internet of Things intrusion detection systems: A comprehensive review and future directions. *Cluster Computing*, *26*(6), 3753–3780. doi:10.1007/s10586-022-03776-z

How, M. L., Cheah, S. M., Chan, Y. J., Khor, A. C., & Say, E. M. P. (2023). Artificial Intelligence for Advancing Sustainable Development Goals (SDGs): An Inclusive Democratized Low-Code Approach. In *The Ethics of Artificial Intelligence for the Sustainable Development Goals* (pp. 145–165). Springer International Publishing. doi:10.1007/978-3-031-21147-8_9

Ibrahim, R. F., Abu Al-Haija, Q., & Ahmad, A. (2022). DDoS attack prevention for Internet of Thing devices using Ethereum blockchain technology. *Sensors (Basel)*, *22*(18), 6806. doi:10.3390/s22186806

Ijemaru, G. K., Ang, L. M., & Seng, K. P. (2023). Swarm Intelligence Internet of Vehicles Approaches for Opportunistic Data Collection and Traffic Engineering in Smart City Waste Management. *Sensors (Basel)*, *23*(5), 2860. doi:10.3390/s23052860

Ioannidis, T., Bolgouras, V., Xenakis, C., & Politis, I. (2023, August). Securing the Flow: Security and Privacy Tools for Flow-based Programming. In *Proceedings of the 18th International Conference on Availability, Reliability and Security* (pp. 1-6). 10.1145/3600160.3605089

Jaihar, J., Lingayat, N., Vijaybhai, P. S., Venkatesh, G., & Upla, K. P. (2020, June). Smart home automation using machine learning algorithms. In 2020 international conference for emerging technology (INCET) (pp. 1-4). IEEE. doi:10.1109/INCET49848.2020.9154007

Jalal, A., Quaid, M. A. K., & Hasan, A. S. (2018, December). Wearable sensor-based human behavior understanding and recognition in daily life for smart environments. In *2018 International Conference on Frontiers of Information Technology (FIT)* (pp. 105-110). IEEE. 10.1109/FIT.2018.00026

James, E., & Rabbi, F. (2023). Fortifying the IoT Landscape: Strategies to Counter Security Risks in Connected Systems. *Tensorgate Journal of Sustainable Technology and Infrastructure for Developing Countries*, *6*(1), 32–46.

Jangjou, M., & Sohrabi, M. K. (2022). A comprehensive survey on security challenges in different network layers in cloud computing. *Archives of Computational Methods in Engineering*, *29*(6), 3587–3608. doi:10.1007/s11831-022-09708-9

Javed, A. R., Ahmed, W., Pandya, S., Maddikunta, P. K. R., Alazab, M., & Gadekallu, T. R. (2023). A survey of explainable artificial intelligence for smart cities. *Electronics (Basel)*, *12*(4), 1020. doi:10.3390/electronics12041020

Jha, R. K. (2023). Cybersecurity and confidentiality in smart grid for enhancing sustainability and reliability. *Recent Research Reviews Journal, 2*(2), 215–241. doi:10.36548/rrrj.2023.2.001

Jmal, R., Ghabri, W., Guesmi, R., Alshammari, B. M., Alshammari, A. S., & Alsaif, H. (2023). Distributed Blockchain-SDN Secure IoT System Based on ANN to Mitigate DDoS Attacks. *Applied Sciences (Basel, Switzerland), 13*(8), 4953. doi:10.3390/app13084953

Joseph, K. J. (2002). *Growth of ICT and ICT for Development: Realities of the Myths of the Indian Experience* (No. 2002/78). WIDER Discussion Paper.

Karie, N. M., Sahri, N. M., Yang, W., Valli, C., & Kebande, V. R. (2021). A review of security standards and frameworks for IoT-based smart environments. *IEEE Access : Practical Innovations, Open Solutions, 9*, 121975–121995. doi:10.1109/ACCESS.2021.3109886

Kaur, T., & Kamboj, S. (2023, January). Descriptive Analysis of the Cloud Computing Services and Deployment Models. In *2023 International Conference for Advancement in Technology (ICONAT)* (pp. 1-6). IEEE. 10.1109/ICONAT57137.2023.10080749

Kaya, M. M., Taşkiran, Y., Kanoğlu, A., Demirtaş, A., Zor, E., Burçak, I., ... Akgül, F. T. (2021). Designing A Smart Home Management System With Artificial Intelligence &. *Machine Learning*.

Khalid, M., Hameed, S., Qadir, A., Shah, S. A., & Draheim, D. (2023). Towards SDN-based smart contract solution for IoT access control. *Computer Communications, 198*, 1–31. doi:10.1016/j.comcom.2022.11.007

Khan, A., Khan, S., Hassan, B., & Zheng, Z. (2022). CNN-based smoker classification and detection in smart city application. *Sensors (Basel), 22*(3), 892. doi:10.3390/s22030892

Kim, H., Choi, H., Kang, H., An, J., Yeom, S., & Hong, T. (2021). A systematic review of the smart energy conservation system: From smart homes to sustainable smart cities. *Renewable & Sustainable Energy Reviews, 140*, 110755. doi:10.1016/j.rser.2021.110755

Kiran, V., Sardana, A., & Kaur, P. (2022, April). Defending Against Ddos Attacks in Rpl Using Subjective Logic Based Trust Approach For IOT. In *2022 2nd International Conference on Advance Computing and Innovative Technologies in Engineering (ICACITE)* (pp. 1656-1660). IEEE. 10.1109/ICACITE53722.2022.9823719

Kodada, B. B., Prasad, G., & Pais, A. R. (2012). Protection against DDoS and data modification attack in computational grid cluster environment. *International Journal of Computer Network and Information Security*, *4*(7), 12–18. doi:10.5815/ijcnis.2012.07.02

Komninos, N., Philippou, E., & Pitsillides, A. (2014). Survey in smart grid and smart home security: Issues, challenges and countermeasures. *IEEE Communications Surveys and Tutorials*, *16*(4), 1933–1954. doi:10.1109/COMST.2014.2320093

Koo, C., Shin, S., Gretzel, U., Hunter, W. C., & Chung, N. (2016). Conceptualization of smart tourism destination competitiveness. *Asia Pacific Journal of Information Systems*, *26*(4), 561–576. doi:10.14329/apjis.2016.26.4.561

Kornaros, G. (2022). Hardware-assisted machine learning in resource-constrained IoT environments for security: Review and future prospective. *IEEE Access : Practical Innovations, Open Solutions*, *10*, 58603–58622. doi:10.1109/ACCESS.2022.3179047

Krishnamoorthy, S., Dua, A., & Gupta, S. (2023). Role of emerging technologies in future IoT-driven Healthcare 4.0 technologies: A survey, current challenges and future directions. *Journal of Ambient Intelligence and Humanized Computing*, *14*(1), 361–407. doi:10.1007/s12652-021-03302-w

Kumar, A., Sharma, S., Goyal, N., Singh, A., Cheng, X., & Singh, P. (2021). Secure and energy-efficient smart building architecture with emerging technology IoT. *Computer Communications*, *176*, 207–217. doi:10.1016/j.comcom.2021.06.003

Kumar, H., Singh, M. K., Gupta, M. P., & Madaan, J. (2020). Moving towards smart cities: Solutions that lead to the Smart City Transformation Framework. *Technological Forecasting and Social Change*, *153*, 119281. doi:10.1016/j.techfore.2018.04.024

Kumar, R., & Venkatesh, K. (2022, April). Centralized and Decentralized Data Backup Approaches. In *Proceedings of International Conference on Deep Learning, Computing and Intelligence: ICDCI 2021* (pp. 687-698). Singapore: Springer Nature Singapore. 10.1007/978-981-16-5652-1_60

Kumar, J. (2020). Implementation of integrations of iot and cloud services in a home automation smart system. *Journal of Critical Reviews, 7*(1).

Kumari, P., & Jain, A. K. (2023). A comprehensive study of DDoS attacks over IoT network and their countermeasures. *Computers & Security*, *127*, 103096. doi:10.1016/j.cose.2023.103096

Lepore, D., Testi, N., & Pasher, E. (2023). Building Inclusive Smart Cities through Innovation Intermediaries. *Sustainability (Basel)*, *15*(5), 4024. doi:10.3390/su15054024

Li, W., Yigitcanlar, T., Erol, I., & Liu, A. (2021). Motivations, barriers and risks of smart home adoption: From systematic literature review to conceptual framework. *Energy Research & Social Science*, *80*, 102211. doi:10.1016/j.erss.2021.102211

Lin, Y., Kämäräinen, T., Di Francesco, M., & Ylä-Jääski, A. (2015). Performance evaluation of remote display access for mobile cloud computing. *Computer Communications*, *72*, 17–25. doi:10.1016/j.comcom.2015.05.006

Liu, H., Ning, H., Mu, Q., Zheng, Y., Zeng, J., Yang, L. T., Huang, R., & Ma, J. (2019). A review of the smart world. *Future Generation Computer Systems*, *96*, 678–691. doi:10.1016/j.future.2017.09.010

Liu, P., & Peng, Z. (2013). China's smart city pilots: A progress report. *Computer*, *47*(10), 72–81. doi:10.1109/MC.2013.149

Lokshina, I. V., Greguš, M., & Thomas, W. L. (2019). Application of integrated building information modeling, IoT and blockchain technologies in system design of a smart building. *Procedia Computer Science*, *160*, 497–502. doi:10.1016/j.procs.2019.11.058

. Madhvan, R., & Zolkipli, M. F. (2023). An Overview of Malware Injection Attacks: Techniques, Impacts, and Countermeasures. *Borneo International Journal, 6*(3), 22-30.

Maggi, F., Volpatto, A., Gasparini, S., Boracchi, G., & Zanero, S. (2011, December). A fast eavesdropping attack against touchscreens. In *2011 7th International Conference on Information Assurance and Security (IAS)* (pp. 320-325). IEEE. 10.1109/ISIAS.2011.6122840

Mahbub, M. (2020). Progressive researches on IoT security: An exhaustive analysis from the perspective of protocols, vulnerabilities, and preemptive architectonics. *Journal of Network and Computer Applications*, *168*, 102761. doi:10.1016/j.jnca.2020.102761

Malviya, S., & Lohiya, H. (2022). An analysis of authentication attacks with countermeasures and various authentication methods in a distributed environment. *International Research Journal of Modernization in Engineering Technology and Science*, *4*(12), 1–8.

Maniatis, P. (n.d.). *Comparison of Public, Private, Hybrid, and Community Cloud Computing in Terms of Purchasing and Supply Management: A Quantitative Approach*. Academic Press.

Mao, J., Lin, Q., Zhu, S., Ma, L., & Liu, J. (2023). SmartTracer: Anomaly-Driven Provenance Analysis Based on Device Correlation in Smart Home Systems. *IEEE Internet of Things Journal*.

Marian, M., Ganea, E., Popescu, D., Stîngă, F., Cusman, A., & Ionică, D. (2022, May). Analysis of Different SaaS Architectures from a Trust Service Provider Perspective. In *2022 23rd International Carpathian Control Conference (ICCC)* (pp. 360-365). IEEE. 10.1109/ICCC54292.2022.9805947

Martins, F., Patrão, C., Moura, P., & de Almeida, A. T. (2021). A Review of Energy Modeling Tools for Energy Efficiency in Smart Cities. *Smart Cities*, *4*(4), 1420–1436. doi:10.3390/smartcities4040075

Mehrabi, N., Morstatter, F., Saxena, N., Lerman, K., & Galstyan, A. (2021). A survey on bias and fairness in machine learning. *ACM Computing Surveys*, *54*(6), 1–35. doi:10.1145/3457607

Mei, Y., Ma, Y., Ma, J., Moeller, L., & Federici, J. F. (2021). Eavesdropping risk evaluation on terahertz wireless channels in atmospheric turbulence. *IEEE Access : Practical Innovations, Open Solutions*, *9*, 101916–101923. doi:10.1109/ACCESS.2021.3098016

Mishra, P., & Singh, G. (2023). Energy management systems in sustainable smart cities based on the internet of energy: A technical review. *Energies*, *16*(19), 6903. doi:10.3390/en16196903

Motevalli, S., & Seyedian, S. A. (2023). Evaluation of Environmental Factors in Smart Homes Implementation (case Study: Cities of Mazandaran Province in Iran). *ISPRS Annals of the Photogrammetry, Remote Sensing and Spatial Information Sciences*, *10*, 563–568. doi:10.5194/isprs-annals-X-4-W1-2022-563-2023

Mukherjee, S. (2020). Emerging frontiers in smart environment and healthcare–A vision. *Information Systems Frontiers*, *22*(1), 23–27. doi:10.1007/s10796-019-09965-3

Nasrullayev, N., Muminova, S., Istamovich, D. K., & Boltaeva, M. (2023, July). Providing IoT Security in Industry 4.0 using Web Application Firewall. In *2023 4th International Conference on Electronics and Sustainable Communication Systems (ICESC)* (pp. 1788-1792). IEEE.

Nassiri Abrishamchi, M. A., Zainal, A., Ghaleb, F. A., Qasem, S. N., & Albarrak, A. M. (2022). Smart home privacy protection methods against a passive wireless Snooping side-channel attack. *Sensors (Basel), 22*(21), 8564. doi:10.3390/s22218564

Navarro, M., Liang, Y., & Zhong, X. (2022). Energy-efficient and balanced routing in low-power wireless sensor networks for data collection. *Ad Hoc Networks, 127,* 102766. doi:10.1016/j.adhoc.2021.102766

Neshenko, N., Bou-Harb, E., Crichigno, J., Kaddoum, G., & Ghani, N. (2019). Demystifying IoT security: An exhaustive survey on IoT vulnerabilities and a first empirical look on Internet-scale IoT exploitations. *IEEE Communications Surveys and Tutorials, 21*(3), 2702–2733. doi:10.1109/COMST.2019.2910750

Olaniyi, O., Okunleye, O. J., & Olabanji, S. O. (2023). Advancing data-driven decision-making in smart cities through big data analytics: A comprehensive review of existing literature. *Current Journal of Applied Science and Technology, 42*(25), 10–18. doi:10.9734/cjast/2023/v42i254181

Ortega-Fernández, A., Martín-Rojas, R., & García-Morales, V. J. (2020). Artificial intelligence in the urban environment: Smart cities as models for developing innovation and sustainability. *Sustainability (Basel), 12*(19), 7860. doi:10.3390/su12197860

Ortega-Fernandez, I., & Liberati, F. (2023). A Review of Denial of Service Attack and Mitigation in the Smart Grid Using Reinforcement Learning. *Energies, 16*(2), 635. doi:10.3390/en16020635

Otuoze, A. O., Mustafa, M. W., & Larik, R. M. (2018). Smart grids security challenges: Classification by sources of threats. *Journal of Electrical Systems and Information Technology, 5*(3), 468–483. doi:10.1016/j.jesit.2018.01.001

Ozer, E., Malekloo, A., Ramadan, W., Tran, T. T., & Di, X. (2023). Systemic reliability of bridge networks with mobile sensing-based model updating for postevent transportation decisions. *Computer-Aided Civil and Infrastructure Engineering, 38*(8), 975–999. doi:10.1111/mice.12892

Patel, B., Patel, K., & Patel, A. (2023). Internet of Thing (IoT): Routing Protocol Classification and Possible Attack on IOT Environment. In *Modern Electronics Devices and Communication Systems: Select Proceedings of MEDCOM 2021* (pp. 543-551). Singapore: Springer Nature Singapore.

Pathak, A. K., Saguna, S., Mitra, K., & Åhlund, C. (2021, June). Anomaly detection using machine learning to discover sensor tampering in IoT systems. In *ICC 2021-IEEE International Conference on Communications* (pp. 1-6). IEEE. 10.1109/ICC42927.2021.9500825

Patra, B. K., Mishra, S., & Patra, S. K. (2022). Genetic Algorithm-Based Energy-Efficient Clustering with Adaptive Grey Wolf Optimization-Based Multipath Routing in Wireless Sensor Network to Increase Network Life Time. *Intelligent Systems Proceedings of ICMIB*, *2021*, 499–512. doi:10.1007/978-981-19-0901-6_44

Pentikousis, K. (2000). TCP in wired-cum-wireless environments. *IEEE Communications Surveys and Tutorials*, *3*(4), 2–14. doi:10.1109/COMST.2000.5340805

Petrovska, I., & Kuchuk, H. (2022). *Static allocation method in a cloud environment with a service model IAAS*. Academic Press.

Prabhavathy, M., & Umamaheswari, S. (2022). Prevention of Runtime Malware Injection Attack in Cloud Using Unsupervised Learning. *Intelligent Automation & Soft Computing*, *32*(1), 101–114. doi:10.32604/iasc.2022.018257

Rahman, A., & Subriadi, A. P. (2022, January). Software as a service (SaaS) adoption factors: individual and organizational perspective. In *2022 2nd International Conference on Information Technology and Education (ICIT&E)* (pp. 31-36). IEEE.

Ramesh, S., Nirmalraj, S., Murugan, S., Manikandan, R., & Al-Turjman, F. (2023). Optimization of energy and security in mobile sensor network using classification based signal processing in heterogeneous network. *Journal of Signal Processing Systems for Signal, Image, and Video Technology*, *95*(2-3), 153–160. doi:10.1007/s11265-021-01690-y

Ramirez, M. A., Kim, S. K., Hamadi, H. A., Damiani, E., Byon, Y. J., Kim, T. Y., . . . Yeun, C. Y. (2022). Poisoning attacks and defenses on artificial intelligence: A survey. *arXiv preprint arXiv:2202.10276*.

Rana, M., Mamun, Q., & Islam, R. (2022). Lightweight cryptography in IoT networks: A survey. *Future Generation Computer Systems*, *129*, 77–89. doi:10.1016/j.future.2021.11.011

Rehman, A., & Farrakh, A. (2022). An Intelligent Approach for Smart Home Energy Management System Empowered with Machine learning Techniques. *International Journal of Computational and Innovative Sciences*, *1*(4), 7–14.

Roberts, M. K., & Ramasamy, P. (2023). An improved high performance clustering based routing protocol for wireless sensor networks in IoT. *Telecommunication Systems*, *82*(1), 45–59. doi:10.1007/s11235-022-00968-1

Rokis, K., & Kirikova, M. (2022, September). Challenges of low-code/no-code software development: A literature review. In *International Conference on Business Informatics Research*(pp. 3-17). Cham: Springer International Publishing. 10.1007/978-3-031-16947-2_1

Saad, R. M., Soufy, K. A. A., & Shaheen, S. I. (2023). Security in smart home environment: Issues, challenges, and countermeasures-a survey. *International Journal of Security and Networks*, *18*(1), 1–9. doi:10.1504/IJSN.2023.129887

Sadeeq, M. M., Abdulkareem, N. M., Zeebaree, S. R., Ahmed, D. M., Sami, A. S., & Zebari, R. R. (2021). IoT and Cloud computing issues, challenges and opportunities: A review. *Qubahan Academic Journal*, *1*(2), 1–7. doi:10.48161/qaj.v1n2a36

Sahu, M. K., & Pattanaik, S. R. (2023). Modified Limited memory BFGS with displacement aggregation. *arXiv preprint arXiv:2301.05447.*

Salah, S., & Amro, B. M. (2022). Big picture: Analysis of DDoS attacks map-systems and network, cloud computing, SCADA systems, and IoT. *International Journal of Internet Technology and Secured Transactions*, *12*(6), 543–565. doi:10.1504/IJITST.2022.126468

Saritepeci, M. (2020). Predictors of cyberloafing among high school students: Unauthorized access to school network, metacognitive awareness and smartphone addiction. *Education and Information Technologies*, *25*(3), 2201–2219. doi:10.1007/s10639-019-10042-0

Sarker, I. H., Khan, A. I., Abushark, Y. B., & Alsolami, F. (2023). Internet of things (iot) security intelligence: A comprehensive overview, machine learning solutions and research directions. *Mobile Networks and Applications*, *28*(1), 296–312. doi:10.1007/s11036-022-01937-3

Sergi, I., Montanaro, T., Benvenuto, F. L., & Patrono, L. (2021). A smart and secure logistics system based on IoT and cloud technologies. *Sensors (Basel)*, *21*(6), 2231. doi:10.3390/s21062231

Shaaban, A. M., Chlup, S., El-Araby, N., & Schmittner, C. (2022). Towards Optimized Security Attributes for IoT Devices in Smart Agriculture Based on the IEC 62443 Security Standard. *Applied Sciences (Basel, Switzerland)*, *12*(11), 5653. doi:10.3390/app12115653

Sharadqh, A. A., Hatamleh, H. A. M., Saloum, S. S., & Alawneh, T. A. (2023). Hybrid Chain: Blockchain Enabled Framework for Bi-Level Intrusion Detection and Graph-Based Mitigation for Security Provisioning in Edge Assisted IoT Environment. *IEEE Access: Practical Innovations, Open Solutions*, *11*, 27433–27449. doi:10.1109/ACCESS.2023.3256277

Shiny, X. A., Ravikumar, D., Chinnasamy, A., & Hemavathi, S. (2023, May). Cloud Computing based Smart Traffic Management System with Priority Switching for Health Care Services. In *2023 2nd International Conference on Applied Artificial Intelligence and Computing (ICAAIC)* (pp. 1159-1163). IEEE. 10.1109/ICAAIC56838.2023.10140942

Shrivastava, R. K., Singh, S. P., Hasan, M. K., Islam, S., Abdullah, S., & Aman, A. H. M. (2022). Securing Internet of Things devices against code tampering attacks using Return Oriented Programming. *Computer Communications*, *193*, 38–46. doi:10.1016/j.comcom.2022.06.033

. Siji, F. G., & Uche, O. P. (2023). An improved model for comparing different endpoint detection and response tools for mitigating insider threat. *Indian Journal of Engineering, 20.*

Singh, M., Fuenmayor, E., Hinchy, E. P., Qiao, Y., Murray, N., & Devine, D. (2021). Digital twin: Origin to future. *Applied System Innovation*, *4*(2), 36. doi:10.3390/asi4020036

Sleem, A., & Elhenawy, I. (2023). Survey of Artificial Intelligence of Things for Smart Buildings: A closer outlook. *Journal of Intelligent Systems & Internet of Things*, *8*(2), 63–71. doi:10.54216/JISIoT.080206

Steingartner, W., Galinec, D., & Kozina, A. (2021). Threat defense: Cyber deception approach and education for resilience in hybrid threats model. *Symmetry*, *13*(4), 597. doi:10.3390/sym13040597

Sundberg, S., Brunstrom, A., Ferlin-Reiter, S., Høiland-Jørgensen, T., & Brouer, J. D. (2023, March). Efficient continuous latency monitoring with eBPF. In *International Conference on Passive and Active Network Measurement* (pp. 191-208). Cham: Springer Nature Switzerland. 10.1007/978-3-031-28486-1_9

Swain, S. R., Saxena, D., Kumar, J., Singh, A. K., & Lee, C. N. (2023). An ai-driven intelligent traffic management model for 6g cloud radio access networks. *IEEE Wireless Communications Letters*, *12*(6), 1056–1060. doi:10.1109/LWC.2023.3259942

Swessi, D., & Idoudi, H. (2022). A survey on internet-of-things security: Threats and emerging countermeasures. *Wireless Personal Communications*, *124*(2), 1557–1592. doi:10.1007/s11277-021-09420-0

Syamala, M., Komala, C. R., Pramila, P. V., Dash, S., Meenakshi, S., & Boopathi, S. (2023). Machine learning-integrated IoT-based smart home energy management System. In *Handbook of Research on Deep Learning Techniques for Cloud-Based Industrial IoT* (pp. 219–235). IGI Global. doi:10.4018/978-1-6684-8098-4.ch013

Taiwo, O., Ezugwu, A. E., Oyelade, O. N., & Almutairi, M. S. (2022). Enhanced intelligent smart home control and security system based on deep learning model. *Wireless Communications and Mobile Computing*, *2022*, 1–22. doi:10.1155/2022/9307961

Tanveer, M., Bashir, A. K., Alzahrani, B. A., Albeshri, A., Alsubhi, K., & Chaudhry, S. A. (2023). CADF-CSE: Chaotic map-based authenticated data access/sharing framework for IoT-enabled cloud storage environment. *Physical Communication*, *59*, 102087. doi:10.1016/j.phycom.2023.102087

Tavares, R., Sousa, H., & Ribeiro, J. (2021). O uso de smart speakers na incapacidade: Uma scoping review. *New Trends in Qualitative Research*, *8*, 392–403. doi:10.36367/ntqr.8.2021.392-403

Tawalbeh, L. A., Muheidat, F., Tawalbeh, M., & Quwaider, M. (2020). IoT Privacy and security: Challenges and solutions. *Applied Sciences (Basel, Switzerland)*, *10*(12), 4102. doi:10.3390/app10124102

Thomaz, G. A., Guerra, M. B., Sammarco, M., Detyniecki, M., & Campista, M. E. M. (2023). Tamper-proof access control for IoT clouds using enclaves. *Ad Hoc Networks*, *147*, 103191. doi:10.1016/j.adhoc.2023.103191

. Thyagaturu, A. S., Nguyen, G., Rimal, B. P., & Reisslein, M. (2022). Ubi-Flex-Cloud: ubiquitous flexible cloud computing: status quo and research imperatives. *Applied Computing and Informatics*.

Tippins, N. T., Oswald, F. L., & McPhail, S. M. (2021). Scientific, legal, and ethical concerns about AI-based personnel selection tools: A call to action. *Personnel Assessment and Decisions*, *7*(2), 1. doi:10.25035/pad.2021.02.001

Tiwari, P. (2022). *Top 5 IoT Security Testing Companies to Rely on*. https://kratikal.com/blog/top-5-iot-security-testing-companies-to-rely-on/

Tripuraneni, N., Jordan, M., & Jin, C. (2020). On the theory of transfer learning: The importance of task diversity. *Advances in Neural Information Processing Systems*, *33*, 7852–7862.

Tushir, B., Dalal, Y., Dezfouli, B., & Liu, Y. (2020). A quantitative study of ddos and e-ddos attacks on wifi smart home devices. *IEEE Internet of Things Journal*, *8*(8), 6282–6292. doi:10.1109/JIOT.2020.3026023

Venkata Subhadu, V. (2023). *IT methods for operational near applications: low-code and no-code as rapid-application-development tools* [Doctoral dissertation]. Technische Hochschule Ingolstadt.

Vuković, O., & Dán, G. (2014). Security of fully distributed power system state estimation: Detection and mitigation of data integrity attacks. *IEEE Journal on Selected Areas in Communications*, *32*(7), 1500–1508. doi:10.1109/JSAC.2014.2332106

Wang, H., Ju, Y., Zhang, N., Pei, Q., Liu, L., Dong, M., & Leung, V. C. (2022). Resisting malicious eavesdropping: Physical layer security of mm wave mimo communications in presence of random blockage. *IEEE Internet of Things Journal*, *9*(17), 16372–16385. doi:10.1109/JIOT.2022.3153054

Wang, Z. H., Jin, M. H., Jiang, L., Feng, C. J., Cao, J. Y., & Yun, Z. (2023, July). Secure access method of power internet of things based on zero trust architecture. In *International Conference on Swarm Intelligence* (pp. 386-399). Cham: Springer Nature Switzerland. 10.1007/978-3-031-36625-3_31

Weingärtner, T. (2019, November). Tokenization of physical assets and the impact of IoT and AI. In *European Union Blockchain Observatory and Forum* (Vol. 10, pp. 1-16). Academic Press.

Whaiduzzaman, M., Barros, A., Chanda, M., Barman, S., Sultana, T., Rahman, M. S., Roy, S., & Fidge, C. (2022). A review of emerging technologies for IoT-based smart cities. *Sensors (Basel)*, *22*(23), 9271. doi:10.3390/s22239271

Wibawa, S. (2023). Analysis of Adversarial Attacks on AI-based With Fast Gradient Sign Method. *International Journal of Engineering Continuity*, *2*(2), 72–79. doi:10.58291/ijec.v2i2.120

Wiyatno, R., & Xu, A. (2018). Maximal jacobian-based saliency map attack. *arXiv preprint arXiv:1808.07945*.

Wu, C., Yang, L., Cai, M., Zhao, X., & Sun, Q. (2023). Security Evaluation Method of Smart Home Cloud Platform. *International Journal of Pattern Recognition and Artificial Intelligence*, *37*(06), 2351012. doi:10.1142/S0218001423510126

Xia, Q., Sifah, E. B., Agyekum, K. O. B. O., Xia, H., Acheampong, K. N., Smahi, A., ... Guizani, M. (2019). Secured fine-grained selective access to outsourced cloud data in IoT environments. *IEEE Internet of Things Journal*, *6*(6), 10749–10762. doi:10.1109/JIOT.2019.2941638

Xu, M., Yu, J., Zhang, X., Wang, C., Zhang, S., Wu, H., & Han, W. (2023). Improving real-world password guessing attacks via bi-directional transformers. In *32nd USENIX Security Symposium (USENIX Security 23)* (pp. 1001-1018). USENIX.

Yang, S., & Wang, X. (2023). Analysis of Computer Network Security and Prevention Technology. *Journal of Electronics and Information Science, 8*(2), 20–24.

Yang, X., Shu, L., Liu, Y., Hancke, G. P., Ferrag, M. A., & Huang, K. (2022). Physical security and safety of iot equipment: A survey of recent advances and opportunities. *IEEE Transactions on Industrial Informatics, 18*(7), 4319–4330. doi:10.1109/TII.2022.3141408

Younes, D., Alghannam, E., Tan, Y., & Lu, H. (2020). Enhancement in quality estimation of resistance spot welding using vision system and fuzzy support vector machine. *Symmetry, 12*(8), 1380. doi:10.3390/sym12081380

Youssef, A. E. (2020). A framework for cloud security risk management based on the business objectives of organizations. *arXiv preprint arXiv:2001.08993.*

Zakzouk, A., Menzel, K., & Hamdy, M. (2023, September). Brain-Computer-Interface (BCI) Based Smart Home Control Using EEG Mental Commands. In *Working Conference on Virtual Enterprises* (pp. 720-732). Cham: Springer Nature Switzerland. 10.1007/978-3-031-42622-3_51

Zawaideh, F. H., Ghanem, W. A. H., Yusoff, M. H., Saany, S. I. A., Jusoh, J. A., & El-Ebiary, Y. A. B. (2022). The Layers of Cloud Computing Infrastructure and Security Attacking Issues. *Journal of Pharmaceutical Negative Results*, 792–800.

Zheng, Y., Li, Z., Xu, X., & Zhao, Q. (2022). Dynamic defenses in cyber security: Techniques, methods and challenges. *Digital Communications and Networks, 8*(4), 422–435. doi:10.1016/j.dcan.2021.07.006

Compilation of References

. Madhvan, R., & Zolkipli, M. F. (2023). An Overview of Malware Injection Attacks: Techniques, Impacts, and Countermeasures. *Borneo International Journal, 6*(3), 22-30.

. Siji, F. G., & Uche, O. P. (2023). An improved model for comparing different endpoint detection and response tools for mitigating insider threat. *Indian Journal of Engineering, 20.*

. Srivastava, S., & Pathak, D. M. (2023). AIoT-based smart education and online teaching. AIoT Technologies and Applications for Smart Environments, 229.

. Thyagaturu, A. S., Nguyen, G., Rimal, B. P., & Reisslein, M. (2022). Ubi-Flex-Cloud: ubiquitous flexible cloud computing: status quo and research imperatives. *Applied Computing and Informatics.*

A., N. (2017). Automation and inequality The changing world of work in the global South Green economy. *International Institute for Environment and Development.* http://pubs.iied.org/11506IIED

Abbes, W., Kechaou, Z., Hussain, A., Qahtani, A. M., Almutiry, O., Dhahri, H., & Alimi, A. M. (2023). An Enhanced Binary Particle Swarm Optimization (E-BPSO) algorithm for service placement in hybrid cloud platforms. *Neural Computing & Applications, 35*(2), 1343–1361. doi:10.1007/s00521-022-07839-5

Abdelmaboud, A., Ahmed, A. I. A., Abaker, M., Eisa, T. E., Albasheer, H., Ghorashi, S., & Karim, F. K. (2022). Blockchain for IoT Applications: Taxonomy, Platforms, Recent Advances, Challenges and Future Research Directions. *Electronics (Basel), 11*(4), 630. doi:10.3390/electronics11040630

Abdel-Rahman, M. (2023). Advanced cybersecurity measures in IT service operations and their crucial role in safeguarding enterprise data in a connected world. *Eigenpub Review of Science and Technology, 7*(1), 138–158.

Abdollahi, A., Pradhan, B., & Alamri, A. (2021). RoadVecNet: A new approach for simultaneous road network segmentation and vectorization from aerial and google earth imagery in a complex urban set-up. *GIScience & Remote Sensing, 58*(7), 1151–1174. doi:10.1080/15481603.2021.1 972713

Abdollahpouri, H., Burke, R., & Mobasher, B. (2017). Context-aware event recommendation in event-based social networks. *User Modeling and User-Adapted Interaction, 27*(4), 715–751.

Abdullahi, M., Baashar, Y., Alhussian, H., Alwadain, A., Aziz, N., Capretz, L. F., & Abdulkadir, S. J. (2022). Detecting cybersecurity attacks in Internet of Things Using Artificial intelligence Methods: A Systematic Literature review. *Electronics (Basel)*, *11*(2), 198. doi:10.3390/electronics11020198

Abidin, A., Senin, N., & Manaf, A. A. A. (2021). A preliminary study of low-code/no-code ecosystem practices: translating design student views on crafting interactive design. *NVEO-Natural Volatiles & Essential Oils Journal*, 10244-10258.

Abid, M. A., Afaqui, N., Khan, M. A., Akhtar, M. W., Malik, A. W., Munir, A., Ahmad, J., & Shabir, B. (2022). Evolution towards Smart and Software-Defined Internet of Things. *AI*, *3*(1), 100–123. doi:10.3390/ai3010007

Accenture. (n.d.). Where AI is Aiding Productivity. *Frontier Economics*. https://www.statista.com/chart/23779/ai-productivity-increas e/

Adi, E., Anwar, A., Baig, Z. A., & Zeadally, S. (2020). Machine learning and data analytics for the IoT. *Neural Computing & Applications*, *32*(20), 16205–16233. doi:10.1007/s00521-020-04874-y

Adir, O., Poley, M., Chen, G., Froim, S., Krinsky, N., Shklover, J., Shainsky-Roitman, J., Lammers, T., & Schroeder, A. (2019). Integrating Artificial Intelligence and Nanotechnology for Precision Cancer Medicine. *Advanced Materials*, *32*(13), 1901989. Advance online publication. doi:10.1002/adma.201901989 PMID:31286573

Adli, H. K., Remli, M. A., Wong, K. N. S. W. S., Ismail, N. A., González-Briones, A., Corchado, J. M., & Mohamad, M. S. (2023). Recent Advancements and Challenges of AIoT Application in Smart Agriculture: A review. *Sensors (Basel)*, *23*(7), 3752. doi:10.3390/s23073752 PMID:37050812

Aggarwal, R. (2020, March). *Organisational Change*. Retrieved from https://www.arsdcollege.ac.in/wp-content/uploads/2020/03/OB-Change.pdf

Agiollo, A., Conti, M., Kaliyar, P., Lin, T. N., & Pajola, L. (2021). DETONAR: Detection of routing attacks in RPL-based IoT. *IEEE Transactions on Network and Service Management*, *18*(2), 1178–1190. doi:10.1109/TNSM.2021.3075496

Aguilar, J., Garces-Jimenez, A., R-Moreno, M. D., & García, R. (2021). A systematic literature review on the use of artificial intelligence in energy self-management in smart buildings. *Renewable & Sustainable Energy Reviews*, *151*, 111530. doi:10.1016/j.rser.2021.111530

Ahmetoglu, S., Cob, Z. C., & Ali, N. (2022). A Systematic Review of Internet of Things adoption in Organizations: Taxonomy, benefits, challenges and Critical factors. *Applied Sciences (Basel, Switzerland)*, *12*(9), 4117. doi:10.3390/app12094117

Alahi, E. E., Sukkuea, A., Tina, F. W., Nag, A., Kurdthongmee, W., Suwannarat, K., & Mukhopadhyay, S. C. (2023). Integration of IoT-Enabled Technologies and Artificial Intelligence (AI) for Smart City Scenario: Recent Advancements and Future Trends. *Sensors (Basel)*, *23*(11), 5206. doi:10.3390/s23115206 PMID:37299934

Albalawi, A. M., & Almaiah, M. A. (2022). Assessing and reviewing of cyber-security threats, attacks, mitigation techniques in IoT environment. *Journal of Theoretical and Applied Information Technology*, *100*, 2988–3011.

Alexakis, G., Panagiotakis, S., Fragkakis, A., Markakis, E., & Vassilakis, K. (2019). Control of smart home operations using natural language processing, voice recognition and IoT technologies in a multi-tier architecture. *Designs*, *3*(3), 32. doi:10.3390/designs3030032

Al-Fuqaha, A., Guizani, M., Mohammadi, M., Aledhari, M., & Ayyash, M. (2015). Internet of Things: A Survey on Enabling Technologies, Protocols, and Applications. *IEEE Communications Surveys and Tutorials*, *17*(4), 2347–2376. doi:10.1109/COMST.2015.2444095

Ali, I., Sabir, S., & Ullah, Z. (2019). Internet of things security, device authentication and access control: a review. *arXiv preprint arXiv:1901.07309.*

Ali, M. L., Ismat, S., Thakur, K., Kamruzzaman, A., Lue, Z., & Thakur, H. N. (2023, March). Network Packet Sniffing and Defense. In *2023 IEEE 13th Annual Computing and Communication Workshop and Conference (CCWC)* (pp. 499-503). IEEE. 10.1109/CCWC57344.2023.10099148

Ali, M., Naeem, F., Adam, N., Kaddoum, G., Adnan, M., & Tariq, M. (2023). Integration of Data Driven Technologies in Smart Grids for Resilient and Sustainable Smart Cities: A Comprehensive Review. *arXiv preprint arXiv:2301.08814.*

Ali, A., Smartt, C., Lester, E., Williams, O., & Greedy, S. (2023). High capacity chipless RFID tags for biomass tracking application. *International Journal of Microwave and Wireless Technologies*, *15*(5), 742–752. doi:10.1017/S1759078722000745

Aliahmadi, A., & Nozari, H. (2022). The Neutrosophic decision-making method evaluates security metrics in AIoT and blockchain-based supply chains. *Supply Chain Forum, 24*(1), 31–42. 10.1080/16258312.2022.2101898

Aliero, M. S., Qureshi, K. N., Pasha, M. F., Ghani, I., & Yauri, R. A. (2021). Systematic mapping study on energy optimization solutions in smart building structure: Opportunities and challenges. *Wireless Personal Communications*, *119*(3), 2017–2053. doi:10.1007/s11277-021-08316-3

Aljumah, A., & Ahanger, T. A. (2020). Cyber security threats, challenges and defence mechanisms in cloud computing. *IET Communications*, *14*(7), 1185–1191. doi:10.1049/iet-com.2019.0040

Allioui, H., & Mourdi, Y. (2023). Exploring the full potentials of IoT for better financial growth and stability: A comprehensive survey. *Sensors (Basel)*, *23*(19), 8015. doi:10.3390/s23198015 PMID:37836845

Alon-Barkat, S., & Busuioc, M. (2023). Human–AI Interactions in Public Sector Decision Making: "Automation Bias" and "Selective Adherence" to Algorithmic Advice. *Journal of Public Administration: Research and Theory*, *33*(1), 153–169. doi:10.1093/jopart/muac007

Alotaibi, Y., & Ilyas, M. (2022). Security risks in internet of things (IoT): a brief survey. In *Proceedings of the 26th World Multi-Conference on Systemics, Cybernetics and Informatics (WMSCI 2022)* (pp. 1-5). 10.54808/WMSCI2022.01.6

AlRikabi, H. T. S., & Hazim, H. T. (2021). Enhanced data security of communication system using combined encryption and steganography. *iJIM, 15*(16), 145.

Al-Shaer, E., Assi, C., & Zhang, N. (2018). Artificial intelligence for wireless communication systems: A comprehensive survey. *IEEE Communications Surveys and Tutorials, 20*(4), 2759–2788.

Alsmadi, I., Tawalbeh, L. A., Alshraideh, M., & Jararweh, Y. (2014). A review of quality of service in mobile computing environments. *The Scientific World Journal, 2014*, 16.

Andrade, T., & Bastos, D. (2019, June). Extended reality in iot scenarios: Concepts, applications and future trends. In *2019 5th Experiment International Conference (exp. at'19)* (pp. 107-112). IEEE.

Andreotta, A. J., Kirkham, N., & Rizzi, M. (2021). AI, big data, and the future of consent. *AI & Society, 37*(4), 1715–1728. doi:10.1007/s00146-021-01262-5 PMID:34483498

Apthorpe, N., Reisman, D., & Feamster, N. (2017). Closing the blinds: Four strategies for protecting smart home privacy from network observers. *arXiv preprint arXiv:1705.06809.*

Araujo, T., Helberger, N., Kruikemeier, S., & de Vreese, C. H. (2020). In AI we trust? Perceptions about automated decision-making by artificial intelligence. *AI & Society, 35*(3), 611–623. doi:10.1007/s00146-019-00931-w

Archi, Y. E., & Benbba, B. (2023). The Applications of Technology Acceptance Models in Tourism and Hospitality Research: A Systematic Literature Review. *Journal of Environmental Management and Tourism, 14*(2), 379. doi:10.14505/jemt.v14.2(66).08

Arévalo-Ascanio, J., & Estrada López, H. (2017). Decisionmaking. A review of the topic. In *Gerencia de las organizaciones: Un enfoque empresarial* (pp. 249–278). Ediciones Universidad Simón Bolivar.

Arinez, J. F., Chang, Q., Gao, R. X., Xu, C., & Zhang, J. (2020). Artificial Intelligence in Advanced Manufacturing: Current Status and Future Outlook. *Journal of Manufacturing Science and Engineering, 142*(11), 110804. Advance online publication. doi:10.1115/1.4047855

Aroba, O. J., Xulu, T., Msani, N. N., Mohlakoana, T. T., Ndlovu, E. E., & Mthethwa, S. M. (2023, March). The adoption of an intelligent waste collection system in a smart city. In *2023 Conference on Information Communications Technology and Society (ICTAS)* (pp. 1-6). IEEE. 10.1109/ICTAS56421.2023.10082750

Arul, E., & Punidha, A. (2022, February). Artificial Intelligence to Protect Cyber Security Attack on Cloud E-Learning Tools (AIPCE). In *International Conference on Computing, Communication, Electrical and Biomedical Systems* (pp. 29-37). Cham: Springer International Publishing. 10.1007/978-3-030-86165-0_3

Asil, A. L. K. A. Y. A. A., & Erdem Hepaktan, A. C. (2003). Organizational Change. *Yönetm ve Ekonom, 10*(1), 31-58. doi:https://dergipark.org.tr/tr/download/article-file/145822

Astarilla, L. (2018). University students' perception towards the use of duolingo application in learning english. *Prosiding CELSciTech, 3*, 1–9.

Atzori, L., Iera, A., & Morabito, G. (2010). The Internet of Things: A survey. *Computer Networks*, *54*(15), 2787–2805. doi:10.1016/j.comnet.2010.05.010

Barik, R. K., Gudey, S. K., Reddy, G. G., Pant, M., Dubey, H., Mankodiya, K., & Kumar, V. (2017, December). FogGrid: Leveraging fog computing for enhanced smart grid network. In *2017 14th IEEE India Council International Conference (INDICON)* (pp. 1-6). IEEE.

Bartolomeo, J. (2015, April 29). *Tipo de decisor según su postura ante el riesgo*. Retrieved from https://administrarconsultora.wordpress.com: https://administrarconsultora.wordpress.com/2015/04/29/tipo-de-decisor-segun-su-postura-ante-el-riesgo/

Bartoň, M., Budjač, R., Tanuška, P., Gašpar, G., & Schreiber, P. (2022). Identification overview of industry 4.0 essential attributes and resource-limited embedded artificial-intelligence-of-things devices for small and medium-sized enterprises. *Applied Sciences (Basel, Switzerland)*, *12*(11), 5672. doi:10.3390/app12115672

Baskerville, R., Bunker, D., Olaisen, J., Pries-Heje, J., Larsen, T. J., & Swanson, E. B. (2014). Diffusion and Innovation Theory: past, present, and future contributions to academia and practice. In Springer eBooks (pp. 295–300). doi:10.1007/978-3-662-43459-8_18

Batra, R., Song, L., & Ramprasad, R. (2021). Emerging materials intelligence ecosystems propelled by machine learning. *Nature Reviews. Materials*, *6*(8), 655–678. doi:10.1038/s41578-020-00255-y

Battina, D. S. (2017). The Role of Machine Learning in Clinical Research: Transforming the Future of Evidence Generation. *Future*, *4*(12).

Bekri, W., Layeb, T., Rihab, J. M. A. L., & Fourati, L. C. (2022, May). Intelligent IoT Systems: Security issues, attacks, and countermeasures. In 2022 International Wireless Communications and Mobile Computing (IWCMC) (pp. 231-236). IEEE.

Belli, L., Cilfone, A., Davoli, L., Ferrari, G., Adorni, P., Di Nocera, F., Dall'Olio, A., Pellegrini, C., Mordacci, M., & Bertolotti, E. (2020). IoT-enabled smart sustainable cities: Challenges and approaches. *Smart Cities*, *3*(3), 1039–1071. doi:10.3390/smartcities3030052

Bentley, J. W. (2018). *Decreasing Operational Distortion and Surrogation through Narrative Reporting* (*SSRN* Scholarly Paper 2924726). doi:10.2139/ssrn.2924726

Benzidia, S., Bentahar, O., Husson, J., & Makaoui, N. (2023). Big data analytics capability in healthcare operations and supply chain management: The role of green process innovation. *Annals of Operations Research*. Advance online publication. doi:10.1007/s10479-022-05157-6 PMID:36687515

Bhanpurawala, A., El-Fakih, K., & Zualkernan, I. (2022). A Formal Assisted Approach for Modeling and Testing Security Attacks in IoT Edge Devices. *arXiv preprint arXiv:2210.05623*.

Bibri, S. E., Allam, Z., & Krogstie, J. (2022). The Metaverse as a Virtual Form of Data-Driven Smart Urbanism: Platformization and Its Underlying Processes, Institutional Dimensions, and Disruptive Impacts. *Computational Urban Science*, 2(1), 24. Advance online publication. doi:10.1007/s43762-022-00051-0 PMID:35974838

Bibri, S. E., & Jagatheesaperumal, S. K. (2023). Harnessing the potential of the metaverse and artificial intelligence for the internet of city things: Cost-Effective XReality and synergistic AIOT technologies. *Smart Cities*, 6(5), 2397–2429. doi:10.3390/smartcities6050109

BillyM. (2020). The influence of dynamic organizations and the application of digital innovations to educational institutions in the world during the COVID-19 pandemic. *Available at* SSRN 3588233. doi:10.2139/ssrn.3588233

Björkman, M. (1992). What is productivity? *IFAC Proceedings Volumes, 25*(8). 10.1016/S1474-6670(17)54065-3

Black, P. W., Meservy, T. O., Tayler, W. B., & Williams, J. O. (2022). Surrogation Fundamentals: Measurement and Cognition. *Journal of Management Accounting Research*, 34(1), 9–29. doi:10.2308/JMAR-2020-071

Bonnett, A. (2023). *What is geography?* Rowman & Littlefield.

Braun, T., Fung, B. C., Iqbal, F., & Shah, B. (2018). Security and privacy challenges in smart cities. *Sustainable Cities and Society*, 39, 499–507. doi:10.1016/j.scs.2018.02.039

Braun, V., & Clarke, V. (2006). Using thematic analysis in psychology. *Qualitative Research in Psychology*, 3(2), 77–101. doi:10.1191/1478088706qp063oa

Brous, P., Janssen, M., & Krans, R. (2020). Data governance as a success factor for data science. In Lecture Notes in Computer Science (pp. 431–442). doi:10.1007/978-3-030-44999-5_36

Brundage, M., Avin, S., Clark, J., Toner, H., Eckersley, P., Garfinkel, B., . . . Amodei, D. (2018). The malicious use of artificial intelligence: Forecasting, prevention, and mitigation. *arXiv preprint arXiv:1802.07228.*

Buçinca, Z., Malaya, M. B., & Gajos, K. Z. (2021). To Trust or to Think: Cognitive Forcing Functions Can Reduce Overreliance on AI in AI-assisted Decision-making. *Proceedings of the ACM on Human-Computer Interaction, 5*(CSCW1), 188:1-188:21. 10.1145/3449287

Bughin, J., Seong, J., Manyika, J., Chui, M., & Joshi, R. (2018). Notes from the AI frontier: Modeling the impact of AI on the world economy. *McKinsey Global Institute, 4.*

Canós, L., Pons, C., Valero, M., & Maheut, J. P. (n.d.). *Toma de decisiones en la empresa: proceso y clasificación.* Retrieved from https://riunet.upv.es/bitstream/handle/10251/16502/TomaDecisiones.pdf

Cao, K., Liu, Y., Meng, G., & Sun, Q. (2020). An overview of edge computing research. *IEEE Access : Practical Innovations, Open Solutions*, 8, 85714–85728. doi:10.1109/ACCESS.2020.2991734

Cepymesnews. (2023, May 10). *Estadísticas de inteligencia artificial.* Retrieved from https://cepymenews.es/estadisticas-inteligencia-artificial-para-2023

Chacín, L. (2010). Information technology as support to decision making management process in organizations in electrical sector in Venezuela. *Espacios, 31*(2), 11-13. Retrieved from https://www.revistaespacios.com/a10v31n02/10310233.html

Chaganti, R., Bhushan, B., & Ravi, V. (2023). A survey on Blockchain solutions in DDoS attacks mitigation: Techniques, open challenges and future directions. *Computer Communications, 197,* 96–112. doi:10.1016/j.comcom.2022.10.026

Chang, Y.-W., Hsu, P.-Y., & Wu, Z.-Y. (2015). Exploring managers' intention to use business intelligence: The role of motivations. *Behaviour & Information Technology, 34*(3), 273–285. doi:10.1080/0144929X.2014.968208

Chang, Z., Liu, S., Xiong, X., Cai, Z., & Tu, G. (2021). A Survey of Recent Advances in Edge-Computing-Powered Artificial Intelligence of Things. *IEEE Internet of Things Journal, 8*(18), 13849–13875. doi:10.1109/JIOT.2021.3088875

Chataut, R., & Phoummalayvane, A. (2023). *Unleashing the Power of IoT: A Comprehensive Review of IoT Applications, Advancements, and Future Prospects in Healthcare, Agriculture, Smart Homes, Smart Cities, and Industry 4.0.* Academic Press.

Chen, D. Q., Preston, D. S., & Swink, M. (2016). How the use of Big Data analytics Affects value creation in supply chain management. *Journal of Management Information Systems, 32*(4), 4–39. doi:10.1080/07421222.2015.1138364

Chen, D., Xue, Y., & Yao, L. (2018). A survey of big data architectures and machine learning algorithms in large scale data processing. *Journal of King Saud University. Computer and Information Sciences.*

Chen, L., Chen, P., & Lin, Z. (2020). Artificial intelligence in education: A review. *IEEE Access : Practical Innovations, Open Solutions, 8,* 75264–75278. doi:10.1109/ACCESS.2020.2988510

Chen, M., Saad, W., & Yin, C. (2018). Virtual Reality over Wireless Networks: Quality-of-Service Model and Learning-Based Resource Management. *IEEE Transactions on Communications, 66*(11), 5621–5635. doi:10.1109/TCOMM.2018.2850303

Chen, M., & Wu, H. (2021). Real-time intelligent image processing for the Internet of Things. *Journal of Real-Time Image Processing, 18*(4), 997–998. doi:10.1007/s11554-021-01149-0

Chen, R., Wu, X., & Liu, X. (2023). RSETP: A Reliable Security Education and Training Platform Based on the Alliance Blockchain. *Electronics (Basel), 12*(6), 1427. doi:10.3390/electronics12061427

Chen, S., Gu, X., Wang, J., & Zhu, H. (2021). AIOT Used for COVID-19 Pandemic Prevention and Control. *Contrast Media & Molecular Imaging, 2021,* 1–23. doi:10.1155/2021/8922504 PMID:34729056

Chinchero, H. F., & Alonso, J. M. (2020, June). A review on energy management methodologies for LED lighting systems in smart buildings. In *2020 IEEE International Conference on Environment and Electrical Engineering and 2020 IEEE Industrial and Commercial Power Systems Europe (EEEIC/I&CPS Europe)* (pp. 1-6). IEEE. 10.1109/EEEIC/ICPSEurope49358.2020.9160796

Chiu, T. K. (2023). The impact of Generative AI (GenAI) on practices, policies and research direction in education: A case of ChatGPT and Midjourney. *Interactive Learning Environments*, 1–17. doi:10.1080/10494820.2023.2253861

Chiu, T. K., Xia, Q., Zhou, X., Chai, C. S., & Cheng, M. (2023). Systematic literature review on opportunities, challenges, and future research recommendations of artificial intelligence in education. *Computers and Education: Artificial Intelligence*, 4, 100118. doi:10.1016/j.caeai.2022.100118

Chobisai, K., Kaul, V., Kumar, D., & Gupta, K. P. A. V. (2022). *Performance Analysis Of Platform As A Service (Paas)*. For Storage In Multi Cloud Environment.

Choi, J., Hecht, G. W., & Tayler, W. B. (2013). Strategy Selection, Surrogation, and Strategic Performance Measurement Systems. *Journal of Accounting Research*, 51(1), 105–133. doi:10.1111/j.1475-679X.2012.00465.x

Chou, C. M., Shen, T. C., Shen, T. C., & Shen, C. H. (2022). Influencing factors on students' learning effectiveness of AI-based technology application: Mediation variable of the human-computer interaction experience. *Education and Information Technologies*, 27(6), 8723–8750. doi:10.1007/s10639-021-10866-9

Choung, H., David, P., & Ross, A. (2022). Trust and ethics in AI. *AI & Society*, 38(2), 733–745. doi:10.1007/s00146-022-01473-4

Ciacci, A., & Penco, L. (2023). Business model innovation: Harnessing big data analytics and digital transformation in hostile environments. *Journal of Small Business and Enterprise Development*. doi:10.1108/JSBED-10-2022-0424

Cisco. (2019). *Cisco Visual Networking Index: Forecast and Trends, 2017–2022*. Cisco.

Clarín. (2022, June 3). *Una computadora cuántica resuelve en 36 microsegundos un problema de 9.000 años*. Retrieved from https://www.clarin.com/tecnologia/computadora-cuantica-resue lve-36-microsegundos-problema-9-000-anos_0_BzzaX42xoE.html

Cleveland Clinic. (2021). *How Cleveland Clinic's home hospital brings care to you*. Retrieved from https://my.clevelandclinic.org/patient/guides/hospital-at-ho me

Cobb–Douglas production function. (n.d.). Wikipedia. Retrieved June 1, 2023, from https://en.wikipedia.org/wiki/Cobb%E2%80%93Douglas_productio n_function

Corchado, J. M. (2020, November 23). *The role of the AIoT and deepint. net.* Retrieved from IEEE International Conference on Electronics Circuits and Systems: https://gredos.usal.es/bitstream/handle/10366/144254/ICECS.pdf

Costa, I., Riccotta, R., Montini, P., Stefani, E., De Goes, R., Gaspar, M. A., Martins, F. S., Fernandes, A. A., Machado, C., Loçano, R., & Larieira, C. L. C. (2022). The Degree of Contribution of Digital Transformation Technology on Company Sustainability Areas. *Sustainability (Basel)*, *14*(1), 462. doi:10.3390/su14010462

Cottrell, A. (2019). *Economics 207, 2019.* The Cobb–Douglas Production Function. Retrieved June 1, 2023, from http://users.wfu.edu/cottrell/ecn207/cobb-douglas.pdf

Cp, V., Kalaivanan, S., Karthik, R., & Sanjana, A. (2022, February). Blockchain-based IoT Device Security. In *2022 2nd International Conference on Artificial Intelligence and Signal Processing (AISP)* (pp. 1-6). IEEE. 10.1109/AISP53593.2022.9760674

Cromby, J. J., Standen, P. J., & Brown, D. J. (1996). The potentials of virtual environments in the education and training of people with learning disabilities. *Journal of Intellectual Disability Research*, *40*(6), 489–501. doi:10.1111/j.1365-2788.1996.tb00659.x PMID:9004109

Cui, Y., Kara, S., & Chan, K. C. (2020). Manufacturing big data ecosystem: A systematic literature review. *Robotics and Computer-integrated Manufacturing*, *62*, 101861. doi:10.1016/j.rcim.2019.101861

Damioli, G., Roy, V. V., & Vertesy, D. (2021, January 21). The impact of artificial intelligence on labor productivity. *Eurasian Business Review.* doi:10.1007/s40821-020-00172-8

Datision. (2021, October 14). *La inteligencia artificial para la toma de decisiones empresariales.* Retrieved from https://datision.com/blog/inteligencia-artificial-decisiones-empresariales/

Daubner, L., & Považanec, A. (2023, August). Data Loss Prevention Solution for Linux Endpoint Devices. In *Proceedings of the 18th International Conference on Availability, Reliability and Security* (pp. 1-10). 10.1145/3600160.3605036

Davenport, T. H. (2018, January 30). Artificial Intelligence for the Real World. *Harvard Business Review.* https://hbr.org/webinar/2018/02/artificial-intelligence-for-the-real-world

Davenport, T. H., Harris, J., & Shapiro, J. (2010). *Competing on analytics: The new science of winning.* Harvard Business Press.

De Kock, J. H., Latham, H. A., Leslie, S. J., Grindle, M., Munoz, S. A., Ellis, L., Polson, R., & O'Malley, C. M. (2021). A rapid review of the impact of COVID-19 on healthcare workers' mental health: Implications for supporting psychological well-being. *BMC Public Health*, *21*(1). Advance online publication. doi:10.1186/s12889-020-10070-3 PMID:33422039

Dekker, I., De Jong, E. M., Schippers, M. C., De Bruijn-Smolders, M., Alexiou, A., & Giesbers, B. (2020). Optimizing students' mental health and academic performance: AI-Enhanced Life Crafting. *Frontiers in Psychology*, *11*, 1063. Advance online publication. doi:10.3389/fpsyg.2020.01063 PMID:32581935

Deloitte. (2020). *Deloitte Survey on AI Adoption in Manufacturing*. Deloitte China. https://www2.deloitte.com/cn/en/pages/consumer-industrial-products/articles/ai-manufacturing-application-survey.html

Deora, V., Shao, J., Gray, W. A., & Fiddian, N. J. (2003). A Quality of Service Management Framework Based on User Expectations. In Lecture Notes in Computer Science (pp. 104–114). doi:10.1007/978-3-540-24593-3_8

Dhanalakshmi, R., Benjamin, M., Sivaraman, A., Sood, K., & Sreedeep, S. S. (2023). Machine Learning-Based Smart Appliances for Everyday Life. In *Smart Analytics, Artificial Intelligence and Sustainable Performance Management in a Global Digitalised Economy* (pp. 289–301). Emerald Publishing Limited. doi:10.1108/S1569-37592023000110A015

Dias, T., Oliveira, N., Sousa, N., Praça, I., & Sousa, O. (2021, December). A hybrid approach for an interpretable and explainable intrusion detection system. In *International Conference on Intelligent Systems Design and Applications* (pp. 1035-1045). Cham: Springer International Publishing.

Diyer, O., Achtaich, N., & Najib, K. (2020). Artificial Intelligence in Learning Skills Assessment. *Proceedings of the 3rd International Conference on Networking, Information Systems & Security*. 10.1145/3386723.3387901

Dizdarević, J., Carpio, F., Jukan, A., & Masip-Bruin, X. (2019). A survey of communication protocols for the Internet of Things and related challenges of FOG and cloud computing integration. *ACM Computing Surveys*, *51*(6), 1–29. doi:10.1145/3292674

Dong, B., Shi, Q., Yang, Y., Wen, F., Zhang, Z., & Lee, C. (2021). Technology evolution from self-powered sensors to AIoT enabled smart homes. *Nano Energy*, *79*, 105414. doi:10.1016/j.nanoen.2020.105414

Duan, Y., Edwards, J. S., & Dwivedi, Y. K. (2019). Artificial intelligence for decision making in the era of Big Data – evolution, challenges and research agenda. *International Journal of Information Management*, *48*, 63–71. doi:10.1016/j.ijinfomgt.2019.01.021

Duong, L. (2023). *The techniques of IoT and it's applications for smart homes: internet of things techniques and standards for building smart homes*. Academic Press.

Dutta, A. (1996). Integrating AI and Optimization for Decision Support: A Survey. *Decision Support Systems*, *18*(3–4), 217–226. doi:10.1016/0167-9236(96)00026-7

Dwivedi, Y. K., Rana, N. P., Jeyaraj, A., Clement, M., & Williams, M. D. (2017). Re-examining the Unified Theory of Acceptance and Use of Technology (UTAUT): Towards a revised theoretical model. *Information Systems Frontiers*, *21*(3), 719–734. doi:10.1007/s10796-017-9774-y

Dzulhikam, D., & Rana, M. E. (2022, March). A critical review of cloud computing environment for big data analytics. In *2022 International Conference on Decision Aid Sciences and Applications (DASA)* (pp. 76-81). IEEE. 10.1109/DASA54658.2022.9765168

Economic Policy Institute. (n.d.). *The gap between productivity and a typical worker's compensation has increased dramatically since 1979.* Author.

Edu, J. S., Such, J. M., & Suarez-Tangil, G. (2020). Smart home personal assistants: A security and privacy review. *ACM Computing Surveys*, *53*(6), 1–36. doi:10.1145/3412383

El Khatib, M., & Al Falasi, A. (2021). Effects of Artificial Intelligence on Decision Making in Project Management. *American Journal of Industrial and Business Management*, *11*(03), 251–260. doi:10.4236/ajibm.2021.113016

Elmarady, A. A., & Rahouma, K. (2021). Studying cybersecurity in civil aviation, including developing and applying aviation cybersecurity risk assessment. *IEEE Access : Practical Innovations, Open Solutions*, *9*, 143997–144016. doi:10.1109/ACCESS.2021.3121230

Elragal, A., & Klischewski, R. (2017). Theory-driven or process-driven prediction? Epistemological challenges of big data analytics. *Journal of Big Data*, *4*(1), 19. doi:10.1186/s40537-017-0079-2

Engels, A., Reyer, M., Xu, X., Mathar, R., Zhang, J., & Zhuang, H. (2013). Autonomous Self-Optimization of Coverage and Capacity in LTE Cellular Networks. *IEEE Transactions on Vehicular Technology*, *62*(5), 1989–2004. doi:10.1109/TVT.2013.2256441

Fabrègue, B. F., & Bogoni, A. (2023). Privacy and Security Concerns in the Smart City. *Smart Cities*, *6*(1), 586–613. doi:10.3390/smartcities6010027

Favaretto, M., De Clercq, E., & Elger, B. S. (2019). Big Data and discrimination: Perils, promises and solutions. A systematic review. *Journal of Big Data*, *6*(1), 12. doi:10.1186/s40537-019-0177-4

Felix, C. V. (2020). The role of the teacher and AI in education. In *International perspectives on the role of technology in humanizing higher education* (pp. 33–48). Emerald Publishing Limited. doi:10.1108/S2055-364120200000033003

Feraudo, A., Popescu, D. A., Yadav, P., Mortier, R., & Bellavista, P. (2024, January). Mitigating IoT Botnet DDoS Attacks through MUD and eBPF based Traffic Filtering. In *Proceedings of the 25th International Conference on Distributed Computing and Networking* (pp. 164-173). 10.1145/3631461.3631549

Franco, P., Martinez, J. M., Kim, Y. C., & Ahmed, M. A. (2021). IoT based approach for load monitoring and activity recognition in smart homes. *IEEE Access : Practical Innovations, Open Solutions*, *9*, 45325–45339. doi:10.1109/ACCESS.2021.3067029

Frank, M. R., Autor, D., & Bessen, J. E. (2019, April 2). Toward understanding the impact of artificial intelligence on labor. *National Academy of Sciences, 116*(14), 10. https://www.jstor.org/stable/10.2307/26698534

Frei-Landau, R., Muchnik-Rozanov, Y., & Avidov-Ungar, O. (2022). Using Rogers' diffusion of innovation theory to conceptualize the mobile-learning adoption process in teacher education in the COVID-19 era. *Education and Information Technologies*, 27(9), 12811–12838. doi:10.1007/s10639-022-11148-8 PMID:35702319

Frisk, J. E., & Bannister, F. (2017). Improving the use of analytics and big data by changing the decision-making culture: A design approach. *Management Decision*, 55(10), 2074–2088. doi:10.1108/MD-07-2016-0460

Gadatsch, A. (2023). IT Investment Calculation and Total Cost of Ownership Analysis: IT Standards as a tool for IT Controlling. In Springer eBooks (pp. 75–93). doi:10.1007/978-3-658-39270-3_6

Gallagher, B., McLean, T., Phillips, L. A., Warren, M., & Morrow, P. (2019). Predictive maintenance in the industry 4.0 era: A comprehensive survey. *IEEE Transactions on Reliability*, 68(3), 1406–1420.

Ganesh, D. E. (2022). Analysis of wireless sensor networks through secure routing protocols using directed diffusion methods. *International Journal of Wireless Network Security*, 7(1), 28–35.

Gardašević, G., Veletić, M., Maletić, N., Vasiljević, D., Radusinović, I., Tomović, S., & Radonjić, M. (2016). The IoT architectural framework, design issues and application domains. *Wireless Personal Communications*, 92(1), 127–148. doi:10.1007/s11277-016-3842-3

Garmaki, M., Gharib, R. K., & Boughzala, I. (2023). Big data analytics capability and contribution to firm performance: The mediating effect of organizational learning on firm performance. *Journal of Enterprise Information Management*, 36(5), 1161–1184. doi:10.1108/JEIM-06-2021-0247

Gaspars-Wieloch, H. (2017). Project Net Present Value estimation under uncertainty. *Central European Journal of Operations Research*, 27(1), 179–197. doi:10.1007/s10100-017-0500-0

Gatlin, K., Hallock, D., & Cooley, L. (2017). Confirmation Bias among Business Students: The Impact on Decision-Making. *Review of Contemporary Business Research*, 6(1). Advance online publication. doi:10.15640/rcbr.v6n2a2

Gaur, B., Shukla, V. K., & Verma, A. (2019, April). Strengthening people analytics through wearable IOT device for real-time data collection. In 2019 international conference on automation, computational and technology management (ICACTM) (pp. 555-560). IEEE. doi:10.1109/ICACTM.2019.8776776

Gebraeel, N., Lawley, M. A., & Cahoon, P. (2016). Data-driven predictive maintenance using deep learning and embedded system technology. *IISE Transactions*, 48(6), 494–505.

General Electric (GE). (2021). *Predictive maintenance for aviation*. Retrieved from https://www.ge.com/research/projects/predictive-maintenance-aviation

George, A. S., & Sagayarajan, S. (2023). Securing Cloud Application Infrastructure: Understanding the Penetration Testing Challenges of IaaS, PaaS, and SaaS Environments. *Partners Universal International Research Journal*, 2(1), 24–34.

Gesualdo, F., Daverio, M., Palazzani, L., Dimitriou, D., Díez-Domingo, J., Fons-Martínez, J., Jackson, S., Vignally, P., Rizzo, C., & Tozzi, A. E. (2021). Digital tools in the informed consent process: A systematic review. *BMC Medical Ethics*, 22(1), 18. Advance online publication. doi:10.1186/s12910-021-00585-8 PMID:33639926

Ghahremani-Nahr, J., Nozari, H., & Sadeghi, M. E. (2021). Green Supply Chain Based on Artificial Intelligence of Things (AIoT). *International Journal of Innovation in Management Economics and Social Sciences*, 1(2), 56–63. doi:10.52547/ijimes.1.2.56

Ghasemaghaei, M., & Turel, O. (2023). The Duality of Big Data in Explaining Decision-Making Quality. *Journal of Computer Information Systems*, 63(5), 1093–1111. doi:10.1080/08874417.2022.2125103

Gholamzadehmir, M., Del Pero, C., Buffa, S., Fedrizzi, R., & Aste, N. (2020). Adaptive-predictive control strategy for HVAC systems in smart buildings–A review. *Sustainable Cities and Society*, 63, 102480. doi:10.1016/j.scs.2020.102480

Gill, S. S., Xu, M., Ottaviani, C., Patros, P., Bahsoon, R., Shaghaghi, A., Golec, M., Stankovski, V., Wu, H., Abraham, A., Singh, M., Mehta, H., Ghosh, S. K., Baker, T., Parlikad, A. K., Lutfiyya, H., Kanhere, S. S., Sakellariou, R., Dustdar, S., ... Uhlig, S. (2022). AI for next generation computing: Emerging trends and future directions. *Internet of Things : Engineering Cyber Physical Human Systems*, 19, 100514. doi:10.1016/j.iot.2022.100514

Glöckler, J., Sedlmeir, J., Frank, M., & Fridgen, G. (2023). A systematic review of identity and access management requirements in enterprises and potential contributions of self-sovereign identity. *Business & Information Systems Engineering*, 1–20.

Goel, A., & Sahil, G. (2023). Implementing privacy and data confidentiality within the framework of the Internet of Things. *Journal of Data Protection & Privacy*, 5(4), 374–387.

Goethals, T., Volckaert, B., & De Turck, F. (2021). Enabling and Leveraging AI in the Intelligent Edge: A Review of Current Trends and Future Directions. *IEEE Open Journal of the Communications Society*, 2, 2311–2341. doi:10.1109/OJCOMS.2021.3116437

Gokturk-Saglam, A. L., & Sevgi-Sole, E. (Eds.). (2023). *Emerging Practices for Online Language Assessment, Exams, Evaluation, and Feedback*. IGI Global. doi:10.4018/978-1-6684-6227-0

Gómez, C., & Paradells, J. (2010). Wireless Home Automation Networks: A Survey of Architectures and Technologies. *IEEE Communications Magazine*, 48(6), 92–101. doi:10.1109/MCOM.2010.5473869

Gudla, S., & Kuda, N. (2023). A reliable routing mechanism with energy-efficient node selection for data transmission using a genetic algorithm in wireless sensor network. *Facta Universitatis, Series. Electronics and Energetics*, 36(2), 209–226. doi:10.2298/FUEE2302209G

Guilherme, A. (2019). AI and education: The importance of teacher and student relations. *AI & Society*, 34(1), 47–54. doi:10.1007/s00146-017-0693-8

Gunasekaran, A., Yusuf, Y. Y., Adeleye, E. O., Papadopoulos, T., Kovvuri, D., & Geyi, D. G. (2019). Agile manufacturing: An evolutionary review of practices. *International Journal of Production Research*, *57*(15–16), 5154–5174. doi:10.1080/00207543.2018.1530478

Güngör, V. Ç., & Lambert, F. (2006). A Survey on Communication Networks for Electric System Automation. *Computer Networks*, *50*(7), 877–897. doi:10.1016/j.comnet.2006.01.005

Guo, X., Shen, Z., Zhang, Y., & Wu, T. (2019). Review on the application of artificial intelligence in smart homes. *Smart Cities*, *2*(3), 402–420. doi:10.3390/smartcities2030025

Gupta, B. B., Chaudhary, P., Chang, X., & Nedjah, N. (2022). Smart defense against distributed Denial of service attack in IoT networks using supervised learning classifiers. *Computers & Electrical Engineering*, *98*, 107726. doi:10.1016/j.compeleceng.2022.107726

Gupta, M., & George, J. F. (2016). Toward the development of a big data analytics capability. *Information & Management*, *53*(8), 1049–1064. doi:10.1016/j.im.2016.07.004

Hafeez, I., Antikainen, M., & Tarkoma, S. (2019, March). Protecting IoT-environments against traffic analysis attacks with traffic morphing. In *2019 IEEE International Conference on Pervasive Computing and Communications Workshops (PerCom Workshops)* (pp. 196-201). IEEE. 10.1109/PERCOMW.2019.8730787

Hagendorff, T. (2020). The Ethics of AI Ethics: An Evaluation of Guidelines. *Minds and Machines*, *30*(1), 99–120. doi:10.1007/s11023-020-09517-8

Hangl, J., Behrens, V. J., & Krause, S. (2022). Barriers, Drivers, and Social Considerations for AI adoption in Supply Chain Management: A Tertiary study. *Logistics*, *6*(3), 63. doi:10.3390/logistics6030063

Han, Z., Kumar, U., & Singh, M. (2013). An overview of Smart Grid: Concepts, benefits and challenges. *Electric Power Components and Systems*, *41*(14), 1453–1489.

Haselton, M. G., Nettle, D., & Andrews, P. W. (2015). The Evolution of Cognitive Bias. In *The Handbook of Evolutionary Psychology* (pp. 724–746). John Wiley & Sons, Ltd. doi:10.1002/9780470939376.ch25

Hasko, R., Shakhovska, N., Vovk, O., & Holoshchuk, R. (2020). A Mixed Fog/Edge/AIoT/Robotics Education Approach based on Tripled Learning. In COAPSN (pp. 227-236). Academic Press.

Hay, L., Cash, P., & McKilligan, S. (2020). The future of design cognition analysis. Design Science, 6. doi:10.1017/dsj.2020.20

Heidari, A., & Jabraeil Jamali, M. A. (2023). Internet of Things intrusion detection systems: A comprehensive review and future directions. *Cluster Computing*, *26*(6), 3753–3780. doi:10.1007/s10586-022-03776-z

Heinrichs, B. (2021). Discrimination in the age of artificial intelligence. *AI & Society*, *37*(1), 143–154. doi:10.1007/s00146-021-01192-2

Herrera, D., & Rivas-Lalaleo, D. (2020). Artificial intelligence of things (AIoT) applied to optimize wireless sensors networks (WSNs). *Sensors (Basel)*, *20*(19), 5417.

He, S., Ren, W., Zhu, T., & Choo, K. K. R. (2019). BoSMoS: A blockchain-based status monitoring system for defending against unauthorized software updating in industrial Internet of Things. *IEEE Internet of Things Journal*, *7*(2), 948–959. doi:10.1109/JIOT.2019.2947339

Hildt, E., & Laas, K. (2022). Informed consent in digital data management. The International library of ethics, law and technology (pp. 55–81). doi:10.1007/978-3-030-86201-5_4

Hinestroza, D. (2018). *El machine learning a través de los tiempos, y los aportes a la humanidad.* Retrieved from https://repository.unilibre.edu.co/bitstream/handle/10901/17 289/EL%20MACHINE%20LEARNING.pdf

Hirankerd, K., & Kittisunthonphisarn, N. (2020). E-learning management system based on reality technology with AI. *International Journal of Information and Education Technology (IJIET)*, *10*(4), 259–264. doi:10.18178/ijiet.2020.10.4.1373

Hogg, S., Kurose, J., & Faloutsos, M. (2016). Network management protocols. In *Computer Networks and Internets* (6th ed., pp. 644–674). Pearson.

Horita, F. E. A., De Albuquerque, J. P., Marchezini, V., & Mendiondo, E. M. (2017). Bridging the Gap Between Decision-Making and Emerging Big Data Sources: An Application of a Model-Based Framework to Disaster Management in Brazil. *Decision Support Systems*, *97*, 12–22. doi:10.1016/j.dss.2017.03.001

Howes, A. (1995). Cognitive Modelling: Experiences in Human-Computer Interaction. In Springer eBooks (pp. 97–112). doi:10.1007/978-94-011-0103-5_8

How, M. L., Cheah, S. M., Chan, Y. J., Khor, A. C., & Say, E. M. P. (2023). Artificial Intelligence for Advancing Sustainable Development Goals (SDGs): An Inclusive Democratized Low-Code Approach. In *The Ethics of Artificial Intelligence for the Sustainable Development Goals* (pp. 145–165). Springer International Publishing. doi:10.1007/978-3-031-21147-8_9

Hsu, H. (2023). Facing the era of smartness – delivering excellent smart hospitality experiences through cloud computing. *Journal of Hospitality Marketing & Management*, 1–27. doi:10.108 0/19368623.2023.2251144

Huang, M., & Rust, R. T. (2020). Engaged with a robot? The role of AI in service. *Journal of Service Research*, *24*(1), 30–41. doi:10.1177/1094670520902266

Hung, L. (2021). Adaptive devices for AIoT systems. *2021 International Symposium on Intelligent Signal Processing and Communication Systems (ISPACS)*. 10.1109/ISPACS51563.2021.9651095

Huynh, M.-T., Nippa, M., & Aichner, T. (2023). Big data analytics capabilities: Patchwork or progress? A systematic review of the status quo and implications for future research. *Technological Forecasting and Social Change*, *197*, 122884. doi:10.1016/j.techfore.2023.122884

Николов, Н. (2020). Research of MQTT, CoAP, HTTP and XMPP IoT Communication protocols for Embedded Systems. *2020 XXIX International Scientific Conference Electronics (ET)*. 10.1109/ET50336.2020.9238208

Iberdrola. (2022). *Descubre los principales beneficios del 'Machine Learning'*. Retrieved from https://www.iberdrola.com/innovacion/machine-learning-aprend izaje-automatico

Ibrahim, R. F., Abu Al-Haija, Q., & Ahmad, A. (2022). DDoS attack prevention for Internet of Thing devices using Ethereum blockchain technology. *Sensors (Basel)*, *22*(18), 6806. doi:10.3390/s22186806

IEEE Computer Society. (2017). *IoT Technical Committee*. IEEE. Retrieved from https://www.computer.org/technical-committees/internet-of-th ings/

Ijemaru, G. K., Ang, L. M., & Seng, K. P. (2023). Swarm Intelligence Internet of Vehicles Approaches for Opportunistic Data Collection and Traffic Engineering in Smart City Waste Management. *Sensors (Basel)*, *23*(5), 2860. doi:10.3390/s23052860

ILO. (2023). *Overview of national employment policies adoption*. ILO. https://www.ilo.org/wesodata/?chart=Z2VuZGVyPVsiVG90YWwiXSZ1bml0PSJODdW1iZXIiJnNlY3Rvcj1bXSZ5ZWFyRnJvbT0xOTkxJmluY29tZT1bXSZpbmRpY2F0b3I9WyJ1bmVtcGxveW1lbnQiXSZzdGF0dXM9W10mcmVnaW9uPVsiV29ybGQiXSZjb3VudHJ5PVtdJndvcmtpbmdQb3ZlcnR5PVtdJnllYXJbz0yMDI0JnppZXXdG

Ioannidis, T., Bolgouras, V., Xenakis, C., & Politis, I. (2023, August). Securing the Flow: Security and Privacy Tools for Flow-based Programming. In *Proceedings of the 18th International Conference on Availability, Reliability and Security* (pp. 1-6). 10.1145/3600160.3605089

Ishengoma, F., Shao, D., Alexopoulos, C., Saxena, S., & Nikiforova, A. (2022). Integration of artificial intelligence of things (AIoT) in the public sector: Drivers, barriers and future research agenda. *Digital Policy, Regulation & Governance*, *24*(5), 449–462. doi:10.1108/DPRG-06-2022-0067

Ivanova, Y. (2020). The data protection impact assessment is a tool to enforce non-discriminatory AI. In Lecture Notes in Computer Science (pp. 3–24). doi:10.1007/978-3-030-55196-4_1

Jadad, A. R. (2019). Promoting innovation in healthcare: The role of AI, IoT, and blockchain. *Journal of Medical Internet Research*, *21*(2), e16286.

Jaihar, J., Lingayat, N., Vijaybhai, P. S., Venkatesh, G., & Upla, K. P. (2020, June). Smart home automation using machine learning algorithms. In 2020 international conference for emerging technology (INCET) (pp. 1-4). IEEE. doi:10.1109/INCET49848.2020.9154007

Jalal, A., Quaid, M. A. K., & Hasan, A. S. (2018, December). Wearable sensor-based human behavior understanding and recognition in daily life for smart environments. In *2018 International Conference on Frontiers of Information Technology (FIT)* (pp. 105-110). IEEE. 10.1109/FIT.2018.00026

James, E., & Rabbi, F. (2023). Fortifying the IoT Landscape: Strategies to Counter Security Risks in Connected Systems. *Tensorgate Journal of Sustainable Technology and Infrastructure for Developing Countries, 6*(1), 32–46.

Jangjou, M., & Sohrabi, M. K. (2022). A comprehensive survey on security challenges in different network layers in cloud computing. *Archives of Computational Methods in Engineering, 29*(6), 3587–3608. doi:10.1007/s11831-022-09708-9

Janssen, C. P., Donker, S. F., Brumby, D. P., & Kun, A. L. (2019). History and future of human-automation interaction. *International Journal of Human-Computer Studies, 131*, 99–107. doi:10.1016/j.ijhcs.2019.05.006

Javed, A. R., Ahmed, W., Pandya, S., Maddikunta, P. K. R., Alazab, M., & Gadekallu, T. R. (2023). A survey of explainable artificial intelligence for smart cities. *Electronics (Basel), 12*(4), 1020. doi:10.3390/electronics12041020

Jeble, S., Dubey, R., Childe, S. J., Papadopoulos, T., Roubaud, D., & Prakash, A. (2018). Impact of big data and predictive analytics capability on supply chain sustainability. *International Journal of Logistics Management, 29*(2), 513–538. doi:10.1108/IJLM-05-2017-0134

Jha, R. K. (2023). Cybersecurity and confidentiality in smart grid for enhancing sustainability and reliability. *Recent Research Reviews Journal, 2*(2), 215–241. doi:10.36548/rrrj.2023.2.001

Jmal, R., Ghabri, W., Guesmi, R., Alshammari, B. M., Alshammari, A. S., & Alsaif, H. (2023). Distributed Blockchain-SDN Secure IoT System Based on ANN to Mitigate DDoS Attacks. *Applied Sciences (Basel, Switzerland), 13*(8), 4953. doi:10.3390/app13084953

Joakin, I. (2021, September 5). *Aplicación de tecnologías de aprendizaje automático para predecir negocios y tomar decisiones empresariales*. Retrieved from http://sedici.unlp.edu.ar/bitstream/handle/10915/127018/Documento_completo.pdf?sequence=1

Joseph, K. J. (2002). *Growth of ICT and ICT for Development: Realities of the Myths of the Indian Experience* (No. 2002/78). WIDER Discussion Paper.

Jung, J., Gohar, M., & Koh, S. J. (2020). COAP-Based streaming control for IoT applications. *Electronics (Basel), 9*(8), 1320. doi:10.3390/electronics9081320

Kamble, S. S., Oza, P., & Dongare, S. B. (2019). Internet of Things (IoT): A review of enabling technologies, challenges, and open research issues. *Computer Communications.*

Karie, N. M., Sahri, N. M., Yang, W., Valli, C., & Kebande, V. R. (2021). A review of security standards and frameworks for IoT-based smart environments. *IEEE Access: Practical Innovations, Open Solutions, 9*, 121975–121995. doi:10.1109/ACCESS.2021.3109886

Kaur, T., & Kamboj, S. (2023, January). Descriptive Analysis of the Cloud Computing Services and Deployment Models. In *2023 International Conference for Advancement in Technology (ICONAT)* (pp. 1-6). IEEE. 10.1109/ICONAT57137.2023.10080749

Kaya, M. M., Taşkiran, Y., Kanoğlu, A., Demirtaş, A., Zor, E., Burçak, I., ... Akgül, F. T. (2021). Designing A Smart Home Management System With Artificial Intelligence &. *Machine Learning.*

Keane, I. (2023, March 26). *New York Post (AI impact).* Retrieved June 1, 2023, from https://nypost.com/2023/03/26/up-to-80-percent-of-workers-co uld-see-jobs-impacted-by-ai/

Keane, I. (2023, March 26). *Up to 80 percent of workers could see jobs impacted by AI.* New York Post. Retrieved June 1, 2023, from https://nypost.com/2023/03/26/up-to-80-percent-of-workers-co uld-see-jobs-impacted-by-ai/

Keleko, A. T., Kamsu-Foguem, B., Ngouna, R. H., & Tongne, A. (2022). Artificial intelligence and real-time predictive maintenance in industry 4.0: A bibliometric analysis. *AI and Ethics,* 2(4), 553–577. doi:10.1007/s43681-021-00132-6

Khalid, M., Hameed, S., Qadir, A., Shah, S. A., & Draheim, D. (2023). Towards SDN-based smart contract solution for IoT access control. *Computer Communications,* 198, 1–31. doi:10.1016/j. comcom.2022.11.007

Khan, A., Khan, S., Hassan, B., & Zheng, Z. (2022). CNN-based smoker classification and detection in smart city application. *Sensors (Basel),* 22(3), 892. doi:10.3390/s22030892

Kim, H., Choi, H., Kang, H., An, J., Yeom, S., & Hong, T. (2021). A systematic review of the smart energy conservation system: From smart homes to sustainable smart cities. *Renewable & Sustainable Energy Reviews,* 140, 110755. doi:10.1016/j.rser.2021.110755

Kim, J., Kim, I., Kim, J., & Kim, Y. (2020). An efficient deep learning-based network intrusion detection system for smart grids. *IEEE Access : Practical Innovations, Open Solutions,* 8, 120087–120097.

Kim, J., Lee, H., & Cho, Y. H. (2022). Learning design to support student-AI collaboration: Perspectives of leading teachers for AI in education. *Education and Information Technologies,* 27(5), 6069–6104. doi:10.1007/s10639-021-10831-6

Kim, S. W., Kong, J. H., Lee, S. W., & Lee, S. (2022). Recent Advances of Artificial Intelligence in Manufacturing Industrial Sectors: A Review. *International Journal of Precision Engineering and Manufacturing,* 23(1), 111–129. doi:10.1007/s12541-021-00600-3

Kim, Y., Tan, R., & Xu, L. (2017). Network security. In *Network Science and Cybersecurity* (pp. 259–279). Springer.

Kiran, V., Sardana, A., & Kaur, P. (2022, April). Defending Against Ddos Attacks in Rpl Using Subjective Logic Based Trust Approach For IOT. In *2022 2nd International Conference on Advance Computing and Innovative Technologies in Engineering (ICACITE)* (pp. 1656-1660). IEEE. 10.1109/ICACITE53722.2022.9823719

Knox, J., Wang, Y., & Gallagher, M. (2019). Introduction: AI, inclusion, and 'everyone learning everything'. *Artificial intelligence and inclusive education: Speculative futures and emerging practices,* 1-13.

Kodada, B. B., Prasad, G., & Pais, A. R. (2012). Protection against DDoS and data modification attack in computational grid cluster environment. *International Journal of Computer Network and Information Security, 4*(7), 12–18. doi:10.5815/ijcnis.2012.07.02

Kolmar, C., & American Community Survey. (n.d.). *An Interactive Exploration Of Earnings By Hours Worked.* Zippia. Retrieved June 1, 2023, from https://www.zippia.com/research/earnings-vs-hours-worked/#le ssons

Komninos, N., Philippou, E., & Pitsillides, A. (2014). Survey in smart grid and smart home security: Issues, challenges and countermeasures. *IEEE Communications Surveys and Tutorials, 16*(4), 1933–1954. doi:10.1109/COMST.2014.2320093

Koo, C., Shin, S., Gretzel, U., Hunter, W. C., & Chung, N. (2016). Conceptualization of smart tourism destination competitiveness. *Asia Pacific Journal of Information Systems, 26*(4), 561–576. doi:10.14329/apjis.2016.26.4.561

Kornaros, G. (2022). Hardware-assisted machine learning in resource-constrained IoT environments for security: Review and future prospective. *IEEE Access : Practical Innovations, Open Solutions, 10,* 58603–58622. doi:10.1109/ACCESS.2022.3179047

Koutsoyiannis, A. (1979). *Modern Microeconomics.* Macmillan. doi:10.1007/9781349160778

Krishnamoorthy, S., Dua, A., & Gupta, S. (2023). Role of emerging technologies in future IoT-driven Healthcare 4.0 technologies: A survey, current challenges and future directions. *Journal of Ambient Intelligence and Humanized Computing, 14*(1), 361–407. doi:10.1007/s12652-021-03302-w

Krueger, A. B., & Meyer, B. D. (n.d.). *Labour supply.* Retrieved June 1, 2023, from https://scholar.harvard.edu/files/gborjas/files/lechapter2.p df

Kübler, A., Holz, E. M., Riccio, A., Zickler, C., Kaufmann, T., Kleih, S. C., Staiger-SälZer, P., Desideri, L., Hoogerwerf, E. J., & Mattia, D. (2014). The User-Centered Design as Novel perspective for evaluating the usability of BCI-Controlled Applications. *PLoS One, 9*(12), e112392. doi:10.1371/journal.pone.0112392 PMID:25469774

Kuğuoğlu, B., Van Der Voort, H., & Janssen, M. (2021). The Giant Leap for Smart Cities: Scaling Up Smart City Artificial Intelligence of Things (AIoT) Initiatives. *Sustainability (Basel), 13*(21), 12295. doi:10.3390/su132112295

Kuka, L., Hörmann, C., & Sabitzer, B. (2022). Teaching and Learning with AI in Higher Education: A Scoping Review. *Learning with Technologies and Technologies in Learning: Experience, Trends and Challenges in Higher Education,* 551-571.

Kumar, J. (2020). Implementation of integrations of iot and cloud services in a home automation smart system. *Journal of Critical Reviews, 7*(1).

Kumar, R., & Venkatesh, K. (2022, April). Centralized and Decentralized Data Backup Approaches. In *Proceedings of International Conference on Deep Learning, Computing and Intelligence: ICDCI 2021* (pp. 687-698). Singapore: Springer Nature Singapore. 10.1007/978-981-16-5652-1_60

Kumar, A., Sharma, S., Goyal, N., Singh, A., Cheng, X., & Singh, P. (2021). Secure and energy-efficient smart building architecture with emerging technology IoT. *Computer Communications, 176*, 207–217. doi:10.1016/j.comcom.2021.06.003

Kumar, H., Singh, M. K., Gupta, M. P., & Madaan, J. (2020). Moving towards smart cities: Solutions that lead to the Smart City Transformation Framework. *Technological Forecasting and Social Change, 153*, 119281. doi:10.1016/j.techfore.2018.04.024

Kumari, P., & Jain, A. K. (2023). A comprehensive study of DDoS attacks over IoT network and their countermeasures. *Computers & Security, 127*, 103096. doi:10.1016/j.cose.2023.103096

Kurosu, M. (2020). Human-Computer Interaction. Design and user experience. Lecture Notes in Computer Science. doi:10.1007/978-3-030-49059-1

Kuzlu, M., Fair, C., & Güler, Ö. (2021). Role of Artificial Intelligence in the Internet of Things (IoT) cybersecurity. *Discover the Internet of Things, 1*(1), 7. Advance online publication. doi:10.1007/s43926-020-00001-4

Kvale, S. (1996). *InterViews: An Introduction to Qualitative Research Interviewing*. SAGE Publications.

Kwon, O., Lee, N., & Shin, B. (2014). Data quality management, data usage experience and acquisition intention of big data analytics. *International Journal of Information Management, 34*(3), 387–394. doi:10.1016/j.ijinfomgt.2014.02.002

Lai, Y. H., Chen, S. Y., Lai, C. F., Chang, Y. C., & Su, Y. S. (2021). Study on enhancing AIoT computational thinking skills by plot image-based VR. *Interactive Learning Environments, 29*(3), 482–495. doi:10.1080/10494820.2019.1580750

Lea, R., Cater, A., & Vo, Q. (2018). Artificial intelligence and the internet of things for sustainable agriculture: A comprehensive review. *IEEE Access : Practical Innovations, Open Solutions, 6*, 38335–38355.

Leech, B. L. (2002). Asking Questions: Techniques for Semistructured Interviews. *PS, Political Science & Politics, 35*(4), 665–668. doi:10.1017/S1049096502001129

Lemus, L., Félix, J. M., Fides-Valero, Á., Benlloch-Dualde, J., & Martínez-Millana, A. (2022). A Proof-of-Concept IoT system for remote healthcare based on interoperability standards. *Sensors (Basel), 22*(4), 1646. doi:10.3390/s22041646 PMID:35214548

Lepore, D., Testi, N., & Pasher, E. (2023). Building Inclusive Smart Cities through Innovation Intermediaries. *Sustainability (Basel), 15*(5), 4024. doi:10.3390/su15054024

Letaief, K. B., Shi, Y., Lu, J., & Lu, J. (2022). Edge Artificial Intelligence for 6G: Vision, Enabling Technologies, and Applications. *IEEE Journal on Selected Areas in Communications*, *40*(1), 5–36. doi:10.1109/JSAC.2021.3126076

Li, B., Hou, B., Yu, W., Lu, X., & Yang, C. (2017). Applications of artificial intelligence in intelligent manufacturing: A review. *Frontiers of Information Technology & Electronic Engineering*, *18*(1), 86–96. doi:10.1631/FITEE.1601885

Lin, S., Chang, T., Jorswieck, E. A., & Lin, P. (2022). Applications in AIoT: IoT security and secrecy. In Information Theory, Mathematical Optimization, and Their Crossroads in 6G System Design (pp. 331–373). doi:10.1007/978-981-19-2016-5_9

Lin, Y. S., Chen, S. Y., Tsai, C. W., & Lai, Y. H. (2021). Exploring computational thinking skills training through augmented reality and AIoT learning. *Frontiers in Psychology*, *12*, 640115. doi:10.3389/fpsyg.2021.640115 PMID:33708166

Lin, Y., Kämäräinen, T., Di Francesco, M., & Ylä-Jääski, A. (2015). Performance evaluation of remote display access for mobile cloud computing. *Computer Communications*, *72*, 17–25. doi:10.1016/j.comcom.2015.05.006

Lioupa, A., Memos, V. A., Stergiou, C. L., Ishibashi, Y., & Psannis, K. E. (2023, September). The Integration of 6G and Blockchain into an Efficient AIoT-Based Smart Education Model. In *2023 6th World Symposium on Communication Engineering (WSCE)* (pp. 1-5). IEEE.

Lipinski, P., Oliveira, L. B., & Botha, R. A. (2018). AIoT: Artificial intelligence as a service over IoT. *Future Generation Computer Systems*, *86*, 941–951.

Liu, H., Ning, H., Mu, Q., Zheng, Y., Zeng, J., Yang, L. T., Huang, R., & Ma, J. (2019). A review of the smart world. *Future Generation Computer Systems*, *96*, 678–691. doi:10.1016/j.future.2017.09.010

Liu, P., & Peng, Z. (2013). China's smart city pilots: A progress report. *Computer*, *47*(10), 72–81. doi:10.1109/MC.2013.149

Li, W., Yigitcanlar, T., Erol, I., & Liu, A. (2021). Motivations, barriers and risks of smart home adoption: From systematic literature review to conceptual framework. *Energy Research & Social Science*, *80*, 102211. doi:10.1016/j.erss.2021.102211

Lokshina, I. V., Greguš, M., & Thomas, W. L. (2019). Application of integrated building information modeling, IoT and blockchain technologies in system design of a smart building. *Procedia Computer Science*, *160*, 497–502. doi:10.1016/j.procs.2019.11.058

López-Granados, F., Jurado-Expósito, M., & Peña-Barragán, J. M. (2011). Quantifying efficacy and limits of unmanned aerial vehicle (UAV) technology for weed seedling detection as affected by sensor resolution. *Sensors (Basel)*, *11*(1), 119–131.

Lopez, J., Sánchez, H., López, T., & Cerrada, M. (2018). IoT-based system for real-time monitoring and management of photovoltaic facilities. *IEEE Transactions on Industrial Informatics*, *14*(6), 2624–2633.

Lozada, N., Arias-Pérez, J., & Henao-García, E. A. (2023). Unveiling the effects of big data analytics capability on innovation capability through absorptive capacity: Why more and better insights matter. *Journal of Enterprise Information Management, 36*(2), 680–701. doi:10.1108/JEIM-02-2021-0092

Luo, D. (2018). *Guide teaching system based on artificial intelligence.* https://online-journals.org/index.php/i-jet/article/view/905
8

Lu, Z., Qian, P., Bi, D., Ye, Z., He, X., Zhao, Y., Su, L., Li, S., & Zheng-Long, Z. (2021). Application of AI and IoT in Clinical Medicine: Summary and Challenges. *Current Medical Science, 41*(6), 1134–1150. doi:10.1007/s11596-021-2486-z PMID:34939144

Maggi, F., Volpatto, A., Gasparini, S., Boracchi, G., & Zanero, S. (2011, December). A fast eavesdropping attack against touchscreens. In *2011 7th International Conference on Information Assurance and Security (IAS)* (pp. 320-325). IEEE. 10.1109/ISIAS.2011.6122840

Maghsudi, S., Lan, A., Xu, J., & van Der Schaar, M. (2021). Personalized education in the artificial intelligence era: What to expect next. *IEEE Signal Processing Magazine, 38*(3), 37–50. doi:10.1109/MSP.2021.3055032

Mahbub, M. (2020). Progressive researches on IoT security: An exhaustive analysis from the perspective of protocols, vulnerabilities, and preemptive architectonics. *Journal of Network and Computer Applications, 168*, 102761. doi:10.1016/j.jnca.2020.102761

Malviya, S., & Lohiya, H. (2022). An analysis of authentication attacks with countermeasures and various authentication methods in a distributed environment. *International Research Journal of Modernization in Engineering Technology and Science, 4*(12), 1–8.

Maniatis, P. (n.d.). *Comparison of Public, Private, Hybrid, and Community Cloud Computing in Terms of Purchasing and Supply Management: A Quantitative Approach.* Academic Press.

Mao, J., Lin, Q., Zhu, S., Ma, L., & Liu, J. (2023). SmartTracer: Anomaly-Driven Provenance Analysis Based on Device Correlation in Smart Home Systems. *IEEE Internet of Things Journal.*

Marangunić, N., & Granić, A. (2014). Technology acceptance model: A literature review from 1986 to 2013. *Universal Access in the Information Society, 14*(1), 81–95. doi:10.1007/s10209-014-0348-1

Marian, M., Ganea, E., Popescu, D., Stîngă, F., Cusman, A., & Ionică, D. (2022, May). Analysis of Different SaaS Architectures from a Trust Service Provider Perspective. In *2022 23rd International Carpathian Control Conference (ICCC)* (pp. 360-365). IEEE. 10.1109/ICCC54292.2022.9805947

Marikyan, D., Papagiannidis, S., & Alamanos, E. (2020). Cognitive dissonance in technology adoption: A study of smart home users. *Information Systems Frontiers, 25*(3), 1101–1123. doi:10.1007/s10796-020-10042-3 PMID:32837263

Marquetti, A. (2007). A cross-country non parametric estimation of the returns to factors of production and the elasticity of scale. *Nova Economia Belo Horizonte, 4*(C14), 95-126.

Martínez-Comesaña, M., Rigueira-Díaz, X., Larrañaga-Janeiro, A., Martínez, J. M., Ocarranza-Prado, I., & Kreibel, D. (2023). Impact of artificial intelligence on assessment methods in primary and secondary education: systematic literature review. Revista De Psicodidáctica. doi:10.1016/j.psicoe.2023.06.002

Martins, F., Patrão, C., Moura, P., & de Almeida, A. T. (2021). A Review of Energy Modeling Tools for Energy Efficiency in Smart Cities. *Smart Cities*, *4*(4), 1420–1436. doi:10.3390/smartcities4040075

Maslej, N., Fattorini, L., & Stanford University. (2023, April). The AI Index 2023 Annual Report. *Institute for Human-Centered AI*. https://aiindex.stanford.edu/report/

Matin, A., Islam, R., Wang, X., Huo, H., & Xu, G. (2023). AIoT for sustainable manufacturing: Overview, challenges, and opportunities. *Internet of Things : Engineering Cyber Physical Human Systems*, *100901*. Advance online publication. doi:10.1016/j.iot.2023.100901

McKenzie, J. E., Bossuyt, P. M., Boutron, I., Hoffmann, T., Mulrow, C. D., Shamseer, L., Tetzlaff, J., Akl, E. A., Brennan, S., Chou, R., Glanville, J., Grimshaw, J., Hróbjartsson, A., Lalu, M. M., Li, T., Loder, E., Mayo-Wilson, E., McDonald, S., . . . Moher, D. (2021). The PRISMA 2020 Statement: An Updated Guideline for Reporting Systematic Reviews. BMJ, 71. doi:10.1136/bmj.n71

McKeown, N., Anderson, T., Balakrishnan, H., Parulkar, G., Peterson, L., Rexford, J., ... Shenker, S. (2008). OpenFlow: Enabling innovation in campus networks. *Computer Communication Review*, *38*(2), 69–74. doi:10.1145/1355734.1355746

McKinsey. (2018, September). *McKinsey Global Institute*. Retrieved May 31, 2023, from https://www.mckinsey.com/mgi

Mehrabi, N., Morstatter, F., Saxena, N., Lerman, K., & Galstyan, A. (2021). A survey on bias and fairness in machine learning. *ACM Computing Surveys*, *54*(6), 1–35. doi:10.1145/3457607

Mei, Y., Ma, Y., Ma, J., Moeller, L., & Federici, J. F. (2021). Eavesdropping risk evaluation on terahertz wireless channels in atmospheric turbulence. *IEEE Access : Practical Innovations, Open Solutions*, *9*, 101916–101923. doi:10.1109/ACCESS.2021.3098016

Meng, X., & Eneh, T. I. (2021). A survey on the impact of 5G network on edge computing and IoT. *IEEE Access : Practical Innovations, Open Solutions*, *9*, 75590–75606.

Menon, D., Anand, B., & Chowdhary, C. L. (2023). Digital Twin: Exploring the Intersection of Virtual and Physical Worlds. *IEEE Access : Practical Innovations, Open Solutions*, *11*, 75152–75172. doi:10.1109/ACCESS.2023.3294985

Merendino, A., Dibb, S., Meadows, M., Quinn, L., Wilson, D., Simkin, L., & Canhoto, A. (2018). Big data, big decisions: The impact of big data on board level decision-making. *Journal of Business Research*, *93*, 67–78. doi:10.1016/j.jbusres.2018.08.029

Mikalef, P., Pappas, I. O., Krogstie, J., & Giannakos, M. (2018). Big data analytics capabilities: A systematic literature review and research agenda. *Information Systems and e-Business Management*, *16*(3), 547–578. doi:10.1007/s10257-017-0362-y

Milaninia, N. (2021, March 1). *Biases in machine learning models and big data analytics: The international criminal and humanitarian law implications.* International Review of the Red Cross. https://international-review.icrc.org/articles/biases-machine-learning-big-data-analytics-ihl-implications-913

Milne-Ives, M., Selby, E., Inkster, B., Lam, C. S., & Meinert, E. (2022). Artificial intelligence and machine learning in mobile apps for mental health: A scoping review. *PLOS Digital Health*, *1*(8), e0000079. doi:10.1371/journal.pdig.0000079 PMID:36812623

Ming, H. X., Ti, L. R., Chen, H. C., Yu, T. C., Ming, H. Y., & Lai, Y. H. (2022, October). Multicultural Knowledge and Information Literacy Learning Using AIoT Integration Technology. In *2022 IET International Conference on Engineering Technologies and Applications (IET-ICETA)* (pp. 1-2). IEEE. 10.1109/IET-ICETA56553.2022.9971523

Ministerio Cultura Argentina. (2020, June 22). *Alan Turing, el padre de la inteligencia artificial.* Retrieved from https://www.cultura.gob.ar/alan-turing-el-padre-de-la-inteligencia-artificial-9162/

Mishra, P., & Singh, G. (2023). Energy management systems in sustainable smart cities based on the internet of energy: A technical review. *Energies*, *16*(19), 6903. doi:10.3390/en16196903

Mohanty, S. P., Hughes, D. P., & Zhang, H. (2016). Using machine learning for crop yield prediction and climate change adaptation. *Computers and Electronics in Agriculture*, *123*, 234–246.

Moreno, A., Armengol, E., Béjar, J., Belanche, L., Cortés, U., Gavaldà, R., . . . Sànchez, M. (1994). *Aprendizaje automático.* Barcelona-España: Edicions de la Universitat Politècnica de Catalunya (UPC).

Motevalli, S., & Seyedian, S. A. (2023). Evaluation of Environmental Factors in Smart Homes Implementation (case Study: Cities of Mazandaran Province in Iran). *ISPRS Annals of the Photogrammetry, Remote Sensing and Spatial Information Sciences*, *10*, 563–568. doi:10.5194/isprs-annals-X-4-W1-2022-563-2023

Moya Anegón, F. d., Herrero Solana, V., & Guerrero Bote, V. (1998). La aplicación de Redes Neuronales Artificiales (RNA): a la recuperación de la información. *Bibliodoc: anuari de biblioteconomia, documentació i informació*, 147-164.

Mukherjee, S. (2020). Emerging frontiers in smart environment and healthcare–A vision. *Information Systems Frontiers*, *22*(1), 23–27. doi:10.1007/s10796-019-09965-3

Munawar, S., Toor, S. K., Aslam, M., & Hamid, M. (2018). Move to smart learning environment: Exploratory research of challenges in computer laboratory and design intelligent virtual laboratory for eLearning technology. *Eurasia Journal of Mathematics, Science and Technology Education*, *14*(5), 1645–1662. doi:10.29333/ejmste/85036

Murphy, K., Di Ruggiero, E., Upshur, R., Willison, D. J., Malhotra, N., Cai, J., Malhotra, N., Lui, V., & Gibson, J. L. (2021). Artificial intelligence for good health: A scoping review of the ethics literature. *BMC Medical Ethics*, *22*(1), 14. Advance online publication. doi:10.1186/s12910-021-00577-8 PMID:33588803

Nagano, H. (2020). The growth of knowledge through the resource-based view. *Management Decision*, *58*(1), 98–111. doi:10.1108/MD-11-2016-0798

Nasrullayev, N., Muminova, S., Istamovich, D. K., & Boltaeva, M. (2023, July). Providing IoT Security in Industry 4.0 using Web Application Firewall. In *2023 4th International Conference on Electronics and Sustainable Communication Systems (ICESC)* (pp. 1788-1792). IEEE.

Nassar, M. M., Soares, A. L., & Yadav, S. (2017). Internet of Things (IoT) and big data: A review of recent developments. *IEEE Internet of Things Journal*, *4*(5), 1032–1043.

NASSCOM. (2022). *Implications of AI on the Indian Economy*. AI Adoption Index.

Nassiri Abrishamchi, M. A., Zainal, A., Ghaleb, F. A., Qasem, S. N., & Albarrak, A. M. (2022). Smart home privacy protection methods against a passive wireless Snooping side-channel attack. *Sensors (Basel)*, *22*(21), 8564. doi:10.3390/s22218564

Navarro, M., Liang, Y., & Zhong, X. (2022). Energy-efficient and balanced routing in low-power wireless sensor networks for data collection. *Ad Hoc Networks*, *127*, 102766. doi:10.1016/j.adhoc.2021.102766

Neshenko, N., Bou-Harb, E., Crichigno, J., Kaddoum, G., & Ghani, N. (2019). Demystifying IoT security: An exhaustive survey on IoT vulnerabilities and a first empirical look on Internet-scale IoT exploitations. *IEEE Communications Surveys and Tutorials*, *21*(3), 2702–2733. doi:10.1109/COMST.2019.2910750

Netflix. (2021). *How Netflix works: The (hugely simplified) complex stuff that happens every time you hit Play*. Retrieved from https://netflixtechblog.com

News Medical Life Science. (2023). *AI virtual patients' diagnostic application revolutionizes medical education in Hong Kong*. https://www.news-medical.net/news/20231117/AI-virtual-patients-revolutionize-medical-education-in-Hong-Kong.aspx

Nguyen, G., Dlugolinský, Š., Tran, V., & García, Á. L. (2020). Deep Learning for Proactive Network Monitoring and Security Protection. *IEEE Access : Practical Innovations, Open Solutions*, *8*, 19696–19716. doi:10.1109/ACCESS.2020.2968718

Nikitas, A., Michalakopoulou, K., Njoya, E. T., & Karampatzakis, D. (2020). Artificial Intelligence, Transport, and the Smart City: Definitions and Dimensions of a New Mobility Era. *Sustainability (Basel)*, *12*(7), 2789. doi:10.3390/su12072789

Nozari, H., Szmelter-Jarosz, A., & Ghahremani-Nahr, J. (2022). Analysis of artificial intelligence of things (AIOT) challenges for the smart supply chain (Case Study: FMCG Industries). *Sensors (Basel)*, *22*(8), 2931. doi:10.3390/s22082931 PMID:35458916

OECD. (2022, December). *Growth in GDP per capita, productivity and ULC*. OECD Statistics. Retrieved June 1, 2023, from https://stats.oecd.org/Index.aspx?DataSetCode=PDB_GR#

Ogundokun, R. O., Awotunde, J. B., Misra, S., Abikoye, O. C., & Folarin, O. (2021). Application of machine learning for ransomware detection in IoT devices. In Springer eBooks (pp. 393–420). doi:10.1007/978-3-030-72236-4_16

Olaniyi, O., Okunleye, O. J., & Olabanji, S. O. (2023). Advancing data-driven decision-making in smart cities through big data analytics: A comprehensive review of existing literature. *Current Journal of Applied Science and Technology, 42*(25), 10–18. doi:10.9734/cjast/2023/v42i254181

Omonayajo, B., Al-Turjman, F., & Cavus, N. (2022). Interactive and innovative technologies for smart education. *Computer Science and Information Systems, 19*(3), 1549–1564. doi:10.2298/CSIS210817027O

Onyx Soft. (n.d.). *¿Qué son los algoritmos predictivos de planificación de procesos?* Retrieved from https://www.onyxerp.com/blog/algoritmos-predictivos

Ortega, B. (2020). Cost-Benefit analysis. In Springer eBooks (pp. 1–5). doi:10.1007/978-3-319-69909-7_600-2

Ortega-Fernández, A., Martín-Rojas, R., & García-Morales, V. J. (2020). Artificial intelligence in the urban environment: Smart cities as models for developing innovation and sustainability. *Sustainability (Basel), 12*(19), 7860. doi:10.3390/su12197860

Ortega-Fernandez, I., & Liberati, F. (2023). A Review of Denial of Service Attack and Mitigation in the Smart Grid Using Reinforcement Learning. *Energies, 16*(2), 635. doi:10.3390/en16020635

Otasek, D., Morris, J. H., Bouças, J., Pico, A., & Demchak, B. (2019). Cytoscape Automation: Empowering Workflow-Based Network Analysis. *Genome Biology, 20*(1), 185. Advance online publication. doi:10.1186/s13059-019-1758-4 PMID:31477170

Otuoze, A. O., Mustafa, M. W., & Larik, R. M. (2018). Smart grids security challenges: Classification by sources of threats. *Journal of Electrical Systems and Information Technology, 5*(3), 468–483. doi:10.1016/j.jesit.2018.01.001

Ozer, E., Malekloo, A., Ramadan, W., Tran, T. T., & Di, X. (2023). Systemic reliability of bridge networks with mobile sensing-based model updating for postevent transportation decisions. *Computer-Aided Civil and Infrastructure Engineering, 38*(8), 975–999. doi:10.1111/mice.12892

Pan, L., Jiang, M., Wei, D., Chao, H. C., & Zhang, Y. (2018). Artificial intelligence of things (AIoT): A comprehensive survey. *Future Generation Computer Systems, 90*, 128–146.

Parihar, V., Malik, A., Bhawna, B. B., & Chaganti, R. (2023). from smart devices to smarter systems: The evolution of artificial intelligence of things (AIoT) with characteristics, architecture, use cases and challenges. In Springer eBooks (pp. 1–28). doi:10.1007/978-3-031-31952-5_1

Paris, I. L. B. M., Habaebi, M. H., & Zyoud, A. (2023). Implementation of SSL/TLS Security with MQTT Protocol in IoT Environment. *Wireless Personal Communications*, *132*(1), 163–182. doi:10.1007/s11277-023-10605-y

Pascual, M. G. (2022, July 11). *La inteligencia artificial de Google es capaz de aprender como un bebé*. Retrieved from https://elpais.com/tecnologia/2022-07-11/la-inteligencia-art ificial-de-google-es-capaz-de-aprender-como-un-bebe.html

Patel, B., Patel, K., & Patel, A. (2023). Internet of Thing (IoT): Routing Protocol Classification and Possible Attack on IOT Environment. In *Modern Electronics Devices and Communication Systems: Select Proceedings of MEDCOM 2021* (pp. 543-551). Singapore: Springer Nature Singapore.

Pathak, A. K., Saguna, S., Mitra, K., & Åhlund, C. (2021, June). Anomaly detection using machine learning to discover sensor tampering in IoT systems. In *ICC 2021-IEEE International Conference on Communications* (pp. 1-6). IEEE. 10.1109/ICC42927.2021.9500825

Patra, B. K., Mishra, S., & Patra, S. K. (2022). Genetic Algorithm-Based Energy-Efficient Clustering with Adaptive Grey Wolf Optimization-Based Multipath Routing in Wireless Sensor Network to Increase Network Life Time. *Intelligent Systems Proceedings of ICMIB*, *2021*, 499–512. doi:10.1007/978-981-19-0901-6_44

Patton, M. Q. (2002). Qualitative Research & Evaluation Methods. *Sage (Atlanta, Ga.)*.

Pelet, J., Lick, E., & Taieb, B. (2021). The internet of things in upscale hotels: Its impact on guests' sensory experiences and behaviour. *International Journal of Contemporary Hospitality Management*, *33*(11), 4035–4056. doi:10.1108/IJCHM-02-2021-0226

Pentikousis, K. (2000). TCP in wired-cum-wireless environments. *IEEE Communications Surveys and Tutorials*, *3*(4), 2–14. doi:10.1109/COMST.2000.5340805

Pereira, A. (2016). Plant Abiotic Stress Challenges from the Changing Environment. *Frontiers in Plant Science*, *7*. Advance online publication. doi:10.3389/fpls.2016.01123 PMID:27512403

Perera, C., Liu, C. H., & Jayawardena, S. (2017). The emerging internet of things marketplace from an industrial perspective: A survey. *IEEE Transactions on Emerging Topics in Computing*, *5*(2), 150–174.

Pérez, T. E., Raya, R., & Santos, E. (2017, September 29). *Las chicas del ENIAC (1946-1955)*. Retrieved from https://mujeresconciencia.com/2017/09/29/las-chicas-del-enia c-1946-1955

Petrovska, I., & Kuchuk, H. (2022). *Static allocation method in a cloud environment with a service model IAAS*. Academic Press.

Phillips-Wren, G., Power, D. J., & Mora, M. (2019). Cognitive bias, decision styles, and risk attitudes in decision making and DSS. *Journal of Decision Systems*, *28*(2), 63–66. doi:10.1080/12460125.2019.1646509

Pise, A. A., Almuzaini, K. K., Ahanger, T. A., Farouk, A., Pant, K., Pareek, P. K., & Nuagah, S. J. (2022). Enabling Artificial Intelligence of Things (AIOT) Healthcare Architectures and Listing Security Issues. *Computational Intelligence and Neuroscience*, *2022*, 1–14. doi:10.1155/2022/8421434 PMID:36911247

Pise, A. A., Yoon, B., & Singh, S. (2023). Enabling ambient intelligence of things (AIoT) healthcare system architectures. *Computer Communications*, *198*, 186–194. doi:10.1016/j.comcom.2022.10.029

Plathottam, S. J., Rzonca, A., Lakhnori, R., & Iloeje, C. O. (2023). A review of artificial intelligence applications in manufacturing operations. *Journal of Advanced Manufacturing and Processing*, *5*(3), e10159. doi:10.1002/amp2.10159

Ponemon Institute. (2016). *Cost of data center outages*. Ponemon Institute, LLC.

Posavac, S. S., Kardes, F. R., & Joško Brakus, J. (2010). Focus induced tunnel vision in managerial judgment and decision making: The peril and the antidote. *Organizational Behavior and Human Decision Processes*, *113*(2), 102–111. doi:10.1016/j.obhdp.2010.07.002

Prabha, C., Mittal, P., Gahlot, K., & Phul, V. (2023). Smart Healthcare System Based on AIoT Emerging Technologies: A Brief Review. In *Proceedings of International Conference on Recent Innovations in Computing* (pp. 299–311). 10.1007/978-981-99-0601-7_23

Prabhavathy, M., & Umamaheswari, S. (2022). Prevention of Runtime Malware Injection Attack in Cloud Using Unsupervised Learning. *Intelligent Automation & Soft Computing*, *32*(1), 101–114. doi:10.32604/iasc.2022.018257

PwC. (2017). *Human in the loop*. Pwc. www.pwc.com/AI

Qian, K., Zhang, Z., Yamamoto, Y., & Schuller, B. (2021). Artificial intelligence internet of things for the elderly: From assisted living to Healthcare Monitoring. *IEEE Signal Processing Magazine*, *38*(4), 78–88. doi:10.1109/MSP.2021.3057298

Qin, J., Liu, Y., & Yung, M. (2019). A survey of IoT key technology and applications in 5G. *IEEE Access : Practical Innovations, Open Solutions*, *7*, 78714–78727.

Qin, Z., Zhang, H., Meng, S., & Choo, K. R. (2020). Imaging and Fusing Time Series for Wearable Sensor-Based Human Activity Recognition. *Information Fusion*, *53*, 80–87. doi:10.1016/j.inffus.2019.06.014

Qiu, T., Chi, J., Zhou, X., Ning, Z., Atiquzzaman, M., & Wu, D. (2020). Edge Computing in Industrial Internet of Things: Architecture, Advances, and Challenges. *IEEE Communications Surveys and Tutorials*, *22*(4), 2462–2488. doi:10.1109/COMST.2020.3009103

Quy, V. K., Thành, B. T., Chehri, A., Linh, D. M., & Tuan, D. A. (2023). AI and digital transformation in higher education: Vision and approach of a specific university in Vietnam. *Sustainability (Basel)*, *15*(14), 11093. doi:10.3390/su151411093

Rahman, A., & Subriadi, A. P. (2022, January). Software as a service (SaaS) adoption factors: individual and organizational perspective. In *2022 2nd International Conference on Information Technology and Education (ICIT&E)* (pp. 31-36). IEEE.

Rahmani, R., Adabi, S., & Javadi, H. H. S. (2018). Edge computing for the Internet of Things: A case study. *IEEE Internet of Things Journal*, 5(3), 1275–1284.

Rajasegarar, S., Leckie, C., Palaniswami, M., Bezahaf, M., & Bishop, A. (2012). Anomaly detection in wireless sensor networks. *IEEE Wireless Communications*, 19(5), 74–80.

Rajasingham, L. (2009). *The impact of artificial intelligence (AI) systems on future university paradigms*. Victoria University of Wellington Wellington.

Ramesh, S., Nirmalraj, S., Murugan, S., Manikandan, R., & Al-Turjman, F. (2023). Optimization of energy and security in mobile sensor network using classification based signal processing in heterogeneous network. *Journal of Signal Processing Systems for Signal, Image, and Video Technology*, 95(2-3), 153–160. doi:10.1007/s11265-021-01690-y

Ramirez, M. A., Kim, S. K., Hamadi, H. A., Damiani, E., Byon, Y. J., Kim, T. Y., . . . Yeun, C. Y. (2022). Poisoning attacks and defenses on artificial intelligence: A survey. *arXiv preprint arXiv:2202.10276*.

Rana, M., Mamun, Q., & Islam, R. (2022). Lightweight cryptography in IoT networks: A survey. *Future Generation Computer Systems*, 129, 77–89. doi:10.1016/j.future.2021.11.011

Ransbotham, S., Candelon, F., Kiron, D., LaFountain, B., & Khodabandeh, S. (2021). The Cultural Benefits of Artificial Intelligence in the Enterprise. *MIT Sloan Management Review*. https://sloanreview.mit.edu/projects/the-cultural-benefits-of-artificial-intelligence-in-the-enterprise/

Razzaq, A. (2020). A Systematic review of software architectures for IoT systems and future direction to adopting a microservices architecture. *SN Computer Science*, 1(6), 350. Advance online publication. doi:10.1007/s42979-020-00359-w

Redacción España. (2019, November 11). *Origen del concepto de Inteligencia Artificial*. Retrieved from https://agenciab12.com/noticia/origen-concepto-inteligencia-artificial

Rehman, A., & Farrakh, A. (2022). An Intelligent Approach for Smart Home Energy Management System Empowered with Machine learning Techniques. *International Journal of Computational and Innovative Sciences*, 1(4), 7–14.

Reinhardt, K. (2022). Trust and trustworthiness in AI ethics. *AI and Ethics*, 3(3), 735–744. doi:10.1007/s43681-022-00200-5

Ren, W., Xin, T., Du, J., Wang, N., Li, S., Min, G., & Zhao, Z. (2021). Privacy enhancing techniques in the internet of things using data anonymization. *Information Systems Frontiers*. Advance online publication. doi:10.1007/s10796-021-10116-w

Revathy, R., Gopal, M., Selvi, M., & Periasamy, J. (2020, June 4). Analysis of artificial intelligence of things. *International Journal of Electrical Engineering and Technology*, *11*(4), 275–280. doi:10.34218/IJEET.11.4.2020.031

Roberts, M. K., & Ramasamy, P. (2023). An improved high performance clustering based routing protocol for wireless sensor networks in IoT. *Telecommunication Systems*, *82*(1), 45–59. doi:10.1007/s11235-022-00968-1

Rokis, K., & Kirikova, M. (2022, September). Challenges of low-code/no-code software development: A literature review. In *International Conference on Business Informatics Research* (pp. 3-17). Cham: Springer International Publishing. 10.1007/978-3-031-16947-2_1

Rothstein, M. A., & Tovino, S. A. (2019). California takes the lead on data privacy law. *The Hastings Center Report*, *49*(5), 4–5. doi:10.1002/hast.1042 PMID:31581323

Rouhani, S. (2020). *AI and the Future of Universities*. Mehr News Agency. https://www.mehrnews.com/news/5035864

Rueda, J. F. (2022). *Aprendizaje supervisado y no supervisado*. Retrieved from https://healthdataminer.com/data-mining/aprendizaje-supervisado-y-no-supervisado/

Russo, M. (2021, February 2). *¿Cuándo se inventó el primer ordenador?* Retrieved from https://www.info-computer.com/blog/cuando-se-invento-el-primer-ordenador/

Ryan, M. (2020). In AI, we trust ethics, artificial intelligence, and reliability. *Science and Engineering Ethics*, *26*(5), 2749–2767. doi:10.1007/s11948-020-00228-y PMID:32524425

Saadé, R. G., Zhang, J., Wang, X., Liu, H., & Guan, H. (2023). Challenges and Opportunities in the Internet of Intelligence of Things in Higher Education—Towards Bridging Theory and Practice. *IoT*, *4*(3), 430–465. doi:10.3390/iot4030019

Saad, R. M., Soufy, K. A. A., & Shaheen, S. I. (2023). Security in smart home environment: Issues, challenges, and countermeasures-a survey. *International Journal of Security and Networks*, *18*(1), 1–9. doi:10.1504/IJSN.2023.129887

Sabharwal, R., & Miah, S. J. (2021). A new theoretical understanding of big data analytics capabilities in organizations: A thematic analysis. *Journal of Big Data*, *8*(1), 159. doi:10.1186/s40537-021-00543-6

Sachs, G. (2023, April 5). *Generative AI Could Raise Global GDP by 7%*. Goldman Sachs. Retrieved June 1, 2023, from https://www.goldmansachs.com/intelligence/pages/generative-ai-could-raise-global-gdp-by-7-percent.html

Sadeeq, M. M., Abdulkareem, N. M., Zeebaree, S. R., Ahmed, D. M., Sami, A. S., & Zebari, R. R. (2021). IoT and Cloud computing issues, challenges and opportunities: A review. *Qubahan Academic Journal*, *1*(2), 1–7. doi:10.48161/qaj.v1n2a36

Sahu, M. K., & Pattanaik, S. R. (2023). Modified Limited memory BFGS with displacement aggregation. *arXiv preprint arXiv:2301.05447.*

Salah, K., Rehman, M. H. U., Nizamuddin, N., & Al-Fuqaha, A. (2019). Blockchain for AI: Review and Open Research Challenges. *IEEE Access : Practical Innovations, Open Solutions, 7,* 10127–10149. doi:10.1109/ACCESS.2018.2890507

Salah, S., & Amro, B. M. (2022). Big picture: Analysis of DDoS attacks map-systems and network, cloud computing, SCADA systems, and IoT. *International Journal of Internet Technology and Secured Transactions, 12*(6), 543–565. doi:10.1504/IJITST.2022.126468

Saleh, Z. (2019). *Artificial Intelligence Definition, Ethics and Standards.* Retrieved from https://www.researchgate.net/publication/332548325_Artificial_Intelligence_Definition_Ethics_and_Standards

Samson, K., & Kostyszyn, P. (2015). Effects of Cognitive Load on Trusting Behavior – An Experiment Using the Trust Game. *PLoS One, 10*(5), e0127680. doi:10.1371/journal.pone.0127680 PMID:26010489

Santhi, A. R., & Muthuswamy, P. (2023). Industry 5.0 or Industry 4.0S? Introduction to Industry 4.0 and a Peek into the Prospective Industry 5.0 Technologies. *International Journal on Interactive Design and Manufacturing, 17*(2), 947–979. doi:10.1007/s12008-023-01217-8

Saritepeci, M. (2020). Predictors of cyberloafing among high school students: Unauthorized access to school network, metacognitive awareness and smartphone addiction. *Education and Information Technologies, 25*(3), 2201–2219. doi:10.1007/s10639-019-10042-0

Sarker, I. H., Furhad, H., & Nowrozy, R. (2021). AI-Driven Cybersecurity: An overview, security intelligence modelling and research directions. *SN Computer Science, 2*(3), 173. Advance online publication. doi:10.1007/s42979-021-00557-0 PMID:33778771

Sarker, I. H., Khan, A. I., Abushark, Y. B., & Alsolami, F. (2023). Internet of things (iot) security intelligence: A comprehensive overview, machine learning solutions and research directions. *Mobile Networks and Applications, 28*(1), 296–312. doi:10.1007/s11036-022-01937-3

Schürmann, T., & Beckerle, P. (2020). Personalizing Human-Agent interaction through cognitive models. *Frontiers in Psychology, 11,* 561510. Advance online publication. doi:10.3389/fpsyg.2020.561510 PMID:33071887

Sergi, I., Montanaro, T., Benvenuto, F. L., & Patrono, L. (2021). A smart and secure logistics system based on IoT and cloud technologies. *Sensors (Basel), 21*(6), 2231. doi:10.3390/s21062231

Seriki, O. (2020). Resource-Based view. In Springer eBooks (pp. 1–4). doi:10.1007/978-3-030-02006-4_469-1

Shaaban, A. M., Chlup, S., El-Araby, N., & Schmittner, C. (2022). Towards Optimized Security Attributes for IoT Devices in Smart Agriculture Based on the IEC 62443 Security Standard. *Applied Sciences (Basel, Switzerland), 12*(11), 5653. doi:10.3390/app12115653

Sharadqh, A. A., Hatamleh, H. A. M., Saloum, S. S., & Alawneh, T. A. (2023). Hybrid Chain: Blockchain Enabled Framework for Bi-Level Intrusion Detection and Graph-Based Mitigation for Security Provisioning in Edge Assisted IoT Environment. *IEEE Access : Practical Innovations, Open Solutions*, *11*, 27433–27449. doi:10.1109/ACCESS.2023.3256277

Sharma, U., Tomar, P., Bhardwaj, H., & Sakalle, A. (2021). Artificial intelligence and its implications in education. In Impact of AI Technologies on Teaching, Learning, and Research in Higher Education (pp. 222-235). IGI Global. doi:10.4018/978-1-7998-4763-2.ch014

Sharma, V., Sharma, K. K., Vashishth, T. K., Panwar, R., Kumar, B., & Chaudhary, S. (2023, November). Brain-Computer Interface: Bridging the Gap Between Human Brain and Computing Systems. In *2023 International Conference on Research Methodologies in Knowledge Management, Artificial Intelligence and Telecommunication Engineering (RMKMATE)* (pp. 1-5). IEEE. 10.1109/RMKMATE59243.2023.10369702

Shen, M. (2023, August 15). Blockchains for Artificial Intelligence of Things: A Comprehensive survey. *IEEE Journals & Magazine*. https://ieeexplore.ieee.org/abstract/document/10105989/authors#authors

Sheng, J., Amankwah-Amoah, J., & Wang, X. (2017). A multidisciplinary perspective of big data in management research. *International Journal of Production Economics*, *191*, 97–112. doi:10.1016/j.ijpe.2017.06.006

Shiny, X. A., Ravikumar, D., Chinnasamy, A., & Hemavathi, S. (2023, May). Cloud Computing based Smart Traffic Management System with Priority Switching for Health Care Services. In *2023 2nd International Conference on Applied Artificial Intelligence and Computing (ICAAIC)* (pp. 1159-1163). IEEE. 10.1109/ICAAIC56838.2023.10140942

Shi, W., Cao, J., Zhang, Q., Li, Y., & Xu, L. (2016). Edge computing: Vision and challenges. *IEEE Internet of Things Journal*, *3*(5), 637–646. doi:10.1109/JIOT.2016.2579198

Shokouhyar, S., Seddigh, M. R., & Panahifar, F. (2020). Impact of big data analytics capabilities on supply chain sustainability: A case study of Iran. *World Journal of Science. Technology and Sustainable Development*, *17*(1), 33–57. doi:10.1108/WJSTSD-06-2019-0031

Shrivastava, R. K., Singh, S. P., Hasan, M. K., Islam, S., Abdullah, S., & Aman, A. H. M. (2022). Securing Internet of Things devices against code tampering attacks using Return Oriented Programming. *Computer Communications*, *193*, 38–46. doi:10.1016/j.comcom.2022.06.033

Shute, V., & Ventura, M. (2013). *Stealth assessment: Measuring and Supporting learning in Video Games*. The MIT Press. doi:10.7551/mitpress/9589.001.0001

Sibanda, M., Khumalo, N. Z., & Fon, F. N. (2023, November). A review of the implications of artificial intelligence tools in higher education. Should we panic? In *The 10th Focus Conference (TFC 2023)* (pp. 128-145). Atlantis Press.

Siemens. (2021). *Digitalization in building management.* Retrieved from https://new.siemens.com/global/en/products/buildings/topic-areas/intelligent-infrastructure.html

Si-Jing, L., & Wang, L. (2018). Artificial Intelligence Education Ethical Problems and Solutions. *2018 13th International Conference on Computer Science & Education (ICCSE).* 10.1109/ICCSE.2018.8468773

Singh, B., Gupta, V. K., Jain, A. K., Vashishth, T. K., & Sharma, S. (2023). *Transforming education in the digital age: A comprehensive study on the effectiveness of online learning.* Academic Press.

Singh, M., Fuenmayor, E., Hinchy, E. P., Qiao, Y., Murray, N., & Devine, D. (2021). Digital twin: Origin to future. *Applied System Innovation, 4*(2), 36. doi:10.3390/asi4020036

Sleem, A., & Elhenawy, I. (2023). Survey of Artificial Intelligence of Things for Smart Buildings: A closer outlook. *Journal of Intelligent Systems & Internet of Things, 8*(2), 63–71. doi:10.54216/JISIoT.080206

Srivastava, N., Hinton, G., Krizhevsky, A., Sutskever, I., & Salakhutdinov, R. (2015). Dropout: A simple way to prevent neural networks from overfitting. *Journal of Machine Learning Research, 15*(1), 1929–1958.

Stackpole, B. (2023, October 25). *For AI in manufacturing, start with data.* https://mitsloan.mit.edu/ideas-made-to-matter/ai-manufacturing-start-data

Standen, P. J., Brown, D. J., Taheri, M., Galvez Trigo, M. J., Boulton, H., Burton, A., & Hortal, E. (2020). An evaluation of an adaptive learning system based on multimodal affect recognition for learners with intellectual disabilities. *British Journal of Educational Technology, 51*(5), 1748–1765. doi:10.1111/bjet.13010

Steingartner, W., Galinec, D., & Kozina, A. (2021). Threat defense: Cyber deception approach and education for resilience in hybrid threats model. *Symmetry, 13*(4), 597. doi:10.3390/sym13040597

Stewart, J. C., Davis, G. A., & Igoche, D. A. (2020). AI, IoT, and AIoT: Definitions and impacts on the artificial intelligence curriculum. *Issues in Information Systems, 21*(4).

Stoyanova, M., Nikoloudakis, Y., Panagiotakis, S., Pallis, E., & Markakis, E. K. (2020). A Survey on the Internet of Things (IoT) Forensics: Challenges, Approaches, and Open Issues. *IEEE Communications Surveys and Tutorials, 22*(2), 1191–1221. doi:10.1109/COMST.2019.2962586

Sturges, J. E., & Hanrahan, K. J. (2004). Comparing Telephone and Face-to-Face Qualitative Interviewing: A Research Note. *Qualitative Research, 4*(1), 107–118. doi:10.1177/1468794104041110

Sundberg, S., Brunstrom, A., Ferlin-Reiter, S., Høiland-Jørgensen, T., & Brouer, J. D. (2023, March). Efficient continuous latency monitoring with eBPF. In *International Conference on Passive and Active Network Measurement* (pp. 191-208). Cham: Springer Nature Switzerland. 10.1007/978-3-031-28486-1_9

Sun, Y., Shi, W., Xu, D., & Liu, Y. (2016). When mobile blockchain meets edge computing. *IEEE Network*, *30*(2), 96–101.

Sun, Y., Yu, W., Han, Z., & Liu, K. (2006). Information Theoretic Framework of Trust Modeling and Evaluation for Ad Hoc Networks. *IEEE Journal on Selected Areas in Communications*, *24*(2), 305–317. doi:10.1109/JSAC.2005.861389

Support, E. D. U. (2023). *Educational uses for teachers.* https://edusupport.rug.nl/2430042249

Suresh Babu, C. V. N. S., A., P., M. V., & Janapriyan, R. (2023). IoT-Based Smart Accident Detection and Alert System. In P. Swarnalatha & S. Prabu (Eds.), Handbook of Research on Deep Learning Techniques for Cloud-Based Industrial IoT (pp. 322-337). IGI Global. doi:10.4018/978-1-6684-8098-4.ch019

Suresh Babu, C. V., Mahalashmi, J., Vidhya, A., Nila Devagi, S., & Bowshith, G. (2023). Save Soil Through Machine Learning. In M. Habib (Ed.), *Global Perspectives on Robotics and Autonomous Systems: Development and Applications* (pp. 345–362). IGI Global. doi:10.4018/978-1-6684-7791-5.ch016

Suresh Babu, C. V., Monika, R., Dhanusha, T., Vishnuvaradhanan, K., & Harish, A. (2023). Smart Street Lighting System for Smart Cities Using IoT (LoRa). In R. Kumar, A. Abdul Hamid, & N. Binti Ya'akub (Eds.), *Effective AI, Blockchain, and E-Governance Applications for Knowledge Discovery and Management* (pp. 78–96). IGI Global. doi:10.4018/978-1-6684-9151-5.ch006

Suresh Babu, C. V., & Srisakthi, S. (2023). Cyber Physical Systems and Network Security: The Present Scenarios and Its Applications. In R. Thanigaivelan, S. Kaliappan, & C. Jegadheesan (Eds.), *Cyber-Physical Systems and Supporting Technologies for Industrial Automation* (pp. 104–130). IGI Global. doi:10.4018/978-1-6684-9267-3.ch006

Sverko, M., Grbac, T. G., & Mikuc, M. (2022). SCADA Systems With Focus on Continuous Manufacturing and Steel Industry: A Survey on Architectures, Standards, Challenges and Industry 5.0. *IEEE Access : Practical Innovations, Open Solutions*, *10*, 109395–109430. doi:10.1109/ACCESS.2022.3211288

Swain, S. R., Saxena, D., Kumar, J., Singh, A. K., & Lee, C. N. (2023). An ai-driven intelligent traffic management model for 6g cloud radio access networks. *IEEE Wireless Communications Letters*, *12*(6), 1056–1060. doi:10.1109/LWC.2023.3259942

Swessi, D., & Idoudi, H. (2022). A survey on internet-of-things security: Threats and emerging countermeasures. *Wireless Personal Communications*, *124*(2), 1557–1592. doi:10.1007/s11277-021-09420-0

Swiecki, Z., Khosravi, H., Chen, G., Martinez-Maldanado, R., Lodge, J. M., Milligan, S., Selwyn, N., & Gašević, D. (2022). Assessment in the age of artificial intelligence. Computers & Education. *Artificial Intelligence*, *3*, 100075. doi:10.1016/j.caeai.2022.100075

Syamala, M., Komala, C. R., Pramila, P. V., Dash, S., Meenakshi, S., & Boopathi, S. (2023). Machine learning-integrated IoT-based smart home energy management System. In *Handbook of Research on Deep Learning Techniques for Cloud-Based Industrial IoT* (pp. 219–235). IGI Global. doi:10.4018/978-1-6684-8098-4.ch013

Szukits, Á. (2022). The illusion of data-driven decision making – The mediating effect of digital orientation and controllers' added value in explaining organizational implications of advanced analytics. *Journal of Management Control*, *33*(3), 403–446. doi:10.1007/s00187-022-00343-w

Taher, M. (2011). Resource-Based view theory. In Springer eBooks (pp. 151–163). doi:10.1007/978-1-4419-6108-2_8

Taiwo, O., Ezugwu, A. E., Oyelade, O. N., & Almutairi, M. S. (2022). Enhanced intelligent smart home control and security system based on deep learning model. *Wireless Communications and Mobile Computing*, *2022*, 1–22. doi:10.1155/2022/9307961

Talebkhah, M., Sali, A., Marjani, M., Gordan, M., Hashim, S. J., & Rokhani, F. Z. (2020). Edge computing: Architecture, Applications and Future Perspectives. *2020 IEEE 2nd International Conference on Artificial Intelligence in Engineering and Technology (IICAIET)*. 10.1109/IICAIET49801.2020.9257824

Tanveer, M., Bashir, A. K., Alzahrani, B. A., Albeshri, A., Alsubhi, K., & Chaudhry, S. A. (2023). CADF-CSE: Chaotic map-based authenticated data access/sharing framework for IoT-enabled cloud storage environment. *Physical Communication*, *59*, 102087. doi:10.1016/j.phycom.2023.102087

Tao, F., Qi, Q., Liu, A., & Kusiak, A. (2018). Data-driven smart manufacturing. *Journal of Manufacturing Systems*, *48*, 157–169. doi:10.1016/j.jmsy.2018.01.006

Tapalova, O., & Zhiyenbayeva, N. (2022). Artificial Intelligence in Education: AIEd for Personalised Learning Pathways. *Electronic Journal of e-Learning*, *20*(5), 639–653. doi:10.34190/ejel.20.5.2597

Tapscott, D. (1997). The Digital Economy: Promise and Peril. In *The Age of Networked Intelligence*. McGraw-Hill Education.

Taroun, A., & Yang, J.-B. (2011). Dempster-Shafer Theory of Evidence: Potential usage for decision making and risk analysis in construction project management. *The Built & Human Environment Review, 4*.

Tavares, R., Sousa, H., & Ribeiro, J. (2021). O uso de smart speakers na incapacidade: Uma scoping review. *New Trends in Qualitative Research*, *8*, 392–403. doi:10.36367/ntqr.8.2021.392-403

Tawalbeh, L. A., Muheidat, F., Tawalbeh, M., & Quwaider, M. (2020). IoT Privacy and security: Challenges and solutions. *Applied Sciences (Basel, Switzerland)*, *10*(12), 4102. doi:10.3390/app10124102

Tellioğlu, H. (2021). User-Centered design. In Springer eBooks (pp. 1–19). doi:10.1007/978-3-030-05324-6_122-1

Tensorway. (2023). *AI in Higher Education: Impact of AI on Student Assessment.* https://www.tensorway.com/post/ai-for-student-assessment-in-higher-education

Thibaud, M., Chi, H., Zhou, W., & Piramuthu, S. (2018). Internet of Things (IoT) in high-risk Environment, Health and Safety (EHS) industries: A comprehensive review. *Decision Support Systems, 108*, 79–95. doi:10.1016/j.dss.2018.02.005

Thomaz, G. A., Guerra, M. B., Sammarco, M., Detyniecki, M., & Campista, M. E. M. (2023). Tamper-proof access control for IoT clouds using enclaves. *Ad Hoc Networks, 147*, 103191. doi:10.1016/j.adhoc.2023.103191

Tikhamarine, Y., Souag-Gamane, D., Ahmed, A. N., Kisi, Ö., & El-Shafie, A. (2020). Improving artificial intelligence models accuracy for monthly streamflow forecasting using grey Wolf optimization (GWO) algorithm. *Journal of Hydrology (Amsterdam), 582*, 124435. doi:10.1016/j.jhydrol.2019.124435

Tippins, N. T., Oswald, F. L., & McPhail, S. M. (2021). Scientific, legal, and ethical concerns about AI-based personnel selection tools: A call to action. *Personnel Assessment and Decisions, 7*(2), 1. doi:10.25035/pad.2021.02.001

Tiwari, P. (2022). *Top 5 IoT Security Testing Companies to Rely on.* https://kratikal.com/blog/top-5-iot-security-testing-companies-to-rely-on/

Triantafyllou, A., Sarigiannidis, P., & Λάγκας, Θ. (2018). Network Protocols, schemes, and Mechanisms for Internet of Things (IoT): Features, open challenges, and trends. *Wireless Communications and Mobile Computing, 2018*, 1–24. doi:10.1155/2018/5349894

Trigo Aranda, V. (n.d.). Algoritmos. *Acta*, 43-50. Retrieved from https://www.acta.es/medios/articulos/matematicas/035041.pdf

Tripuraneni, N., Jordan, M., & Jin, C. (2020). On the theory of transfer learning: The importance of task diversity. *Advances in Neural Information Processing Systems, 33*, 7852–7862.

Tsai, C. C., Cheng, Y. M., Tsai, Y. S., & Lou, S. J. (2021). Impacts of AIOT implementation course on the learning outcomes of senior high school students. *Education Sciences, 11*(2), 82. doi:10.3390/educsci11020082

Tsai, M., & Li, M. (2021). Attendance Monitoring System based on Artificial Intelligence Facial Recognition Technology. *2021 IEEE International Conference on Consumer Electronics-Taiwan (ICCE-TW).* 10.1109/ICCE-TW52618.2021.9603093

Tushir, B., Dalal, Y., Dezfouli, B., & Liu, Y. (2020). A quantitative study of ddos and e-ddos attacks on wifi smart home devices. *IEEE Internet of Things Journal, 8*(8), 6282–6292. doi:10.1109/JIOT.2020.3026023

U.S. Bureau of Labor Statistics. (2023, March 23). *News releases from the Office of Productivity and Technology.* Bureau of Labor Statistics. Retrieved June 1, 2023, from https://www.bls.gov/productivity/news-releases.htm

Ursavaş, Ö. F. (2022). Technology Acceptance Model: history, theory, and application. In Springer eBooks (pp. 57–91). doi:10.1007/978-3-031-10846-4_4

Vagdatli, T., & Petroutsatou, K. (2022). Modelling Approaches of Life Cycle Cost–Benefit Analysis of Road Infrastructure: A Critical Review and Future Directions. *Buildings*, *13*(1), 94. doi:10.3390/buildings13010094

Van Wynsberghe, A. (2021). Sustainable AI: AI for sustainability and the sustainability of AI. *AI and Ethics*, *1*(3), 213–218. doi:10.1007/s43681-021-00043-6

Vashishth, T. K., Sharma, V., Sharma, K. K., Panwar, R., & Chaudhary, S. (2023). The Impact of Information Technology on the Education Sector: An Analysis of the Advantages, Challenges, and Strategies for Effective Integration. *International Journal of Research and Analytical Review*, *10*(2), 265–274.

Vázquez-Cano, E., Mengual-Andrés, S., & López-Meneses, E. (2021). Chatbot to improve learning punctuation in Spanish and to enhance open and flexible learning environments. *International Journal of Educational Technology in Higher Education*, *18*(1), 1–20. doi:10.1186/s41239-021-00269-8

Venkata Subhadu, V. (2023). *IT methods for operational near applications: low-code and no-code as rapid-application-development tools* [Doctoral dissertation]. Technische Hochschule Ingolstadt.

Verma, N. (2023). *How Effective is AI in Education? 10 Case Studies and Examples.* https://axonpark.com/how-effective-is-ai-in-education-10-case-studies-and-examples/

Vermanen, M., Rantanen, M. M., & Koskinen, J. (2022). Privacy in the internet of things ecosystems – a prerequisite for companies' ethical data collection and use. *IFIP Advances in Information and Communication Technology*, 18–26. doi:10.1007/978-3-031-15688-5_2

Vermesan, O., Bahr, R., Ottella, M., Serrano, M. M., Karlsen, T., Wahlstrøm, T., Sand, H. E., Ashwathnarayan, M., & Gamba, M. T. (2020). Internet of Robotic Things intelligent connectivity and platforms. *Frontiers in Robotics and AI*, *7*, 104. Advance online publication. doi:10.3389/frobt.2020.00104 PMID:33501271

Vuković, O., & Dán, G. (2014). Security of fully distributed power system state estimation: Detection and mitigation of data integrity attacks. *IEEE Journal on Selected Areas in Communications*, *32*(7), 1500–1508. doi:10.1109/JSAC.2014.2332106

Wang, Z. H., Jin, M. H., Jiang, L., Feng, C. J., Cao, J. Y., & Yun, Z. (2023, July). Secure access method of power internet of things based on zero trust architecture. In *International Conference on Swarm Intelligence* (pp. 386-399). Cham: Springer Nature Switzerland. 10.1007/978-3-031-36625-3_31

Wang, C., He, T., Zhou, H., Zhang, Z., & Lee, C. (2023). Artificial intelligence enhanced sensors - enabling technologies to next-generation health-care and biomedical platforms. *Bioelectronic Medicine*, *9*(1), 17. Advance online publication. doi:10.1186/s42234-023-00118-1 PMID:37528436

Wang, H., Ju, Y., Zhang, N., Pei, Q., Liu, L., Dong, M., & Leung, V. C. (2022). Resisting malicious eavesdropping: Physical layer security of mm wave mimo communications in presence of random blockage. *IEEE Internet of Things Journal, 9*(17), 16372–16385. doi:10.1109/JIOT.2022.3153054

Wang, J., Xu, C., Zhang, J., & Zhong, R. (2022). Big data analytics for intelligent manufacturing systems: A review. *Journal of Manufacturing Systems, 62*, 738–752. doi:10.1016/j.jmsy.2021.03.005

Wang, X., & Gill, C. (2018). Integration of blockchain and IoT. In *Blockchain applications* (pp. 39–44). Springer.

Wassouf, W. N., Alkhatib, R., Salloum, K., & Balloul, S. (2020). Predictive analytics using big data for increased customer loyalty: Syriatel Telecom Company case study. *Journal of Big Data, 7*(1), 29. Advance online publication. doi:10.1186/s40537-020-00290-0

Week, S. (2018). *Seldon: AI can replace Ofsted inspectors within a decade.* https://schoolsweek.co.uk/seldon-ai-can-replace-ofsted-inspectors-within-a-decade/

Weingärtner, T. (2019, November). Tokenization of physical assets and the impact of IoT and AI. In *European Union Blockchain Observatory and Forum* (Vol. 10, pp. 1-16). Academic Press.

Wei, W., Zhou, F., Hu, R. Q., & Wang, B. (2020). Energy-Efficient resource allocation for secure NOMA-Enabled mobile edge computing networks. *IEEE Transactions on Communications, 68*(1), 493–505. doi:10.1109/TCOMM.2019.2949994

Wei, X., & Taecharungroj, V. (2022). How to improve learning experience in MOOCs an analysis of online reviews of business courses on Coursera. *International Journal of Management Education, 20*(3), 100675. doi:10.1016/j.ijme.2022.100675

Whaiduzzaman, M., Barros, A., Chanda, M., Barman, S., Sultana, T., Rahman, M. S., Roy, S., & Fidge, C. (2022). A review of emerging technologies for IoT-based smart cities. *Sensors (Basel), 22*(23), 9271. doi:10.3390/s22239271

Wibawa, S. (2023). Analysis of Adversarial Attacks on AI-based With Fast Gradient Sign Method. *International Journal of Engineering Continuity, 2*(2), 72–79. doi:10.58291/ijec.v2i2.120

Wiyatno, R., & Xu, A. (2018). Maximal jacobian-based saliency map attack. *arXiv preprint arXiv:1808.07945.*

Wongvorachan, T., Lai, K. W., Bulut, O., Tsai, Y. S., & Chen, G. (2022). Artificial intelligence: Transforming the future of feedback in education. *Journal of Applied Testing Technology, 23,* 95–116.

World Economic Forum. (2023, January 5). *Future of jobs Report 2023.* Weforum. Retrieved May 31, 2023, from https://www.weforum.org/reports/the-future-of-%20jobs-report-2023/

Wu, T. N., Chin, K. Y., & Lai, Y. C. (2022, July). Applications of Intelligent Environmental IoT Detection Tools in Environmental Education of College Students. In *2022 12th International Congress on Advanced Applied Informatics (IIAI-AAI)* (pp. 240-243). IEEE. 10.1109/IIAIAAI55812.2022.00055

Wu, C., Yang, L., Cai, M., Zhao, X., & Sun, Q. (2023). Security Evaluation Method of Smart Home Cloud Platform. *International Journal of Pattern Recognition and Artificial Intelligence*, *37*(06), 2351012. doi:10.1142/S0218001423510126

Xiang, Z., Zheng, Y., He, M., Shi, L., Wang, D., Deng, S., & Zheng, Z. (2021). Energy-effective artificial internet-of-things application deployment in edge-cloud systems. *Peer-to-Peer Networking and Applications*, *15*(2), 1029–1044. doi:10.1007/s12083-021-01273-5

Xia, Q., Sifah, E. B., Agyekum, K. O. B. O., Xia, H., Acheampong, K. N., Smahi, A., ... Guizani, M. (2019). Secured fine-grained selective access to outsourced cloud data in IoT environments. *IEEE Internet of Things Journal*, *6*(6), 10749–10762. doi:10.1109/JIOT.2019.2941638

Xie, H., Hwang, G. J., & Wong, T. L. (2021). Editorial note: from conventional AI to modern AI in education: reexamining AI and analytic techniques for teaching and learning. *Journal of Educational Technology & Society*, *24*(3).

Xu, M., Yu, J., Zhang, X., Wang, C., Zhang, S., Wu, H., & Han, W. (2023). Improving real-world password guessing attacks via bi-directional transformers. In *32nd USENIX Security Symposium (USENIX Security 23)* (pp. 1001-1018). USENIX.

Yadava, A. K., Chouhan, V., Uke, N., Kumar, S., Banerjee, J., & Benjeed, A. O. S. (2022). A study of the progress, challenges, and opportunities in artificial intelligence of things (AIoT). *International Journal of Health Sciences*. doi:10.53730/ijhs.v6nS1.7224

Yadav, P., & Marwah, C. (2015). The Concept of Productivity. *International Journal of Engineering and Technical Research*, *3*(5), 192–196. https://www.erpublication.org/published_paper/IJETR032199.pdf

Yan, B., Yang, Y., & Huang, R. (2023). Memristive dynamics enabled neuromorphic computing systems. *Science China. Information Sciences*, *66*(10), 200401. Advance online publication. doi:10.1007/s11432-023-3739-0

Yang, C., Chen, H. W., Chang, E. J., Kristiani, E., Nguyen, K. L. P., & Chang, J. S. (2021). Current advances and future challenges of AIoT applications in particulate matter (PM) monitoring and control. *Journal of Hazardous Materials*, *419*, 126442. doi:10.1016/j.jhazmat.2021.126442

Yang, R., & Wibowo, S. (2022). User trust in artificial intelligence: A comprehensive conceptual framework. *Electronic Markets*, *32*(4), 2053–2077. doi:10.1007/s12525-022-00592-6

Yang, S., & Wang, X. (2023). Analysis of Computer Network Security and Prevention Technology. *Journal of Electronics and Information Science*, *8*(2), 20–24.

Yang, X., Shu, L., Liu, Y., Hancke, G. P., Ferrag, M. A., & Huang, K. (2022). Physical security and safety of iot equipment: A survey of recent advances and opportunities. *IEEE Transactions on Industrial Informatics*, *18*(7), 4319–4330. doi:10.1109/TII.2022.3141408

Yaqoob, I., Hashem, I. A. T., Inayat, Z., Mokhtar, S., Gani, A., & Ullah Khan, S. (2019). Internet of things forensics: Recent advances, taxonomy, requirements, and open challenges. *IEEE Internet of Things Journal*, *6*(4), 6358–6370.

Yaqub, M. Z., & Alsabban, A. (2023). Industry-4.0-Enabled Digital Transformation: Prospects, instruments, challenges, and implications for business strategies. *Sustainability (Basel)*, *15*(11), 8553. doi:10.3390/su15118553

Ye, L., Wang, Z., Jia, T., Ma, Y., Shen, L., Zhang, Y., Li, H., Chen, P., Wu, M., Liu, Y., Jing, Y. P., Zhang, H., & Huang, R. (2023). Research progress on the low-power artificial intelligence of things (AIoT) chip design. *Science China. Information Sciences*, *66*(10), 200407. Advance online publication. doi:10.1007/s11432-023-3813-8

Yiğitcanlar, T., Desouza, K. C., Butler, L., & Roozkhosh, F. (2020). Contributions and Risks of Artificial Intelligence (AI) in Building Smarter Cities: Insights from a Systematic Review of the Literature. *Energies*, *13*(6), 1473. doi:10.3390/en13061473

York, A. (2023). *10 Educational AI Tools for Students in 2023*. https://clickup.com/blog/ai-tools-for-students/

Younes, D., Alghannam, E., Tan, Y., & Lu, H. (2020). Enhancement in quality estimation of resistance spot welding using vision system and fuzzy support vector machine. *Symmetry*, *12*(8), 1380. doi:10.3390/sym12081380

Youssef, A. E. (2020). A framework for cloud security risk management based on the business objectives of organizations. *arXiv preprint arXiv:2001.08993*.

Zakzouk, A., Menzel, K., & Hamdy, M. (2023, September). Brain-Computer-Interface (BCI) Based Smart Home Control Using EEG Mental Commands. In *Working Conference on Virtual Enterprises* (pp. 720-732). Cham: Springer Nature Switzerland. 10.1007/978-3-031-42622-3_51

Zameer, A., Anwar, S. M., Warsi, M. B., & Guergachi, A. (2017). Machine learning for Industry 4.0: A comprehensive survey. *Computers & Industrial Engineering*, *136*, 1–17.

Zawaideh, F. H., Ghanem, W. A. H., Yusoff, M. H., Saany, S. I. A., Jusoh, J. A., & El-Ebiary, Y. A. B. (2022). The Layers of Cloud Computing Infrastructure and Security Attacking Issues. *Journal of Pharmaceutical Negative Results*, 792–800.

Zhang, Y., Ning, Y., Li, B., & Jun, Y. (2021, June). The research on talent education for AI-based IoT system development and implementation by the CDIO concept. In *2021 2nd International Conference on Artificial Intelligence and Education (ICAIE)* (pp. 110-114). IEEE. 10.1109/ICAIE53562.2021.00031

Zhang, K., Leng, S., He, Y., Maharjan, S., & Zhang, Y. (2018). Mobile edge computing and networking for Green and Low-Latency Internet of Things. *IEEE Communications Magazine*, *56*(5), 39–45. doi:10.1109/MCOM.2018.1700882

Zhang, X., Cao, Z., & Wei, D. (2020). Overview of edge computing in the agricultural Internet of Things: Key technologies, applications, challenges. *IEEE Access : Practical Innovations, Open Solutions*, *8*, 141748–141761. doi:10.1109/ACCESS.2020.3013005

Zhang, Y., Chen, C., Zhang, Y., Han, Z., Xu, C., & Liu, K. (2014). Edge computing in the Internet of Things: A multidisciplinary survey. *IEEE Access : Practical Innovations, Open Solutions*, *6*, 6900–6944.

Zhang, Y., Yu, H. Z., Zhou, W., & Man, M. (2022). Application and research of IoT architecture for End-Net-Cloud Edge Computing. *Electronics (Basel)*, *12*(1), 1. doi:10.3390/electronics12010001

Zhang, Z., Wen, F., Sun, Z., Guo, X., He, T., & Lee, C. (2022). Artificial intelligence-enabled sensing technologies in the 5G/internet of things era: From virtual reality/augmented reality to the digital twin. *Advanced Intelligent Systems*, *4*(7), 2100228. Advance online publication. doi:10.1002/aisy.202100228

Zhang, Z., Zheng, Y., Sun, D., & Du, X. (2019). A survey of network anomaly detection techniques. *Journal of Network and Computer Applications*, *60*, 19–31.

Zheng, Y., Li, Z., Xu, X., & Zhao, Q. (2022). Dynamic defenses in cyber security: Techniques, methods and challenges. *Digital Communications and Networks*, *8*(4), 422–435. doi:10.1016/j.dcan.2021.07.006

Zhong, R. Y., Xu, X., Klotz, E., & Newman, S. T. (2017). Intelligent Manufacturing in the context of Industry 4.0: A review. *Engineering (Beijing)*, *3*(5), 616–630. doi:10.1016/J.ENG.2017.05.015

About the Contributors

Sajad Rezaei is a senior lecturer in digital marketing at Worcester Business School, University of Worcester. Dr. Rezaei has published over 70 research papers in world-leading publishers and supervised more than 200 students at the undergraduate, postgraduate, executive MBA, DBA, and PhD levels. Dr. Rezaei's research focuses on the intersection of emerging technologies and consumer psychology, with a particular interest in how these technologies can be used for business innovation, sustainability, and productivity. His work has been featured in top academic journals such as the Journal of Business Research, Journal of Retailing and Consumer Services, Marketing Intelligence & Planning, Online Information Review, and Computers in Human Behavior, and he is the winner of the Emerald Literati Award in 2019. Dr. Rezaei has been named in a list of the Stanford University "Top 2% Most Influential Scientists" for the fifth consecutive year (2019 to 2023).

* * *

C. V. Suresh Babu is a pioneer in content development. A true entrepreneur, he founded Anniyappa Publications, a company that is highly active in publishing books related to Computer Science and Management. Dr. C.V. Suresh Babu has also ventured into SB Institute, a center for knowledge transfer. He holds a Ph.D. in Engineering Education from the National Institute of Technical Teachers Training & Research in Chennai, along with seven master's degrees in various disciplines such as Engineering, Computer Applications, Management, Commerce, Economics, Psychology, Law, and Education. Additionally, he has UGC-NET/SET qualifications in the fields of Computer Science, Management, Commerce, and Education. Currently, Dr. C.V. Suresh Babu is a Professor in the Department of Information Technology at the School of Computing Science, Hindustan Institute of Technology and Science (Hindustan University) in Padur, Chennai, Tamil Nadu, India. For more information, you can visit his personal blog.

Sachin Chaudhary completed his Graduation from MJPRU, and Post Graduation from AKTU, Moradabad, U.P. Currently Pursuing his Ph.D. in Computer Science and Engineering from Govt. Recognized University. Presently, he is working as an Assistant Professor in the Department of Computer Science and Applications, IIMT University, Meerut, U.P, India. He has been awarded as Excellence in teaching award 2019. He is the reviewer member of some reputed journals. He has published several book chapters and research papers of national and international reputed journals.

Qasim Hamakhurshid Hamamurad holds a Doctor of Philosophy degree in Information Systems from UTM. He received his Diploma in Technology from Sulaimany technology institute in 1992, BSc in Civil Engineering from the University of Salahadin, in 1996 and MSc degrees in Management Information System 2012 at Lebanese French University (LFU) and 2nd MSc in Computer Science and Communication Technology from the University Amiens Picardy Jules Verne (UPJV) in 2013, respectively, and his PhD degree in Information System from the University Technology of Malaysia. He is the author of over 14 journal papers and over 7 books. His current research interests include Framework for Land-use Spatial data sharing towards smart city application in Malaysia. Qasim is a member of KEU, He is work from 1997 to 1999 With UN organisation, from 1999 to 2008 head of Asia-QASIM company for information and communication technology, same year Lecturer in university of Salahaddin University, Sulaimany university and ISHQ university. From 2009 to 2019 Direct manager of municipality, and international engineering academic trainer, from 2019 still now he is an academic researcher in UTM. 9d23c30b-296f-4e14-bb0b-87499e8f9131

Richard Jones holds an MBA and works as an IT Infrastructure Manager for a global manufacturing organisation. Richard also works as a guest lecturer at the University of Worcester. Richard is qualified in CISCO Network and VMWare Certified Associate, with over 15 years of experience in designing and supporting complex server, network and telecoms environments.

Bhupendra Kumar completed his Graduation and Post Graduation from Chaudhary Charan Singh University, Meerut, U.P. and Ph.D. in Computer Science and Engineering from Mewar University, Hapur. Presently, he is working as a Professor in the Department of Computer Science and Applications, IIMT University, Meerut, U.P. He has been a huge teaching experience of 19 years. He is the reviewer member of some reputed journals. He has published several book chapters and research papers of national and international reputed journals.

Ali Omidi is an Associate Professor of International Relations and Senior Lecturer of Public International Law at the Department of Political Science, University of Isfahan, Isfahan, Iran. His detailed CV and publications are available at and https://scholar.google.com/citations?hl=en&user=F70iCB0AAAAJ&view_op=list_works&sortby=pubdate.5b0f6796-59b1-4d39-8e0a-caed9d799445

Saman Omidi has a BSc in Computer Engineering from the University of Isfahan-Iran. His main focus is Artificial Intelligence and he oftentimes writes on relations between AI and social issues.

Rajneesh Panwar graduated and post graduated in Mathematics and Computer Application from Ch. Charan Singh University, Meerut (U.P.) and received his M. Tech. in Computer Science from Shobhit University, Meerut. Presently, he is working as an Assistant Professor in the School of Computer Science and Application IIMT University, Meerut, U.P. He qualifies GATE 2021 and UGC-NET June 2020 and December 2020. He has published several book chapters and research papers of national and international repute.

V. Santhi has received her Ph.D. in Computer Science and Engineering from VIT University, Vellore, India. She has pursued her M.Tech. in Computer Science and Engineering from Pondicherry University, Puducherry. She has received her B.E. in Computer Science and Engineering from Bharathidasan University, Trichy, India. Currently she is working as Associate Professor in the School of Computing Science and Engineering, VIT University, Vellore, India. She has authored many national and international journal papers and one book. She is currently in the process of editing two books. Also, she has published many chapters in different books published by International publishers. She is senior member of IEEE and she is holding membership in many professional bodies like CSI, ISTE, IACSIT, IEEE and IAENG. Her areas of research include Image Processing, Digital Signal Processing, Digital Watermarking, Data Compression and Computational Intelligence.

Kewal Krishan Sharma is a professor in computer sc. in IIMT University, Meerut, U.P, India. He did his Ph.D. in computer network with this he has MCA, MBA and Law degree also. He did variously certification courses also. He has an overall experience of around 33 year in academic, business and industry. He wrote a number of research papers and books.

Vikas Sharma completed his Graduation and Post Graduation from Chaudhary Charan Singh University, Meerut, U.P. Currently Pursuing his Ph.D. in Computer Science and Engineering from Govt. Recognized University. Presently, he is working as an Assistant Professor in the Department of Computer Science and Applications, IIMT University, Meerut, U.P. He has been awarded as Excellence in teaching award 2019. He is the reviewer member of some reputed journals. He has published several book chapters and research papers of national and international reputed journals.

Ranjit Singha, a doctoral research fellow at Christ University and esteemed member of the APA, he is expertise in Mindfulness, Addiction Psychology. His credentials include mindfulness certifications from IBM and Oxford Mindfulness Centre, along with training in Yoga and CBCT. His educational background spans PGDBA, MBA, MSc in Counselling Psychology, and a Senior Diploma in Tabla. Active in the SEE Learning® program, he focuses on mindfulness interventions and has authored numerous publications. With teaching experience in diverse subjects like Forensic Psychology and Positive Psychology, Mr. Ranjit mentors graduate and undergraduate research projects, emphasizing student well-being. His dedication extends to personal counselling, reflecting a holistic commitment to education and psychology, positioning him as a key figure in the field.

Surjit Singha is an academician with a broad spectrum of interests, including UN Sustainable Development Goals, Organizational Climate, Workforce Diversity, Organizational Culture, HRM, Marketing, Finance, IB, Global Business, Business, AI, K12 & Higher Education. Currently a faculty member at Kristu Jayanti College, Dr. Surjit also serves as an Editor, reviewer, and author for prominent global publications and journals, including being on the Editorial review board of Information Resources Management Journal and contributor to various publications. With over 13 years of experience in Administration, Teaching, and Research, Dr. Surjit is dedicated to imparting knowledge and guiding students in their research pursuits. As a research mentor, Dr. Surjit has nurtured young minds and fostered academic growth. Dr. Surjit has an impressive track record of over 75 publications, including articles, book chapters, and textbooks, holds two US Copyrights, and has successfully completed and published two fully funded minor research projects from Kristu Jayanti College. Dr. Surjit Singha holds a PhD, M.Phil, and MBA from Christ University, as well as an M.Com and BBM from KJC (Bangalore University).) and a UGC-NET Commerce qualified professional. With a broad academic background, he also serves as an Examiner for IB examinations.

Abdulmaten Taroun is as Senior Lecturer at the University of Worcester, specialising in Project Management, Operations Management, Supply Chain Management, Risk and value management and Quality & Performance Management. Abdulmaten also supervises doctoral students and is also a reviewer for several academic journals including the International Journal of Project Management,

Journal of Operations Research Society, Journal of Automation in Construction and the International Journal of Computational Intelligence Systems.

Tarun Kumar Vashishth is an active academician and researcher in the field of computer science with 21 years of experience. He earned Ph.D. Mathematics degree specialized in Operations Research; served several academic positions such as HoD, Dy. Director, Academic Coordinator, Member Secretary of Department Research Committee, Assistant Center superintendent and Head Examiner in university examinations. He is involved in academic development and scholarly activities. He is member of International Association of Engineers, The Society of Digital Information and Wireless Communications, Global Professors Welfare Association, International Association of Academic plus Corporate (IAAC), Computer Science Teachers Association and Internet Society. His research interest includes Cloud Computing, Artificial Intelligence, Machine Learning and Operations Research; published more than 20 research articles with 1 book and 10 book chapters in edited books. He is contributing as member of editorial and reviewers boards in conferences and various computer journals published by CRC Press, Taylor and Francis, Springer, IGI global and other universities.

Helen Watts is a Principal Lecturer in Business at the University of Worcester, and teaches marketing, consumer psychology, consumer behaviour, and research methods (quantitative, qualitative, mixed), as well as supervising and examining research students. Helen holds a Chartership in Occupational Psychology, and a PhD in Customer Psychology. Outside of academia, Helen has conducted numerous contract research projects with high profile organisations, researching value and brand perceptions, satisfaction, and loyalty. As a consultant, Helen has worked with a range of national and international organisations to improve retention, selection and assessment, talent management, and learning and development, using diagnostic tools, techniques and data-driven processes.

Index

 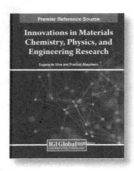  

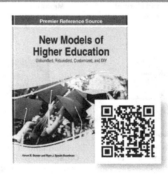

Individual Article & Chapter Downloads

US$ 37.50/each

Easily Identify, Acquire, and Utilize Published Peer-Reviewed Findings in Support of Your Current Research

- Browse Over *170,000+ Articles & Chapters*
- *Accurate & Advanced* Search
- Affordably Acquire *International Research*
- *Instantly Access* Your Content
- Benefit from the *InfoSci® Platform Features*

THE UNIVERSITY
of NORTH CAROLINA
at CHAPEL HILL

Printed in the United States
by Baker & Taylor Publisher Services